Multiscale Modeling in Solid Mechanics

Computational Approaches

Computational and Experimental Methods in Structures

Series Editor: Ferri M. H. Aliabadi *(Imperial College London, UK)*

Vol. 1 Buckling and Postbuckling Structures: Experimental, Analytical and Numerical Studies
edited by B. G. Falzon and M. H. Aliabadi (Imperial College London, UK)

Vol. 2 Advances in Multiphysics Simulation and Experimental Testing of MEMS
edited by A. Frangi, C. Cercignani (Politecnico di Milano, Italy), S. Mukherjee (Cornell University, USA) and N. Aluru (University of Illinois at Urbana Champaign, USA)

Vol. 3 Multiscale Modeling in Solid Mechanics: Computational Approaches
edited by U. Galvanetto and M. H. Aliabadi (Imperial College London, UK)

Computational and Experimental Methods in Structures – Vol. 3

Multiscale Modeling in Solid Mechanics

Computational Approaches

Editors

Ugo Galvanetto

M H Ferri Aliabadi

Imperial College London, UK

ICP

Imperial College Press

Published by

Imperial College Press
57 Shelton Street
Covent Garden
London WC2H 9HE

Distributed by

World Scientific Publishing Co. Pte. Ltd.
5 Toh Tuck Link, Singapore 596224
USA office: 27 Warren Street, Suite 401-402, Hackensack, NJ 07601
UK office: 57 Shelton Street, Covent Garden, London WC2H 9HE

British Library Cataloguing-in-Publication Data
A catalogue record for this book is available from the British Library.

Computational and Experimental Methods in Structures — Vol. 3
MULTISCALE MODELING IN SOLID MECHANICS
Computational Approaches

ISBN-13 978-1-84816-307-2
ISBN-10 1-84816-307-X

Desk Editor: Tjan Kwang Wei

Typeset by Stallion Press
Email: enquiries@stallionpress.com

Printed in Singapore by B & Jo Enterprise Pte Ltd

PREFACE

This unique volume presents the state of the art in the field of multi-scale modelling in solid mechanics with particular emphasis on computational approaches. Contributions from leading experts in the field and younger promising researchers are reunited to give a comprehensive description of recently proposed techniques and of the engineering problems which can be tackled with them. The first four chapters provide a detailed introduction to the theories on which different multi-scale approaches are based, Chapters 5–6 present advanced applications of multi-scale approaches used to investigate the behaviour of non-linear structures. Finally Chapter 7 introduces the novel topic of materials with self-similar structure. All chapters are self-contained and can be read independently.

Chapter 1 by V. G. Kouznetsova, M. G. D. Geers and W. A. M. Brekelmans is a concise but comprehensive introduction to the problem of mechanical multi-scale modelling in the general non-linear environment. This chapter presents a *computational homogenization strategy*, which provides a rigorous approach to determine the macroscopic response of heterogeneous materials with accurate account for microstructural characteristics and evolution. The implementation of the computational homogenization scheme in a Finite Element framework is discussed.

Chapter 2 by Qi-Zhi Xiao and Bhushan Lai Karihaloo is limited to *linear problems: higher order homogenization theory* and corresponding consistent solution strategies are fully described. Modern high performance Finite Element Methods, which are powerful for the solution of sub-problems from homogenization analysis, are also discussed.

Chapter 3 by G. K. Sfantos and M. H. Aliabadi presents a multi-scale modelling of material degradation and fracture based on the use of the *Boundary Element method*. Both micro and macro-scales are being modelled with the boundary element method. Additionally, a scheme for coupling the micro-BEM with a macro-FEM is presented.

Chapter 4 by J. C. Michel and P. Suquet is devoted to the *Nonuniform Transformation Field Analysis* which is a reduction technique introduced in the field of multi-scale problems in Nonlinear Solid Mechanics. The flexibility and accuracy of the method are illustrated by assessing the lifetime of a plate subjected to cyclic four-point bending.

Chapter 5 by M. Lefik, D. Boso, and B. A. Schrefler presents a *multi-scale approach for the thermo-mechanical analysis of hierarchical structures.* Both linear and non-linear material behaviours are considered. The case of composites with periodic microstructure is dealt with in detail and an example shows the capability of the method. It is also shown how *Artificial Neural Networks* can be used either to substitute the overall material relationship or to identify the parameters of the constitutive relation.

Chapter 6 by P. B. Lourenço, on recent advances in *masonry modelling*: micro-modelling and homogenization, addresses the issue of mechanical data necessary for advanced non-linear analysis first, with a set of recommendations. Then, the possibilities of using micro-modelling strategies replicating units and joints are addressed, with a focus on an interface finite element model for cyclic loading and a limit analysis model.

Finally Chapter 7 by R. C. Picu and M. A. Soare deals with the mechanics of materials with self-similar hierarchical microstructure. Many natural materials have hierarchical microstructure that extends over a broad range of length scales. Performing efficient design of structures made from such materials requires the ability to integrate the governing equations of the respective physics on supports with complex geometry.

CONTENTS

Preface v

Contributors ix

Computational Homogenisation for Non-Linear Heterogeneous Solids 1
V. G. Kouznetsova, M. G. D. Geers and W. A. M. Brekelmans

Two-Scale Asymptotic Homogenisation-Based Finite Element Analysis of Composite Materials 43
Qi-Zhi Xiao and Bhushan Lal Karihaloo

Multi-Scale Boundary Element Modelling of Material Degradation and Fracture 101
G. K. Sfantos and M. H. Aliabadi

Non-Uniform Transformation Field Analysis: A Reduced Model for Multiscale Non-Linear Problems in Solid Mechanics 159
Jean-Claude Michel and Pierre Suquet

Multiscale Approach for the Thermomechanical Analysis of Hierarchical Structures 207
Marek J. Lefik, Daniela P. Boso and Bernhard A. Schrefler

Recent Advances in Masonry Modelling: Micromodelling and Homogenisation 251
Paulo B. Lourenço

Mechanics of Materials With Self-Similar Hierarchical Microstructure 295
R. C. Picu and M. A. Soare

Index 333

CONTRIBUTORS

Prof. Wing Kam Liu
Director of NSF Summer Institute on Nano Mechanics and Materials
Northwestern University
Department of Mechanical Engineering
2145 Sheridan Rd., Evanston, IL 60208-3111

Prof. Ian Hutchings
University of Cambridge
Institute for Manufacturing
Mill Lane, Cambridge CB2 1RX

Prof.dr.ir. R. Huiskes (Rik)
Eindhoven University of Technology
Biomedical Engineering, Materials Technology
Eindhoven, The Netherlands

Prof. Manuel Doblare
Structural Mechanics, Department of Mechanical Engineering
Director of the Aragon Institute of Engineering Research
Maria de Luna, Zaragoza (Spain)

Prof. David Hills
Lincoln College, Oxford University
Oxford, UK

Prof. Paulo Sollero
University of Campinas
Sao Paulo, Brasil

Prof. Brian Falzon
Department of Aeronautics, Monash University
Australia

Prof. K. Nikbin
Department of Mechanical Engineering
Imperial College London, UK

Dr. M. Denda
Department of Mechanical and Aerospace Engineering
Rutgers University
New Jersey, USA

Prof. Ramon Abascal
Escuela Superior de Ingenieros
Camino de los descubrimientos s/n
Sevilla, Spain

Prof. K.J. Bathe
Massachusetts Institute of Technology
Boston, USA

Prof. Ugo Galvanetto
Department Engineering, Padova University
Padova, Italy

Prof. J.C.F. Telles
COPPE, Brasil

Prof. M Edirisinghe
Department of Engineering
University College London, UK

Prof. Spiros Pantelakis
Laboratory of Technology and Strength of Materials
Department of Mechanical, Engineering and Aeronautics
University of Patras, Greece

Prof. Eugenio Onate
Director of CIMNE, UPC
Barcelona, Spain

Prof. J. Woody
Department of Civil and Environmental Engineering
UCLA, Los Angeles, CA, 90095-1593, USA

Prof. Jan Sladek
Slovak Academy of Sciences
Slovakia

Prof. Ch-Zhang
University of Siegen
Germany

Prof. Mario Gugalinao
University of Milan
Italy

Prof. B. Abersek
University of Maribor
Slovenia

Dr. P. Dabnichki
Department of Engineering
Queen Mary, University of London

Prof. Alojz Ivankovic
Head of Mechanical Engineering
UCD School of Electrical
Electronic and Mechanical Engineering
University College, Dublin, Ireland

Prof. A. Chan
Department of Engineering
University of Birmingham, UK

Prof. Carmine Pappalettere
Department of Engineering
Bari University, Italy

Prof. H. Espinosa
Mechanical Engineering
Northwestern University, USA

Prof. Kon-Well Wang
The Pennsylvania State University
University Park, PA 16802
USA

Prof. Ole Thybo Thomsen
Department of Mechanical Engineering
Aalborg University
Aalborg East, Denmark

Prof. Herbert A. Mang
Technische Universität Wien
Vienna University of Technology
Institute for Mechanics of Materials
and Structures
Karlsplatz 13/202, 1040, Wien
Austria

Prof. Peter Gudmundson
Department of Solid Mechanics
KTH Engineering Sciences
SE-100 44, Stockholm, Sweden

Prof. Pierre Jacquot
Nanophotonics and Metrology Laboratory
Swiss Federal Institute of Technology Lausanne
CH -1015 Lausanne, Switzerland

Prof. K. Ravi-Chandar
Department of Aerospace Engineering
and Engineering Mechanics
University of Texas at Austin, USA

A. Sellier
LadHyX. Ecole Polytechnique
Palaiseau Cedex, France

COMPUTATIONAL HOMOGENISATION FOR NON-LINEAR HETEROGENEOUS SOLIDS

V. G. Kouznetsova[*,†,‡], M. G. D. Geers[†,§] and W. A. M. Brekelmans[†,¶]

[*]Netherlands Institute for Metals Research, Mekelweg 2
2628 CD Delft, The Netherlands

[†]Eindhoven University of Technology
Department of Mechanical Engineering, P. O. Box 513
5600 MB Eindhoven, The Netherlands
[‡]V.G.Kouznetsova@tue.nl
[§]M.G.D.Geers@tue.nl
[¶]W.A.M.Brekelmans@tue.nl

This chapter presents a computational homogenisation strategy, which provides a rigorous approach to determine the macroscopic response of heterogeneous materials with accurate account for microstructural characteristics and evolution. When using this micro–macro strategy there is no necessity to define homogenised macroscopic constitutive equations, which, in the case of large deformations and complex microstructures, would be generally a hardly feasible task. Instead, the constitutive behaviour at macroscopic integration points is determined by averaging the response of the deforming microstructure. This enables a straightforward application of the method to geometrically and physically non-linear problems, making it a particularly valuable tool for the modelling of evolving non-linear heterogeneous microstructures under complex macroscopic loading paths. In this chapter, the underlying concepts and the details of the computational homogenisation technique are given. Formulation of the microscopic boundary value problem and the consistent micro–macro coupling in a geometrically and physically non-linear framework are elaborated. The implementation of the computational homogenisation scheme in a finite element framework is discussed. Some recent extensions of the computational homogenisation schemes are summarised.

1. Introduction

Industrial and engineering materials, as well as natural materials, are heterogeneous at a certain scale. Typical examples include metal alloy systems, polycrystalline materials, composites, polymer blends, porous and cracked media, biological materials and many functional materials. This heterogeneous nature has a significant impact on the observed macroscopic behaviour of multiphase materials. Various phenomena occurring on the macroscopic level originate from the physics and mechanics of the underlying microstructure. The overall behaviour of micro-heterogeneous materials depends strongly on the size, shape, spatial distribution and properties of the microstructural constituents and their respective interfaces. The microstructural morphology and properties may also evolve under a macroscopic thermo-mechanical loading. Consequently, these microstructural influences are important for the processing and the reliability of the material and resulting products.

Determination of the macroscopic overall characteristics of heterogeneous media is an essential problem in many engineering applications. Studying the relation between microstructural phenomena and the macroscopic behaviour not only allows to predict the behaviour of existing multiphase materials, but also provides a tool to design a material microstructure such that the resulting macroscopic behaviour exhibits the required characteristics. An additional challenge for multiscale modelling is provided by ongoing technological developments, e.g. miniaturisation of products, development of functional and smart materials and increasing complexity of forming operations. In micro and submicron applications the microstructure is no longer negligible with respect to the component size, thus giving rise to a so-called size effect. Functional materials (e.g. as used in flexible electronics) typically involve materials with large thermo-mechanical mismatches combined with highly complex interconnects. Furthermore, advanced forming operations force a material to undergo complex loading paths. This results in varying microstructural responses and easily provokes an evolution of the microstructure, e.g. phase transformations. From an economical (time and costs) point of view, performing straightforward experimental measurements on a number of material samples of different sizes, accounting for various geometrical and physical phase properties, volume fractions and loading paths is a hardly feasible task. Hence, there is a clear need for modelling strategies that provide a better understanding of micro–macro structure–property relations in multiphase materials.

The simplest method leading to homogenised moduli of a heterogeneous material is based on the rule of mixtures. The overall property is then calculated as an average over the respective properties of the constituents, weighted with their volume fractions. This approach takes only one microstructural characteristic, i.e. the volume ratio of the heterogeneities, into consideration and, strictly speaking, denies the influence of other aspects.

A more sophisticated method is the effective medium approximation, as established by Eshelby[1] and further developed by a number of authors.[2–4] Equivalent material properties are derived as a result of the analytical (or semi-analytical) solution of a boundary value problem (BVP) for a spherical or ellipsoidal inclusion of one material in an infinite matrix of another material. An extension of this method is the self-consistent approach, in which a particle of one phase is embedded into an effective material, the properties of which are not known *a priori*.[5,6] These strategies give a reasonable approximation for structures that possess some kind of geometrical regularity, but fail to describe the behaviour of clustered structures. Moreover, high contrasts between the properties of the phases cannot be represented accurately.

Although some work has been done on the extension of the self-consistent approach to non-linear cases (originating from the work by Hill[5] who has proposed an "incremental" version of the self-consistent method), significantly more progress in estimating advanced properties of composites has been achieved by variational bounding methods.[7–10] The variational bounding methods are based on suitable variational (minimum energy) principles and provide upper and lower bounds for the overall composite properties.

Another homogenisation approach is based on the mathematical asymptotic homogenisation theory.[11,12] This method applies an asymptotic expansion of displacement and stress fields on a "natural scale parameter", which is the ratio of a characteristic size of the heterogeneities and a measure of the macrostructure.[13–17] The asymptotic homogenisation approach provides effective overall properties as well as local stress and strain values. However, usually the considerations are restricted to very simple microscopic geometries and simple material models, mostly at small strains. A comprehensive overview of different homogenisation methods may be found in a work done by Nemat-Nasser and Hori.[18]

The increasing complexity of microstructural mechanical and physical behaviour, along with the development of computational methods, made the class of so-called unit cell methods attractive. These approaches have been used in a great number of different applications.[19–26] A selection of examples in the field of metal matrix composites has been collected, for example, in a work done by Suresh *et al.*[27] The unit cell methods serve a twofold purpose: they provide valuable information on the local microstructural fields as well as the effective material properties. These properties are generally determined by fitting the averaged microscopical stress–strain fields, resulting from the analysis of a microstructural representative cell subjected to a certain loading path, on macroscopic closed-form phenomenological constitutive equations in a format established *a priori*. Once the constitutive behaviour becomes non-linear (geometrically, physically or both), it becomes intrinsically difficult to make a well-motivated assumption on a suitable macroscopic constitutive format. For example, McHugh *et al.*[28] have demonstrated that when a composite is characterised by power-law slip system hardening, the power-law hardening behaviour is not preserved at the macroscale. Hence, most of the known homogenisation techniques are not suitable for large deformations nor complex loading paths, neither do they account for the geometrical and physical changes of the microstructure (which is relevant, for example, when dealing with phase transitions).

In recent years, a promising alternative approach for the homogenisation of engineering materials has been developed, i.e. multiscale computational homogenisation, also called global–local analysis or FE^2 in a more particular form. Computational homogenisation is a multiscale technique, which is essentially based on the derivation of the local macroscopic constitutive response (input leading to output, e.g. stress driven by deformation) from the underlying microstructure through the adequate construction and solution of a microstructural BVP.

The basic principles of the classical computational homogenisation have gradually evolved from concepts employed in other homogenisation methods and may be fit into the four-step homogenisation scheme established by Suquet[29]: (i) definition of a microstructural representative volume element (RVE), of which the constitutive behaviour of individual constituents is assumed to be known; (ii) formulation of the microscopic boundary conditions from the macroscopic input variables and their application on the RVE (macro-to-micro transition); (iii) calculation of the macroscopic output variables from the analysis of the deformed microstructural RVE (micro-to-macro transition); (iv) obtaining the (numerical) relation between the macroscopic input and output variables. The main ideas of the computational homogenisation have been established by Suquet[29] and Guedes and Kikuchi[15] and further developed and improved in more recent works.[30–42]

Among several advantageous characteristics of the computational homogenisation technique the following are worth to be mentioned.

Techniques of this type

- do not require any explicit assumptions on the format of the macroscopic local constitutive equations, since the macroscopic constitutive behaviour is obtained from the solution of the associated microscale BVP;
- enable the incorporation of large deformations and rotations on both micro- and macrolevels;
- are suitable for arbitrary material behaviour, including physically non-linear and time dependent;
- provide the possibility to introduce detailed microstructural information, including the physical and geometrical evolution of the microstructure, into the macroscopic analysis;
- allow the use of any modelling technique on the microlevel, e.g. the finite element method (FEM),[33,37,38,40] the boundary element method,[43] the Voronoi cell method,[31,32] a crystal plasticity framework[34,35] or numerical methods based on Fast Fourier Transforms[36,44] and Transformation Field Analysis.[45]

Although the fully coupled micro–macro technique (i.e. the solution of a nested BVP) is still computationally rather expensive, this concern can be overcome by naturally parallelising computations.[37,42] Another option is selective usage, where non-critical regions are modelled by continuum closed-form homogenised constitutive relations or by the constitutive tangents obtained from the microstructural analysis but kept constant in the elastic domain, while in the critical regions the multiscale analysis of the microstructure is fully performed.[39] Despite the required computational efforts the computational homogenisation technique has proven to be a valuable tool to establish non-linear micro–macro structure–property relations, especially in the cases where the complexity of the mechanical and geometrical microstructural properties and the evolving character prohibit the use of other homogenisation methods. Moreover, this direct micro–macro modelling technique is useful for constructing, evaluating and verifying other homogenisation methods or micromechanically based macroscopic constitutive models.

In this chapter a computational homogenisation scheme is presented and details of its numerical implementation are elaborated. After a short summary of the underlying hypotheses and general framework of the computational homogenisation in Sec. 2, the microstructural BVP is stated and different types of boundary conditions are discussed in Sec. 3. Section 4 summarises the averaging theorems providing the basis for the micro–macro coupling. Several types of boundary conditions are shown to automatically satisfy these theorems. Next, in Sec. 5, implementation issues are discussed, whereby special attention is given to the imposition of the periodic boundary conditions and extraction of the overall stress tensor and the consistent tangent operator. The coupled nested solution scheme is summarised in Sec. 6 followed by a simple illustrative example. A general concept of an RVE in the computational homogenisation context is discussed in Sec. 8. Finally, in Sec. 9 several extensions of the classical computational homogenisation scheme are outlined, i.e. homogenisation towards second gradient continuum, computational homogenisation for beams and shells and computational homogenisation for heat conduction problems.

Cartesian tensors and tensor products are used throughout the chapter: $\mathbf{a}$, $\mathbf{A}$ and ${}^n\mathbf{A}$ denote, respectively, a vector, a second-order tensor and a nth-order tensor, respectively. The following notation for vector and tensor operations is employed: the dyadic product $\mathbf{ab} = a_i b_j \mathbf{e}_i \mathbf{e}_j$ and the scalar products $\mathbf{A} \cdot \mathbf{B} = A_{ij} B_{jk} \mathbf{e}_i \mathbf{e}_k$, $\mathbf{A} : \mathbf{B} = A_{ij} B_{ji}$, with $\mathbf{e}_i, i = 1, 2, 3$ the unit vectors of a Cartesian basis; conjugation $A^c_{ij} = A_{ji}$. A matrix and a column are denoted by $\underline{A}$ and $\underset{\sim}{a}$, respectively. The subscript "M" refers to a macroscopic quantity, whereas the subscript "m" denotes a microscopic quantity.

2. Basic Hypotheses

The material configuration to be considered is assumed to be macroscopically sufficiently homogeneous, but microscopically heterogeneous (the morphology consists of distinguishable components as, e.g. inclusions, grains, interfaces, cavities). This is schematically illustrated in Fig. 1. The

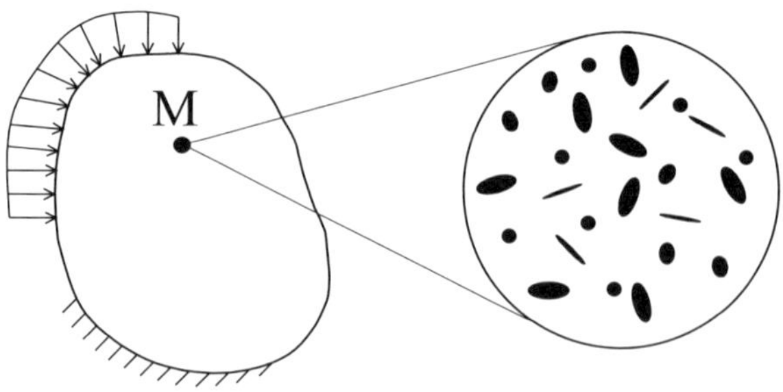

Fig. 1. Continuum macrostructure and heterogeneous microstructure associated with the macroscopic point M.

microscopic length scale is much larger than the molecular dimensions, so that a continuum approach is justified for every constituent. At the same time, in the context of the principle of separation of scales, the microscopic length scale should be much smaller than the characteristic size of the macroscopic sample or the wave length of the macroscopic loading.

Most of the homogenisation approaches make an assumption on global periodicity of the microstructure, suggesting that the whole macroscopic specimen consists of spatially repeated unit cells. In the computational homogenisation approach, a more realistic assumption on local periodicity is proposed, i.e. the microstructure can have different morphologies corresponding to different macroscopic points, while it repeats itself in a small vicinity of each individual macroscopic point. The concept of local and global periodicity is schematically illustrated in Fig. 2. The assumption of local periodicity adopted in the computational homogenisation allows the modelling of the effects of a non-uniform distribution of the microstructure on the macroscopic response (e.g. in functionally graded materials).

In the classical computational homogenisation procedure, a macroscopic deformation (gradient) tensor $\mathbf{F}_{\mathrm{M}}$ is calculated for every material point of the macrostructure (e.g. the integration points of the macroscopic mesh within a finite element (FE) environment). The deformation tensor $\mathbf{F}_{\mathrm{M}}$ for a macroscopic point is next used to formulate the boundary conditions to be imposed on the RVE that is assigned to this point. Upon the solution of the BVP for the RVE, the macroscopic stress tensor $\mathbf{P}_{\mathrm{M}}$ is obtained by averaging the resulting RVE stress field over the volume of the RVE. As a

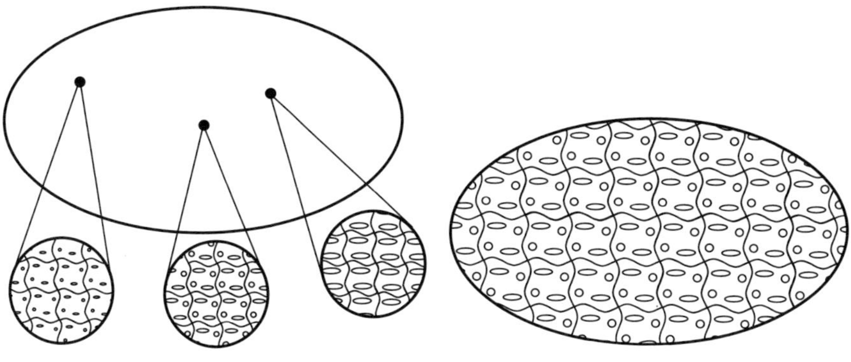

(a) (b)

Fig. 2. Schematic representation of a macrostructure with (a) a locally and (b) a globally periodic microstructure.

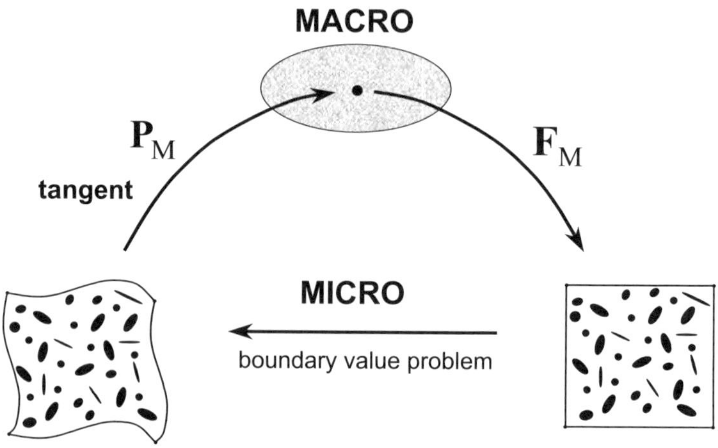

Fig. 3. Computational homogenisation scheme.

result, the (numerical) stress–deformation relationship at the macroscopic point is readily available. Additionally, the local macroscopic consistent tangent is derived from the microstructural stiffness. This framework is schematically illustrated in Fig. 3. This computational homogenisation technique is built entirely within a standard local continuum mechanics concept, where the response at a (macroscopic) material point depends only on the first gradient of the displacement field. Thus, this computational homogenisation framework is sometimes referred to as the "first-order".

The micro–macro procedure outlined here is "deformation driven", i.e. on the local macroscopic level the problem is formulated as follows: given a macroscopic deformation gradient tensor $\mathbf{F}_M$, determine the stress $\mathbf{P}_M$ and the constitutive tangent, based on the response of the underlying microstructure. A "stress-driven" procedure (given a local macroscopic stress, obtain the deformation) is also possible. However, such a procedure does not directly fit into the standard displacement-based FE framework, which is usually employed for the solution of macroscopic BVPs. Moreover, in the case of large deformations, the macroscopic rotational effects have to be added to the stress tensor in order to uniquely determine the deformation gradient tensor, thus complicating the implementation. Therefore, the "stress-driven" approach, which is often used in the analysis of single unit cells, is generally not adopted in coupled micro–macro computational homogenisation strategies.

In the subsequent sections, the essential steps of the first-order computational homogenisation process are discussed in more detail. First the problem on the microlevel is defined, then the aspects of the coupling between micro- and macrolevel are considered and finally the realisation of the whole procedure within an FE context is explained.

3. Definition of the Problem on the Microlevel

The physical and geometrical properties of the microstructure are identified by an RVE. An example of a typical two-dimensional RVE is depicted in Fig. 4. The actual choice of the RVE is a rather delicate task. The RVE should be large enough to represent the microstructure, without introducing non-existing properties (e.g. undesired anisotropy) and at the same time it should be small enough to allow efficient computational modelling. Some issues related to the concept of a representative cell are discussed in Sec. 8. Here it is supposed that an appropriate RVE has been already selected. Then the problem on the RVE level can be formulated as a standard problem in quasi-static continuum solid mechanics.

The RVE deformation field in a point with the initial position vector $\mathbf{X}$ (in the reference domain V_0) and the actual position vector $\mathbf{x}$ (in the current domain V) is described by the microstructural deformation gradient tensor $\mathbf{F}_{\mathrm{m}} = (\nabla_{0\mathrm{m}}\mathbf{x})^c$, where the gradient operator $\nabla_{0\mathrm{m}}$ is taken with respect to the reference microstructural configuration.

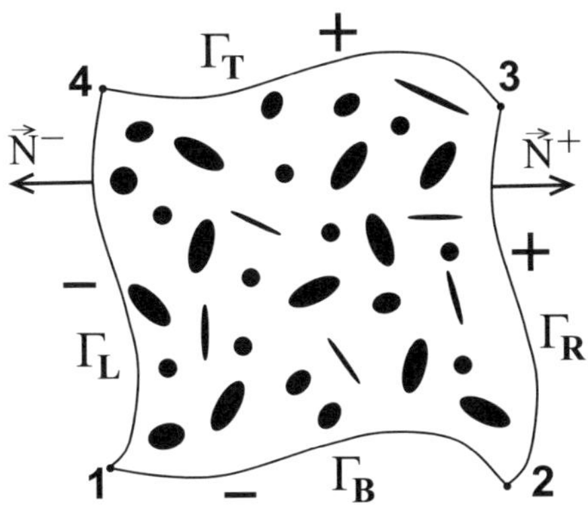

Fig. 4. Schematic representation of a typical two-dimensional representative volume element (RVE).

The RVE is in a state of equilibrium. This is mathematically reflected by the equilibrium equation in terms of the Cauchy stress tensor $\boldsymbol{\sigma}_{\mathrm{m}}$ or, alternatively, in terms of the first Piola–Kirchhoff stress tensor $\mathbf{P}_{\mathrm{m}} = \det(\mathbf{F}_{\mathrm{m}})\boldsymbol{\sigma}_{\mathrm{m}} \cdot (\mathbf{F}_{\mathrm{m}}^{c})^{-1}$ according to (in the absence of body forces)

$$\nabla_{\mathrm{m}} \cdot \boldsymbol{\sigma}_{\mathrm{m}} = \mathbf{0} \text{ in } V \quad \text{or} \quad \nabla_{0\mathrm{m}} \cdot \mathbf{P}_{\mathrm{m}}^{c} = \mathbf{0} \text{ in } V_0, \tag{1}$$

where ∇_{m} is the the gradient operator with respect to the current configuration of the microstructural cell.

The mechanical characterisations of the microstructural components are described by certain constitutive laws, specifying a time- and history-dependent stress–deformation relationship for every microstructural constituent

$$\begin{aligned} \boldsymbol{\sigma}_{\mathrm{m}}^{(\alpha)}(t) &= \mathcal{F}_{\sigma}^{(\alpha)}\{\mathbf{F}_{\mathrm{m}}^{(\alpha)}(\tau), \tau \in [0,t]\} \quad \text{or} \\ \mathbf{P}_{\mathrm{m}}^{(\alpha)}(t) &= \mathcal{F}_{P}^{(\alpha)}\{\mathbf{F}_{\mathrm{m}}^{(\alpha)}(\tau), \tau \in [0,t]\}, \end{aligned} \tag{2}$$

where t denotes the current time; $\alpha = \overline{1,N}$, with N being the number of microstructural constituents to be distinguished (e.g. matrix, inclusion, etc.).

The actual macro-to-micro transition is performed by imposing the macroscopic deformation gradient tensor $\mathbf{F}_{\mathrm{M}}$ on the microstructural RVE through a specific approach. Probably the simplest way is to assume that all the microstructural constituents undergo a constant deformation identical to the macroscopic one. In the literature this is called the Taylor (or Voigt) assumption. Another simple strategy is to assume an identical constant stress (and additionally identical rotation) in all the components. This is called the Sachs (or Reuss) assumption. Also some intermediate procedures are possible, where the Taylor and Sachs assumptions are applied only to certain components of the deformation and stress tensors. All these simplified procedures do not really require a detailed microstructural modelling. Accordingly, they generally provide very rough estimates of the overall material properties and are hardly suitable in the non-linear deformation regimes. The Taylor assumption usually overestimates the overall stiffness, whereas the Sachs assumption leads to an underestimation of the stiffness. Nevertheless, the Taylor and Sachs averaging procedures are sometimes used to quickly obtain a first estimate of the composite's overall stiffness. The Taylor assumption and some intermediate procedures are often employed in multicrystal plasticity modelling.

More accurate averaging strategies that do require the solution of the detailed microstructural BVP transfer the given macroscopic variables to the microstructural RVE via the boundary conditions. Classically three types of RVE boundary conditions are used, i.e. prescribed displacements, prescribed tractions and prescribed periodicity.

In the case of prescribed displacement boundary conditions, the position vector of a point on the RVE boundary in the deformed state is given by

$$\mathbf{x} = \mathbf{F}_{\mathrm{M}} \cdot \mathbf{X} \quad \text{with } \mathbf{X} \text{ on } \Gamma_0, \tag{3}$$

where Γ_0 denotes the undeformed boundary of the RVE. This condition prescribes a linear mapping of the RVE boundary.

For the traction boundary conditions, it is prescribed

$$\mathbf{t} = \mathbf{n} \cdot \boldsymbol{\sigma}_{\mathrm{M}} \text{ on } \Gamma \quad \text{or} \quad \mathbf{p} = \mathbf{N} \cdot \mathbf{P}^c_{\mathrm{M}} \text{ on } \Gamma_0, \tag{4}$$

where $\mathbf{n}$ and $\mathbf{N}$ are the normals to the current (Γ) and initial (Γ_0) RVE boundaries, respectively. However, the traction boundary conditions (4) do not completely define the microstructural BVP, as discussed at the end of Sec. 2. Moreover, they are not appropriate in the deformation driven procedure to be pursued in the present computational homogenisation scheme. Therefore, the RVE traction boundary conditions are not used in the actual implementation of the coupled computational homogenisation scheme; they were presented here for the sake of generality only.

Based on the assumption of microstructural periodicity presented in Sec. 2, periodic boundary conditions are introduced. The periodicity conditions for the microstructural RVE are written in a general format as

$$\mathbf{x}^+ - \mathbf{x}^- = \mathbf{F}_{\mathrm{M}} \cdot (\mathbf{X}^+ - \mathbf{X}^-), \tag{5}$$

$$\mathbf{p}^+ = -\mathbf{p}^-, \tag{6}$$

representing periodic deformations (5) and antiperiodic tractions (6) on the boundary of the RVE. Here the (opposite) parts of the RVE boundary Γ_0^- and Γ_0^+ are defined such that $\mathbf{N}^- = -\mathbf{N}^+$ at corresponding points on Γ_0^- and Γ_0^+, see Fig. 4. The periodicity condition (5), being prescribed on an initially periodic RVE, preserves the periodicity of the RVE in the deformed state. Also it should be mentioned that, as has been observed by several authors,[46,47] the periodic boundary conditions provide a better estimation of the overall properties than the prescribed displacement or prescribed traction boundary conditions (see also the discussion in Sec. 8).

Other types of RVE boundary conditions are possible. The only general requirement is that they should be consistent with the so-called averaging theorems. The averaging theorems, dealing with the coupling between the micro- and macrolevels in an energetically consistent way, will be presented in the following section. The consistency of the three types of boundary conditions presented above with these averaging theorems will be verified.

4. Coupling of the Macroscopic and Microscopic Levels

The actual coupling between the macroscopic and microscopic levels is based on averaging theorems. The integral averaging expressions have been initially proposed by Hill[48] for small deformations and later extended to a large deformation framework.[49,50]

4.1. *Deformation*

The first of the averaging relations concerns the micro–macro coupling of kinematic quantities. It is postulated that the macroscopic deformation gradient tensor $\mathbf{F}_{\mathrm{M}}$ is the volume average of the microstructural deformation gradient tensor $\mathbf{F}_{\mathrm{m}}$

$$\mathbf{F}_{\mathrm{M}} = \frac{1}{V_0}\int_{V_0} \mathbf{F}_{\mathrm{m}}\,\mathrm{d}V_0 = \frac{1}{V_0}\int_{\Gamma_0} \mathbf{x}\mathbf{N}\,\mathrm{d}\Gamma_0, \tag{7}$$

where the divergence theorem has been used to transform the integral over the undeformed volume V_0 of the RVE to a surface integral.

Verification that the use of the prescribed displacement boundary conditions (3) indeed leads to satisfaction of (7) is rather trivial. Substitution of (3) into (7) and use of the divergence theorem with account for $\nabla_{0\mathrm{m}}\mathbf{X} = \mathbf{I}$ give

$$\begin{aligned}\mathbf{F}_{\mathrm{M}} &= \frac{1}{V_0}\int_{\Gamma_0} (\mathbf{F}_{\mathrm{M}}\cdot\mathbf{X})\mathbf{N}\,\mathrm{d}\Gamma_0\\ &= \frac{1}{V_0}\mathbf{F}_{\mathrm{M}}\cdot\int_{\Gamma_0}\mathbf{X}\mathbf{N}\,\mathrm{d}\Gamma_0 = \frac{1}{V_0}\mathbf{F}_{\mathrm{M}}\cdot\int_{V_0}(\nabla_{0\mathrm{m}}\mathbf{X})^c\,\mathrm{d}V_0 = \mathbf{F}_{\mathrm{M}}.\end{aligned} \tag{8}$$

The validation for the periodic boundary conditions (5) follows the same lines except that the RVE boundary is split into the parts Γ_0^+ and Γ_0^-

$$\begin{aligned}\mathbf{F}_{\mathrm{M}} &= \frac{1}{V_0}\left\{\int_{\Gamma_0^+}\mathbf{x}^+\mathbf{N}^+\,\mathrm{d}\Gamma_0 + \int_{\Gamma_0^-}\mathbf{x}^-\mathbf{N}^-\,\mathrm{d}\Gamma_0\right\}\\ &= \frac{1}{V_0}\int_{\Gamma_0^+}(\mathbf{x}^+ - \mathbf{x}^-)\mathbf{N}^+\,\mathrm{d}\Gamma_0\\ &= \frac{1}{V_0}\mathbf{F}_{\mathrm{M}}\cdot\int_{\Gamma_0^+}(\mathbf{X}^+ - \mathbf{X}^-)\mathbf{N}^+\,\mathrm{d}\Gamma_0 = \frac{1}{V_0}\mathbf{F}_{\mathrm{M}}\cdot\int_{\Gamma_0}\mathbf{X}\mathbf{N}\,\mathrm{d}\Gamma_0 = \mathbf{F}_{\mathrm{M}}.\end{aligned} \tag{9}$$

In the general case of large strains and large rotations, attention should be given to the fact that due to the non-linear character of the relations between different kinematic measures not all macroscopic kinematic quantities may be obtained as the volume average of their microstructural counterparts. For example, the volume average of the Green–Lagrange strain tensor

$$\mathbf{E}_{\mathrm{M}}^{*} = \frac{1}{2V_0}\int_{V_0} (\mathbf{F}_{\mathrm{m}}^{c} \cdot \mathbf{F}_{\mathrm{m}} - \mathbf{I})\,\mathrm{d}V_0 \tag{10}$$

is in general not equal to the macroscopic Green–Lagrange strain obtained according to

$$\mathbf{E}_{\mathrm{M}} = \frac{1}{2}(\mathbf{F}_{\mathrm{M}}^{c} \cdot \mathbf{F}_{\mathrm{M}} - \mathbf{I}). \tag{11}$$

4.2. *Stress*

Similarly, the averaging relation for the first Piola–Kirchhoff stress tensor is established as

$$\mathbf{P}_{\mathrm{M}} = \frac{1}{V_0}\int_{V_0} \mathbf{P}_{\mathrm{m}}\,\mathrm{d}V_0. \tag{12}$$

In order to express the macroscopic first Piola–Kirchhoff stress tensor $\mathbf{P}_{\mathrm{M}}$ in the microstructural quantities defined on the RVE surface, the following relation is used (with account for microscopic equilibrium $\nabla_{0\mathrm{m}} \cdot \mathbf{P}_{\mathrm{m}}^{c} = \mathbf{0}$ and the equality $\nabla_{0\mathrm{m}}\mathbf{X} = \mathbf{I}$):

$$\mathbf{P}_{\mathrm{m}} = (\nabla_{0\mathrm{m}} \cdot \mathbf{P}_{\mathrm{m}}^{c})\mathbf{X} + \mathbf{P}_{\mathrm{m}} \cdot (\nabla_{0\mathrm{m}}\mathbf{X}) = \nabla_{0\mathrm{m}} \cdot (\mathbf{P}_{\mathrm{m}}^{c}\mathbf{X}). \tag{13}$$

Substitution of (13) into (12), application of the divergence theorem, and the definition of the first Piola–Kirchhoff stress vector $\mathbf{p} = \mathbf{N} \cdot \mathbf{P}_{\mathrm{m}}^{c}$ give

$$\mathbf{P}_{\mathrm{M}} = \frac{1}{V_0}\int_{V_0} \nabla_{0\mathrm{m}} \cdot (\mathbf{P}_{\mathrm{m}}^{c}\mathbf{X})\,\mathrm{d}V_0 = \frac{1}{V_0}\int_{\Gamma_0} \mathbf{N} \cdot \mathbf{P}_{\mathrm{m}}^{c}\mathbf{X}\,\mathrm{d}\Gamma_0 = \frac{1}{V_0}\int_{\Gamma_0} \mathbf{p}\mathbf{X}\,\mathrm{d}\Gamma_0. \tag{14}$$

Now it is a trivial task to validate that substitution of the traction boundary conditions (4_2) into this equation leads to an identity.

The volume average of the microscopic Cauchy stress tensor $\boldsymbol{\sigma}_{\mathrm{m}}$ over the current RVE volume V can be elaborated similarly to (14)

$$\boldsymbol{\sigma}_{\mathrm{M}}^{*} = \frac{1}{V}\int_{V} \boldsymbol{\sigma}_{\mathrm{m}}\,\mathrm{d}V = \frac{1}{V}\int_{\Gamma} \mathbf{t}\mathbf{x}\,\mathrm{d}\Gamma. \tag{15}$$

Just as it is the case for kinematic quantities, the usual continuum mechanics relation between stress measures (e.g. the Cauchy and the first Piola–Kirchhoff stress tensors) is, in general, not valid for the volume averages of the microstructural counterparts $\boldsymbol{\sigma}^*_{\mathrm{M}} \neq \mathbf{P}_{\mathrm{M}} \cdot \mathbf{F}^c_{\mathrm{M}}/\det(\mathbf{F}_{\mathrm{M}})$. However, the Cauchy stress tensor on the macrolevel should be defined as

$$\boldsymbol{\sigma}_{\mathrm{M}} = \frac{1}{\det(\mathbf{F}_{\mathrm{M}})} \mathbf{P}_{\mathrm{M}} \cdot \mathbf{F}^c_{\mathrm{M}}. \tag{16}$$

Clearly, there is some arbitrariness in the choice of associated deformation and stress quantities, whose macroscopic measures are obtained as a volume average of their microscopic counterparts. The remaining macroscopic measures are then expressed in terms of these averaged quantities using the standard continuum mechanics relations. The specific selection should be made with care and based on experimental results and convenience of the implementation. The actual choice of the "primary" averaging measures: the deformation gradient tensor $\mathbf{F}$ and the first Piola–Kirchhoff stress tensor $\mathbf{P}$ (and their rates) has been advocated in the literature[34,49,50] (in the last two references the nominal stress $\mathbf{S}_N = \det(\mathbf{F})\mathbf{F}^{-1} \cdot \boldsymbol{\sigma} = \mathbf{P}^c$ has been used). This particular choice is motivated by the fact that these two measures are work conjugated, combined with the observation that their volume averages can exclusively be defined in terms of the microstructural quantities on the RVE boundary only. This feature will be used in the following section, where the averaging theorem for the micro–macro energy transition is discussed.

4.3. *Internal work*

The energy averaging theorem, known in the literature as the Hill–Mandel condition or macrohomogeneity condition,[29,48] requires that the macroscopic volume average of the variation of work performed on the RVE is equal to the local variation of the work on the macroscale. Formulated in terms of a work conjugated set, i.e. the deformation gradient tensor and the first Piola–Kirchhoff stress tensor, the Hill–Mandel condition reads

$$\frac{1}{V_0} \int_{V_0} \mathbf{P}_{\mathrm{m}} : \delta \mathbf{F}^c_{\mathrm{m}} \, \mathrm{d}V_0 = \mathbf{P}_{\mathrm{M}} : \delta \mathbf{F}^c_{\mathrm{M}} \quad \forall \delta \mathbf{x}. \tag{17}$$

The averaged microstructural work in the left-hand side of (17) may be expressed in terms of RVE surface quantities

$$\delta W_{0\mathrm{M}} = \frac{1}{V_0} \int_{V_0} \mathbf{P}_{\mathrm{m}} : \delta \mathbf{F}^c_{\mathrm{m}} \, \mathrm{d}V_0 = \frac{1}{V_0} \int_{\Gamma_0} \mathbf{p} \cdot \delta \mathbf{x} \, \mathrm{d}\Gamma_0, \tag{18}$$

where the relation (with account for microstructural equilibrium)

$$\mathbf{P}_{\mathrm{m}} : \nabla_{0\mathrm{m}} \delta \mathbf{x} = \nabla_{0\mathrm{m}} \cdot (\mathbf{P}^c_{\mathrm{m}} \cdot \delta \mathbf{x}) - (\nabla_{0\mathrm{m}} \cdot \mathbf{P}^c_{\mathrm{m}}) \cdot \delta \mathbf{x} = \nabla_{0\mathrm{m}} \cdot (\mathbf{P}^c_{\mathrm{m}} \cdot \delta \mathbf{x}),$$

and the divergence theorem have been used.

Now it is easy to verify that the three types of boundary conditions: prescribed displacements (3), prescribed tractions (4), or the periodicity conditions (5) and (6) all satisfy the Hill–Mandel condition *a priori*, if the averaging relations for the deformation gradient tensor (7) and for the first Piola–Kirchhoff stress tensor (12) are adopted. In the case of the prescribed displacements (3), substitution of the variation of the boundary position vectors $\delta \mathbf{x} = \delta \mathbf{F}_{\mathrm{M}} \cdot \mathbf{X}$ into the expression for the averaged microwork (18) with incorporation of (14) gives

$$\delta W_{0\mathrm{M}} = \frac{1}{V_0} \int_{\Gamma_0} \mathbf{p} \cdot (\delta \mathbf{F}_{\mathrm{M}} \cdot \mathbf{X}) \, \mathrm{d}\Gamma_0 = \frac{1}{V_0} \int_{\Gamma_0} \mathbf{p} \mathbf{X} \, \mathrm{d}\Gamma_0 : \delta \mathbf{F}^c_{\mathrm{M}} = \mathbf{P}_{\mathrm{M}} : \delta \mathbf{F}^c_{\mathrm{M}}. \tag{19}$$

Similarly, substitution of the traction boundary condition (4) into (18), with account for the variation of the macroscopic deformation gradient tensor obtained by varying relation (7), leads to

$$\delta W_{0\mathrm{M}} = \frac{1}{V_0} \int_{\Gamma_0} (\mathbf{N} \cdot \mathbf{P}^c_{\mathrm{M}}) \cdot \delta \mathbf{x} \, \mathrm{d}\Gamma_0 = \mathbf{P}_{\mathrm{M}} : \frac{1}{V_0} \int_{\Gamma_0} \mathbf{N} \delta \mathbf{x} \, \mathrm{d}\Gamma_0 = \mathbf{P}_{\mathrm{M}} : \delta \mathbf{F}^c_{\mathrm{M}}. \tag{20}$$

Finally, for the periodic boundary conditions (5) and (6),

$$\begin{aligned}
\delta W_{0\mathrm{M}} &= \frac{1}{V_0} \left\{ \int_{\Gamma_0^+} \mathbf{p}^+ \cdot \delta \mathbf{x}^+ \, \mathrm{d}\Gamma_0 + \int_{\Gamma_0^-} \mathbf{p}^- \cdot \delta \mathbf{x}^- \, \mathrm{d}\Gamma_0 \right\} \\
&= \frac{1}{V_0} \int_{\Gamma_0^+} \mathbf{p}^+ \cdot (\delta \mathbf{x}^+ - \delta \mathbf{x}^-) \, \mathrm{d}\Gamma_0 \\
&= \frac{1}{V_0} \int_{\Gamma_0} \mathbf{p}^+ (\mathbf{X}^+ - \mathbf{X}^-) \, \mathrm{d}\Gamma_0 : \delta \mathbf{F}^c_{\mathrm{M}} \\
&= \frac{1}{V_0} \int_{\Gamma_0} \mathbf{p} \mathbf{X} \, \mathrm{d}\Gamma_0 : \delta \mathbf{F}^c_{\mathrm{M}} = \mathbf{P}_{\mathrm{M}} : \delta \mathbf{F}^c_{\mathrm{M}}.
\end{aligned} \tag{21}$$

5. FE Implementation

5.1. *RVE boundary value problem*

The RVE problem to be solved is a standard non-linear quasi-static BVP with kinematic boundary conditions.[a] Thus, any numerical technique suitable for solution of this type of problems may be used. In the following, the FEM will be adopted. Following the standard FE procedure for the microlevel RVE, after discretisation, the weak form of equilibrium (1) with account for the constitutive relations (2) leads to a system of non-linear algebraic equations in the unknown nodal displacements $\underset{\sim}{u}$:

$$\underset{\sim}{f}_{\text{int}}(\underset{\sim}{u}) = \underset{\sim}{f}_{\text{ext}}, \tag{22}$$

expressing the balance of internal and external nodal forces. This system has to be completed by boundary conditions. Hence, the earlier introduced boundary conditions (3) or (5) have to be elaborated in more detail.

5.1.1. *Fully prescribed boundary displacements*

In the case of the fully prescribed displacement boundary conditions (3), the displacements of all nodes on the boundary are simply given by

$$\mathbf{u}_p = (\mathbf{F}_{\text{M}} - \mathbf{I}) \cdot \mathbf{X}_p, \qquad p = \overline{1, N_p} \tag{23}$$

where N_p is the number of prescribed nodes, which in this case equals the number of boundary nodes. The boundary conditions (23) are added to the system (22) in a standard manner by static condensation, Lagrange multipliers, or penalty functions.

5.1.2. *Periodic boundary conditions*

Before application of the periodic boundary conditions (5), they have to be rewritten into a format more suitable for the FE framework. Consider a two-dimensional periodic RVE schematically depicted in Fig. 4. The boundary of this RVE can be split into four parts, here denoted as "T" top, "B" bottom, "R" right and "L" left. For the following it is supposed that the FE discretization is performed such that the distribution of nodes on opposite RVE edges is equal. During the initial periodicity of the RVE, for every respective pair of nodes on the top–bottom and right–left boundaries it is

[a] The traction boundary conditions are not considered in the following, as they do not fit into the deformation-driven procedure, as has been discussed above.

valid in the reference configuration

$$\begin{aligned} \mathbf{X}_{\mathrm{T}} - \mathbf{X}_{\mathrm{B}} &= \mathbf{X}_4 - \mathbf{X}_1, \\ \mathbf{X}_{\mathrm{R}} - \mathbf{X}_{\mathrm{L}} &= \mathbf{X}_2 - \mathbf{X}_1, \end{aligned} \tag{24}$$

where $\mathbf{X}_p, p = 1, 2, 4$ are the position vectors of the corner nodes 1, 2, and 4 in the undeformed state. Then by considering pairs of corresponding nodes on the opposite boundaries, (5) can be written as

$$\begin{aligned} \mathbf{x}_{\mathrm{T}} - \mathbf{x}_{\mathrm{B}} &= \mathbf{F}_{\mathrm{M}} \cdot (\mathbf{X}_4 - \mathbf{X}_1), \\ \mathbf{x}_{\mathrm{R}} - \mathbf{x}_{\mathrm{L}} &= \mathbf{F}_{\mathrm{M}} \cdot (\mathbf{X}_2 - \mathbf{X}_1). \end{aligned} \tag{25}$$

Now if the position vectors of the corner nodes in the deformed state are prescribed according to

$$\mathbf{x}_p = \mathbf{F}_{\mathrm{M}} \cdot \mathbf{X}_p, \quad p = 1, 2, 4 \tag{26}$$

then the periodic boundary conditions may be rewritten as

$$\begin{aligned} \mathbf{x}_{\mathrm{T}} &= \mathbf{x}_{\mathrm{B}} + \mathbf{x}_4 - \mathbf{x}_1, \\ \mathbf{x}_{\mathrm{R}} &= \mathbf{x}_{\mathrm{L}} + \mathbf{x}_2 - \mathbf{x}_1. \end{aligned} \tag{27}$$

Since these conditions are trivially satisfied in the undeformed configuration (cf. relation (24)), they may be formulated in terms of displacements

$$\begin{aligned} \mathbf{u}_{\mathrm{T}} &= \mathbf{u}_{\mathrm{B}} + \mathbf{u}_4 - \mathbf{u}_1, \\ \mathbf{u}_{\mathrm{R}} &= \mathbf{u}_{\mathrm{L}} + \mathbf{u}_2 - \mathbf{u}_1, \end{aligned} \tag{28}$$

and

$$\mathbf{u}_p = (\mathbf{F}_{\mathrm{M}} - \mathbf{I}) \cdot \mathbf{X}_p, \quad p = 1, 2, 4. \tag{29}$$

In a discretised format the relations (28) lead to a set of homogeneous constraints of the type

$$\underset{\sim}{C}_a \underset{\sim}{u}_a = \underset{\sim}{0}, \tag{30}$$

where $\underset{\sim}{C}_a$ is a matrix containing coefficients in the constraint relations and $\underset{\sim}{u}_a$ is a column with the degrees of freedom involved in the constraints. Procedures for imposing constraints (30) include the direct elimination of the dependent degrees of freedom from the system of equations, or the use of Lagrange multipliers or penalty functions. In the following, constraints (30) are enforced by elimination of the dependent degrees of freedom. Although such a procedure may be found in many works on finite elements,[51] here it is summarised for the sake of completeness and also in the context of the derivation of the macroscopic tangent stiffness, which will be presented in Sec. 5.3.

First, (30) is partitioned according to

$$[\underline{C}_i \quad \underline{C}_d]\begin{bmatrix} \underset{\sim}{u}_i \\ \underset{\sim}{u}_d \end{bmatrix} = \underset{\sim}{0}, \tag{31}$$

where $\underset{\sim}{u}_i$ are the independent degrees of freedom (to be retained in the system) and $\underset{\sim}{u}_d$ are the dependent degrees of freedom (to be eliminated from the system). Because there are as many dependent degrees of freedom $\underset{\sim}{u}_d$ as there are independent constraint equations in (31), matrix $\underline{C}_d$ is square and non-singular. Solution for $\underset{\sim}{u}_d$ yields

$$\underset{\sim}{u}_d = \underline{C}_{di}\underset{\sim}{u}_i, \quad \text{with } \underline{C}_{di} = -\underline{C}_d^{-1}\underline{C}_i. \tag{32}$$

This relation may be further rewritten as

$$\begin{bmatrix} \underset{\sim}{u}_i \\ \underset{\sim}{u}_d \end{bmatrix} = \underline{T}\underset{\sim}{u}_i, \quad \text{with } \underline{T} = \begin{bmatrix} \underline{I} \\ \underline{C}_{di} \end{bmatrix}, \tag{33}$$

where $\underline{I}$ is a unit matrix of size $[N_i \times N_i]$, with N_i being the number of independent degrees of freedom.

With the transformation matrix $\underline{T}$ defined such that $\underset{\sim}{d} = \underline{T}\underset{\sim}{d}'$, the common transformations $\underset{\sim}{r}' = \underline{T}^T\underset{\sim}{r}$ and $\underline{K}' = \underline{T}^T\underline{K}\underline{T}$ can be applied to a linear system of equations of the form $\underline{K}\underset{\sim}{d} = \underset{\sim}{r}$, leading to a new system $\underline{K}'\underset{\sim}{d}' = \underset{\sim}{r}'$.

The standard linearisation of the non-linear system of equations (22) leads to a linear system in the iterative corrections $\delta\underset{\sim}{u}$ to the current estimate $\underset{\sim}{u}$. This system may be partitioned as

$$\begin{bmatrix} \underline{K}_{ii} & \underline{K}_{id} \\ \underline{K}_{di} & \underline{K}_{dd} \end{bmatrix}\begin{bmatrix} \delta\underset{\sim}{u}_i \\ \delta\underset{\sim}{u}_d \end{bmatrix} = \begin{bmatrix} \delta\underset{\sim}{r}_i \\ \delta\underset{\sim}{r}_d \end{bmatrix}, \tag{34}$$

with the residual nodal forces at the right-hand side. Noting that all constraint equations considered above are linear, and thus their linearisation is straightforward, application of the transformation (33) to the system (34) gives

$$[\underline{K}_{ii} + \underline{K}_{id}\underline{C}_{di} + \underline{C}_{di}^{\mathrm{T}}\underline{K}_{di} + \underline{C}_{di}^{\mathrm{T}}\underline{K}_{dd}\underline{C}_{di}]\delta\underset{\sim}{u}_i = [\delta\underset{\sim}{r}_i + \underline{C}_{di}^{\mathrm{T}}\delta\underset{\sim}{r}_d]. \tag{35}$$

Note that the boundary conditions (29) prescribing displacements of the corner nodes have not yet been applied. The column of "independent" degrees of freedom $\underset{\sim}{u}_i$ includes the prescribed corner nodes $\underset{\sim}{u}_p$ among other nodes. The boundary conditions (29) should be applied to the system (35) in a standard manner.

The condition of antiperiodic tractions (6) will be addressed in Sec. 5.2.2.

5.2. *Calculation of the macroscopic stress*

After the analysis of a microstructural RVE is completed, the RVE-averaged stresses have to be extracted. The macroscopic stress tensor can be calculated by numerically evaluating the volume integral (12). However, it is computationally more efficient to compute the surface integral (14), which can be further simplified for the case of the periodic boundary conditions.

5.2.1. *Fully prescribed boundary displacements*

For the case of prescribed displacement boundary conditions, the surface integral (14) simply leads to

$$\mathbf{P}_{\mathrm{M}} = \frac{1}{V_0} \sum_{p=1}^{N_p} \mathbf{f}_p \mathbf{X}_p, \tag{36}$$

where $\mathbf{f}_p$ are the resulting external forces at the boundary nodes; $\mathbf{X}_p$ are the position vectors of these nodes in the undeformed state; and N_p is the number of the nodes on the boundary.

5.2.2. *Periodic boundary conditions*

In order to simplify the surface integral (14) for the case of periodic boundary conditions, consider all forces acting on the RVE boundary subjected to the boundary conditions according to (28) and (29). At the three prescribed corner nodes, the resulting external forces $\mathbf{f}_p^{\mathrm{e}}, p = 1, 2, 4$ act. Additionally, there are forces involved in every constraint (tying) relation (28). For example, for each constraint relation between pairs of the nodes on the bottom–top boundaries there are a tying force at the node on the bottom boundary $\mathbf{p}_{\mathrm{B}}^{\mathrm{t}}$, a tying force at the node on the top boundary $\mathbf{p}_{\mathrm{T}}^{\mathrm{t}}$, and tying forces at the corner nodes 1 and 4, $\mathbf{p}_1^{\mathrm{t_B}}$ and $\mathbf{p}_4^{\mathrm{t_B}}$, respectively. Similarly, there are forces $\mathbf{p}_{\mathrm{L}}^{\mathrm{t}}$, $\mathbf{p}_{\mathrm{R}}^{\mathrm{t}}$, $\mathbf{p}_1^{\mathrm{t_L}}$ and $\mathbf{p}_2^{\mathrm{t_L}}$ corresponding to the left–right constraints. All these forces are schematically shown in Fig. 5.

Each constraint relation satisfies the condition of zero virtual work; thus,

$$\begin{aligned} \mathbf{p}_{\mathrm{B}}^{\mathrm{t}} \cdot \delta\mathbf{x}_{\mathrm{B}} + \mathbf{p}_{\mathrm{T}}^{\mathrm{t}} \cdot \delta\mathbf{x}_{\mathrm{T}} + \mathbf{p}_1^{\mathrm{t_B}} \cdot \delta\mathbf{x}_1 + \mathbf{p}_4^{\mathrm{t_B}} \cdot \delta\mathbf{x}_4 = 0, \\ \mathbf{p}_{\mathrm{L}}^{\mathrm{t}} \cdot \delta\mathbf{x}_{\mathrm{L}} + \mathbf{p}_{\mathrm{R}}^{\mathrm{t}} \cdot \delta\mathbf{x}_{\mathrm{R}} + \mathbf{p}_1^{\mathrm{t_L}} \cdot \delta\mathbf{x}_1 + \mathbf{p}_2^{\mathrm{t_L}} \cdot \delta\mathbf{x}_2 = 0. \end{aligned} \tag{37}$$

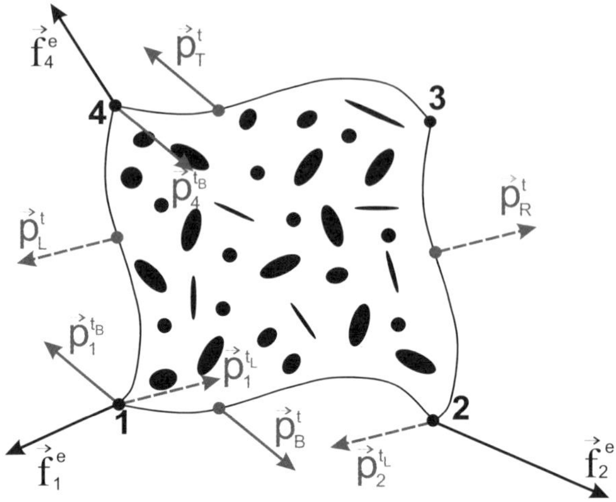

Fig. 5. Schematic representation of the forces acting on the boundary of a two-dimensional RVE subjected to periodic boundary conditions.

Substitution of the variation of the constraints (27) into (37) gives

$$\begin{aligned}
&(\mathbf{p}_\mathrm{B}^\mathrm{t} + \mathbf{p}_\mathrm{T}^\mathrm{t}) \cdot \delta\mathbf{x}_\mathrm{B} + (\mathbf{p}_1^{\mathrm{t_B}} - \mathbf{p}_\mathrm{T}^\mathrm{t}) \cdot \delta\mathbf{x}_1 + (\mathbf{p}_\mathrm{T}^\mathrm{t} + \mathbf{p}_4^{\mathrm{t_B}}) \cdot \delta\mathbf{x}_4 = 0,\\
&(\mathbf{p}_\mathrm{L}^\mathrm{t} + \mathbf{p}_\mathrm{R}^\mathrm{t}) \cdot \delta\mathbf{x}_\mathrm{L} + (\mathbf{p}_1^{\mathrm{t_L}} - \mathbf{p}_\mathrm{R}^\mathrm{t}) \cdot \delta\mathbf{x}_1 + (\mathbf{p}_\mathrm{R}^\mathrm{t} + \mathbf{p}_2^{\mathrm{t_L}}) \cdot \delta\mathbf{x}_2 = 0.
\end{aligned} \tag{38}$$

These relations should hold for any $\delta\mathbf{x}_\mathrm{B}$, $\delta\mathbf{x}_\mathrm{L}$, $\delta\mathbf{x}_1$, $\delta\mathbf{x}_2$, $\delta\mathbf{x}_4$; therefore,

$$\begin{aligned}
\mathbf{p}_\mathrm{B}^\mathrm{t} &= -\mathbf{p}_\mathrm{T}^\mathrm{t} = -\mathbf{p}_1^{\mathrm{t_B}} = \mathbf{p}_4^{\mathrm{t_B}},\\
\mathbf{p}_\mathrm{L}^\mathrm{t} &= -\mathbf{p}_\mathrm{R}^\mathrm{t} = -\mathbf{p}_1^{\mathrm{t_L}} = \mathbf{p}_2^{\mathrm{t_L}}.
\end{aligned} \tag{39}$$

Note that (39) reflects antiperiodicity of tying forces on the opposite boundaries, thus, the condition (6) is indeed satisfied.

With account for all forces acting on the RVE boundary, the surface integral (14) is written as

$$\begin{aligned}
\mathbf{P}_\mathrm{M} = \frac{1}{V_0}\Bigg(&\mathbf{f}_1^\mathrm{e}\mathbf{X}_1 + \mathbf{f}_2^\mathrm{e}\mathbf{X}_2 + \mathbf{f}_4^\mathrm{e}\mathbf{X}_4 + \int_{\Gamma_{0\mathrm{B}}} \mathbf{p}_\mathrm{B}^\mathrm{t}\mathbf{X}_\mathrm{B}\,\mathrm{d}\Gamma_0 + \int_{\Gamma_{0\mathrm{T}}} \mathbf{p}_\mathrm{T}^\mathrm{t}\mathbf{X}_\mathrm{T}\,\mathrm{d}\Gamma_0\\
&+ \int_{\Gamma_{0\mathrm{L}}} \mathbf{p}_\mathrm{L}^\mathrm{t}\mathbf{X}_\mathrm{L}\,\mathrm{d}\Gamma_0 + \int_{\Gamma_{0\mathrm{R}}} \mathbf{p}_\mathrm{R}^\mathrm{t}\mathbf{X}_\mathrm{R}\,\mathrm{d}\Gamma_0 + \left(\int_{\Gamma_{0\mathrm{B}}} \mathbf{p}_1^{\mathrm{t_B}}\,\mathrm{d}\Gamma_0\right)\mathbf{X}_1\\
&+ \left(\int_{\Gamma_{0\mathrm{L}}} \mathbf{p}_1^{\mathrm{t_L}}\,\mathrm{d}\Gamma_0\right)\mathbf{X}_1 + \left(\int_{\Gamma_{0\mathrm{B}}} \mathbf{p}_4^{\mathrm{t_B}}\,\mathrm{d}\Gamma_0\right)\mathbf{X}_4 + \left(\int_{\Gamma_{0\mathrm{L}}} \mathbf{p}_2^{\mathrm{t_L}}\,\mathrm{d}\Gamma_0\right)\mathbf{X}_2\Bigg).
\end{aligned} \tag{40}$$

Making use of the relation between tying forces (39) gives

$$\mathbf{P}_{\mathrm{M}} = \frac{1}{V_0}\Bigg(\sum_{p=1,2,4} \mathbf{f}_p^{\mathrm{e}}\mathbf{X}_p + \int_{\Gamma_{0\mathrm{B}}} \mathbf{p}_{\mathrm{B}}^{\mathrm{t}}(\mathbf{X}_{\mathrm{B}} - \mathbf{X}_{\mathrm{T}})\,\mathrm{d}\Gamma_0 + \int_{\Gamma_{0\mathrm{L}}} \mathbf{p}_{\mathrm{L}}^{\mathrm{t}}(\mathbf{X}_{\mathrm{L}} - \mathbf{X}_{\mathrm{R}})\,\mathrm{d}\Gamma_0$$
$$+ \left(\int_{\Gamma_{0\mathrm{B}}} \mathbf{p}_1^{\mathrm{t_B}}\,\mathrm{d}\Gamma_0\right)\mathbf{X}_1 + \left(\int_{\Gamma_{0\mathrm{L}}} \mathbf{p}_1^{\mathrm{t_L}}\,\mathrm{d}\Gamma_0\right)\mathbf{X}_1$$
$$+ \left(\int_{\Gamma_{0\mathrm{B}}} \mathbf{p}_4^{\mathrm{t_B}}\,\mathrm{d}\Gamma_0\right)\mathbf{X}_4 + \left(\int_{\Gamma_{0\mathrm{L}}} \mathbf{p}_2^{\mathrm{t_L}}\,\mathrm{d}\Gamma_0\right)\mathbf{X}_2\Bigg). \tag{41}$$

Inserting the conditions of the initial periodicity of the RVE (24) results in

$$\mathbf{P}_{\mathrm{M}} = \frac{1}{V_0}\Bigg(\sum_{p=1,2,4} \mathbf{f}_p^{\mathrm{e}}\mathbf{X}_p + \int_{\Gamma_{0\mathrm{B}}} (\mathbf{p}_{\mathrm{B}}^{\mathrm{t}} + \mathbf{p}_1^{\mathrm{t_B}})\mathbf{X}_1\,\mathrm{d}\Gamma_0 + \int_{\Gamma_{0\mathrm{L}}} (\mathbf{p}_{\mathrm{L}}^{\mathrm{t}} + \mathbf{p}_1^{\mathrm{t_L}})\mathbf{X}_1\,\mathrm{d}\Gamma_0$$
$$+ \int_{\Gamma_{0\mathrm{B}}} (\mathbf{p}_4^{\mathrm{t_B}} - \mathbf{p}_{\mathrm{B}}^{\mathrm{t}})\mathbf{X}_4\,\mathrm{d}\Gamma_0 + \int_{\Gamma_{0\mathrm{L}}} (\mathbf{p}_2^{\mathrm{t_L}} - \mathbf{p}_{\mathrm{L}}^{\mathrm{t}})\mathbf{X}_2\,\mathrm{d}\Gamma_0\Bigg), \tag{42}$$

which after substitution of the remaining relations between tying forces (39) gives

$$\mathbf{P}_{\mathrm{M}} = \frac{1}{V_0}\sum_{p=1,2,4} \mathbf{f}_p^{\mathrm{e}}\mathbf{X}_p. \tag{43}$$

Therefore, when the periodic boundary conditions are used, all terms with forces involved in the periodicity constraints cancel out from the boundary integral (14) and the only contribution left is by the external forces at the three prescribed corner nodes.

5.3. *Macroscopic tangent stiffness*

When the micro–macro approach is implemented within the framework of a non-linear FE code, the stiffness matrix at every macroscopic integration point is required. Because in the computational homogenisation approach there is no explicit form of the constitutive behaviour on the macrolevel assumed *a priori*, the stiffness matrix has to be determined numerically from the relation between variations of the macroscopic stress and variations of the macroscopic deformation at such a point. This may be realised by numerical differentiation of the numerical macroscopic stress–strain relation, for example, using a forward difference approximation.[52] Another approach is to condense the microstructural stiffness to the local macroscopic stiffness. This is achieved by reducing the total RVE system of equations to the relation between the forces acting on the RVE boundary and the associated boundary displacements. Elaboration of such a procedure in combination with the Lagrange multiplier method to impose boundary constraints can be found in the literature.[41] Here an alternative scheme,[40,42] which employs the direct condensation of the constrained degrees of freedom, will be considered. After the condensed microscopic stiffness relating the prescribed displacement and force variations is obtained, it needs to be transformed to arrive at an expression relating variations of the macroscopic stress and deformation tensors, typically used in the FE codes. These two steps are elaborated in the following.

5.3.1. *Condensation of the microscopic stiffness: Fully prescribed boundary displacements*

First the total microstructural system of equations (in its linearised form) is partitioned as

$$\begin{bmatrix} \underline{K}_{pp} & \underline{K}_{pf} \\ \underline{K}_{fp} & \underline{K}_{ff} \end{bmatrix} \begin{bmatrix} \delta \underset{\sim}{u}_p \\ \delta \underset{\sim}{u}_f \end{bmatrix} = \begin{bmatrix} \delta \underset{\sim}{f}_p \\ \underset{\sim}{0} \end{bmatrix}, \tag{44}$$

where $\delta \underset{\sim}{u}_p$ and $\delta \underset{\sim}{f}_p$ are the columns with iterative displacements and external forces of the boundary nodes, respectively; $\delta \underset{\sim}{u}_f$ is the column with the iterative displacements of the remaining (interior) nodes; and $\underline{K}_{pp}$, $\underline{K}_{pf}$, $\underline{K}_{fp}$ and $\underline{K}_{ff}$ are the corresponding partitions of the total RVE stiffness matrix. The stiffness matrix in the formulation (44) is taken at the end of a microstructural increment, where a converged state is reached. Elimination of $\delta \underset{\sim}{u}_f$ from (44) leads to the reduced stiffness matrix $\underline{K}_{\mathrm{M}}$ relating boundary displacement variations to boundary force variations

$$\underline{K}_{\mathrm{M}} \delta \underset{\sim}{u}_p = \delta \underset{\sim}{f}_p \quad \text{with } \underline{K}_{\mathrm{M}} = \underline{K}_{pp} - \underline{K}_{pf} (\underline{K}_{ff})^{-1} \underline{K}_{fp}. \tag{45}$$

5.3.2. *Condensation of the microscopic stiffness: Periodic boundary conditions*

In the case of the periodic boundary conditions, the point of departure is the microscopic system of equations (35) from which the dependent degrees of freedom have been eliminated (as described in Sec. 5.1.2)

$$\underline{K}^{\star} \delta \underset{\sim}{u}_i = \delta \underset{\sim}{r}^{\star}, \tag{46}$$

with

$$\begin{aligned} \underline{K}^{\star} &= \underline{K}_{ii} + \underline{K}_{id} \underline{C}_{di} + \underline{C}_{di}^T \underline{K}_{di} + \underline{C}_{di}^T \underline{K}_{dd} \underline{C}_{di}, \\ \delta \underset{\sim}{r}^{\star} &= \delta \underset{\sim}{r}_i + \underline{C}_{di}^T \delta \underset{\sim}{r}_d. \end{aligned}$$

Next, system (46) is further split, similarly to (44), into the parts corresponding to the variations of the prescribed degrees of freedom $\delta \underset{\sim}{u}_p$ (which in this case are the varied positions of the three corner nodes prescribed according to (29)), variations of the external forces at these prescribed nodes denoted by $\delta \underset{\sim}{f}_p^{\star}$, and the remaining (free) displacement variations $\delta \underset{\sim}{u}_f$:

$$\begin{bmatrix} \underline{K}_{pp}^{\star} & \underline{K}_{pf}^{\star} \\ \underline{K}_{fp}^{\star} & \underline{K}_{ff}^{\star} \end{bmatrix} \begin{bmatrix} \delta \underset{\sim}{u}_p \\ \delta \underset{\sim}{u}_f \end{bmatrix} = \begin{bmatrix} \delta \underset{\sim}{f}_p^{\star} \\ \underset{\sim}{0} \end{bmatrix}. \tag{47}$$

Then the reduced stiffness matrix $\underline{K}_{\mathrm{M}}^{\star}$ in the case of periodic boundary conditions is obtained as

$$\underline{K}_{\mathrm{M}}^{\star} \delta \underset{\sim}{u}_p = \delta \underset{\sim}{f}_p^{\star}, \quad \text{with } \underline{K}_{\mathrm{M}}^{\star} = \underline{K}_{pp}^{\star} - \underline{K}_{pf}^{\star} (\underline{K}_{ff}^{\star})^{-1} \underline{K}_{fp}^{\star}. \tag{48}$$

Note that in the two-dimensional case $\underline{K}_{\mathrm{M}}^{\star}$ is only $[6 \times 6]$ matrix and in three-dimensional case $[12 \times 12]$ matrix.

5.3.3. *Macroscopic tangent*

Finally, the resulting relation between displacement and force variations (relation (45) if prescribed displacement boundary conditions are used, or relation (48) if periodicity conditions are employed) needs to be transformed to arrive at an expression relating variations of the macroscopic stress and deformation tensors:

$$\delta \mathbf{P}_{\mathrm{M}} = {}^4\mathbf{C}_{\mathrm{M}}^{\mathrm{P}} : \delta \mathbf{F}_{\mathrm{M}}^{c}, \tag{49}$$

where the fourth-order tensor ${}^4\mathbf{C}_{\mathrm{M}}^{\mathrm{P}}$ represents the required consistent tangent stiffness at the macroscopic integration point level.

In order to obtain this constitutive tangent from the reduced stiffness matrix $\underline{K}_{\mathrm{M}}$ (or $\underline{K}_{\mathrm{M}}^{\star}$), first relations (45) and (48) are rewritten in a specific vector/tensor format

$$\sum_j \mathbf{K}_{\mathrm{M}}^{(ij)} \cdot \delta \mathbf{u}_{(j)} = \delta \mathbf{f}_{(i)}, \tag{50}$$

where indices i and j take the values $i, j = \overline{1, N_p}$ for prescribed displacement boundary conditions (N_p is the number of boundary nodes) and $i, j = 1, 2, 4$ for the periodic boundary conditions. In (50) the components of the tensors $\mathbf{K}_{\mathrm{M}}^{(ij)}$ are simply found in the tangent matrix $\underline{K}_{\mathrm{M}}$ (for displacement boundary conditions) or in the matrix $\underline{K}_{\mathrm{M}}^{\star}$ (for periodic boundary conditions) at the rows and columns of the degrees of freedom in the nodes i and j.

For example, for the case of the periodic boundary conditions the total matrix $\underline{K}^{\star}_{\mathrm{M}}$ has the format

$$\underline{K}^{\star}_{\mathrm{M}} = \begin{bmatrix} \begin{bmatrix} K^{(11)}_{11} & K^{(11)}_{12} \\ K^{(11)}_{21} & K^{(11)}_{22} \end{bmatrix} & \begin{bmatrix} K^{(12)}_{11} & K^{(12)}_{12} \\ K^{(12)}_{21} & K^{(12)}_{22} \end{bmatrix} & \begin{bmatrix} K^{(14)}_{11} & K^{(14)}_{12} \\ K^{(14)}_{21} & K^{(14)}_{22} \end{bmatrix} \\ \begin{bmatrix} K^{(21)}_{11} & K^{(21)}_{12} \\ K^{(21)}_{21} & K^{(21)}_{22} \end{bmatrix} & \begin{bmatrix} K^{(22)}_{11} & K^{(22)}_{12} \\ K^{(22)}_{21} & K^{(22)}_{22} \end{bmatrix} & \begin{bmatrix} K^{(24)}_{11} & K^{(24)}_{12} \\ K^{(24)}_{21} & K^{(24)}_{22} \end{bmatrix} \\ \begin{bmatrix} K^{(41)}_{11} & K^{(41)}_{12} \\ K^{(41)}_{21} & K^{(41)}_{22} \end{bmatrix} & \begin{bmatrix} K^{(42)}_{11} & K^{(42)}_{12} \\ K^{(42)}_{21} & K^{(42)}_{22} \end{bmatrix} & \begin{bmatrix} K^{(44)}_{11} & K^{(44)}_{12} \\ K^{(44)}_{21} & K^{(44)}_{22} \end{bmatrix} \end{bmatrix}, \tag{51}$$

where the superscripts in round brackets refer to the nodes and the subscripts to the degrees of freedom at those nodes. Then each submatrix in (51) may be considered as the representation of a second-order tensor $\mathbf{K}^{(ij)}_{\mathrm{M}}$.

Next, the expression for the variation of the nodal forces (50) is substituted into the relation for the variation of the macroscopic stress following from (36) or (43)

$$\delta\mathbf{P}_{\mathrm{M}} = \frac{1}{V_0}\sum_i\sum_j(\mathbf{K}^{(ij)}_{\mathrm{M}} \cdot \delta\mathbf{u}_{(j)})\mathbf{X}_{(i)}. \tag{52}$$

Substitution of the equation $\delta\mathbf{u}_{(j)} = \mathbf{X}_{(j)} \cdot \delta\mathbf{F}^c_{\mathrm{M}}$ into (52) gives

$$\delta\mathbf{P}_{\mathrm{M}} = \frac{1}{V_0}\sum_i\sum_j(\mathbf{X}_{(i)}\mathbf{K}^{(ij)}_{\mathrm{M}}\mathbf{X}_{(j)})^{\mathrm{LC}} : \delta\mathbf{F}^c_{\mathrm{M}}, \tag{53}$$

where the superscript LC denotes left conjugation, which for a fourth-order tensor ${}^4\mathbf{T}$ is defined as $T^{\mathrm{LC}}_{ijkl} = T_{jikl}$. Finally, by comparing (53) with (49) the consistent constitutive tangent is identified as

$${}^4\mathbf{C}^{\mathrm{P}}_{\mathrm{M}} = \frac{1}{V_0}\sum_i\sum_j(\mathbf{X}_{(i)}\mathbf{K}^{(ij)}_{\mathrm{M}}\mathbf{X}_{(j)})^{\mathrm{LC}}. \tag{54}$$

If the macroscopic FE scheme requires the constitutive tangent relating the variation of the macroscopic Cauchy stress to the variation of the macroscopic deformation gradient tensor according to

$$\delta\boldsymbol{\sigma}_{\mathrm{M}} = {}^4\mathbf{C}^{\sigma}_{\mathrm{M}} : \delta\mathbf{F}^c_{\mathrm{M}}, \tag{55}$$

this tangent may be obtained by varying the definition equation of the macroscopic Cauchy stress tensor (16), followed by substitution of (36) (or

(43)) and (53). This gives

$$\delta\boldsymbol{\sigma}_{\mathrm{M}} = \left[\frac{1}{V}\sum_i\sum_j(\mathbf{x}_{(i)}\mathbf{K}_{\mathrm{M}}^{(ij)}\mathbf{X}_{(j)})^{\mathrm{LC}} + \frac{1}{V}\sum_i \mathbf{f}_{(i)}\mathbf{I}\mathbf{X}_{(i)} - \boldsymbol{\sigma}_{\mathrm{M}}\mathbf{F}_{\mathrm{M}}^{-c}\right] : \delta\mathbf{F}_{\mathrm{M}}^{c}, \tag{56}$$

where the expression in square brackets is identified as the required tangent stiffness tensor ${}^4\mathbf{C}_{\mathrm{M}}^{\sigma}$. In the derivation of (56) it has been used that in the case of prescribed displacements of the RVE boundary (3) or of periodic boundary conditions (5), the initial and current volumes of an RVE are related according to $J_{\mathrm{M}} = \det(\mathbf{F}_{\mathrm{M}}) = V/V_0$.

6. Nested Solution Scheme

Based on the above developments, the actual implementation of the computational homogenisation strategy may be described by the following subsequent steps.

The macroscopic structure to be analysed is discretised by FEs. The external load is applied by an incremental procedure. Increments can be associated with discrete time steps. The solution of the macroscopic non-linear system of equations is performed in a standard iterative manner. To each macroscopic integration point, a discretised RVE is assigned. The geometry of the RVE is based on the microstructural morphology of the material under consideration.

For each macroscopic integration point, the local macroscopic deformation gradient tensor $\mathbf{F}_{\mathrm{M}}$ is computed from the iterative macroscopic nodal displacements (during the initialisation step, zero deformation is assumed throughout the macroscopic structure, i.e. $\mathbf{F}_{\mathrm{M}} = \mathbf{I}$, which allows to obtain the initial macroscopic constitutive tangent). The macroscopic deformation gradient tensor is used to formulate the boundary conditions according to (23) or (28) and (29) to be applied on the corresponding representative cell.

The solution of the RVE BVP employing a fine-scale FE procedure provides the resulting stress and strain distributions in the microstructural cell. Using the resulting forces at the prescribed nodes, the RVE averaged first Piola–Kirchhoff stress tensor $\mathbf{P}_{\mathrm{M}}$ is computed according to (36) or (43) and returned to the macroscopic integration point as a local macroscopic stress. From the global RVE stiffness matrix, the local macroscopic consistent tangent ${}^4\mathbf{C}_{\mathrm{M}}^{\mathrm{P}}$ is obtained according to (54).

When the analysis of all microstructural RVEs is finished, the stress tensor is available at every macroscopic integration point. Thus, the internal macroscopic forces can be calculated. If these forces are in balance with the external load, incremental convergence has been achieved and the next time increment can be evaluated. If there is no convergence, the procedure is continued to achieve an updated estimation of the macroscopic nodal displacements. The macroscopic stiffness matrix is assembled using the constitutive tangents available at every macroscopic integration point from the RVE analysis. The solution of the macroscopic system of equations leads to an updated estimation of the macroscopic displacement field. The solution scheme is summarised in Table 1. It is remarked that the two-level scheme outlined above can be used selectively depending on the macroscopic deformation, e.g. in the elastic domain the macroscopic constitutive tangents do not have to be updated at every macroscopic loading step.

7. Computational Example

As an example, the computational homogenisation approach is applied to pure bending of a rectangular strip under plane strain conditions. Both the length and the height of the sample equal 0.2 m, the thickness is taken 1 m. The macromesh is composed of five quadrilateral eight node plane strain reduced integration elements. The undeformed and deformed geometries of the macromesh are schematically depicted in Fig. 6. At the left side the strip is fixed in axial (horizontal) direction, the displacement in transverse (vertical) direction is left free. At the right side the rotation of the cross section is prescribed. As pure bending is considered the behaviour of the strip is uniform in axial direction and, therefore, a single layer of elements on the macrolevel suffices to simulate the situation.

In this example two heterogeneous microstructures consisting of a homogeneous matrix material with initially 12% and 30% volume fractions of voids are studied. The microstructural cells used in the calculations are presented in Fig. 7. It is worth mentioning that the absolute size of the microstructure is irrelevant for the first-order computational homogenisation analysis (see also discussion in Sec. 9.1).

The matrix material behaviour has been described by a modified elasto-visco-plastic Bodner–Partom model.[53] This choice is motivated by the intention to demonstrate that the method is well suited for complex

Table 1. Incremental-iterative nested multiscale solution scheme for the computational homogenisation.

Macro		Micro
1. Initialisation		
• initialise the macroscopic model		
• assign an RVE to every integration point		
• loop over all integration points		Initialisation RVE analysis
set $\mathbf{F}_\mathrm{M} = \mathbf{I}$	$\xrightarrow{\mathbf{F}_\mathrm{M}}$	
		• prescribe boundary conditions
		• assemble the RVE stiffness
	$\xleftarrow{\text{tangent}}$	• calculate the tangent ${}^4\mathbf{C}^\mathrm{P}_\mathrm{M}$
store the tangent		
• end integration point loop		
2. Next increment		
• apply increment of the macro load		
3. Next iteration		
• assemble the macroscopic tangent stiffness		
• solve the macroscopic system		
• loop over all integration points		RVE analysis
calculate $\mathbf{F}_\mathrm{M}$	$\xrightarrow{\mathbf{F}_\mathrm{M}}$	
		• prescribe boundary conditions
		• assemble the RVE stiffness
		• solve the RVE problem
	$\xleftarrow{\mathbf{P}_\mathrm{M}}$	• calculate $\mathbf{P}_\mathrm{M}$
store $\mathbf{P}_\mathrm{M}$		
	$\xleftarrow{\text{tangent}}$	• calculate the tangent ${}^4\mathbf{C}^\mathrm{P}_\mathrm{M}$
store the tangent		
• end integration point loop		
• assemble the macroscopic internal forces		
4. Check for convergence		
• if not converged $\Rightarrow$ step 3		
• else $\Rightarrow$ step 2		

microstructural material behaviour, e.g. non-linear history and strain rate dependent at large strains. The material parameters for annealed aluminum AA 1050 have been used[53]; elastic parameters: shear modulus $G = 2.6 \times 10^4\,\mathrm{MPa}$, bulk modulus $K = 7.8 \times 10^4\,\mathrm{MPa}$ and viscosity parameters: $\Gamma_0 = 10^8\,\mathrm{s}^{-2}$, $m = 13.8$, $n = 3.4$, $Z_0 = 81.4\,\mathrm{MPa}$, $Z_1 = 170\,\mathrm{MPa}$.

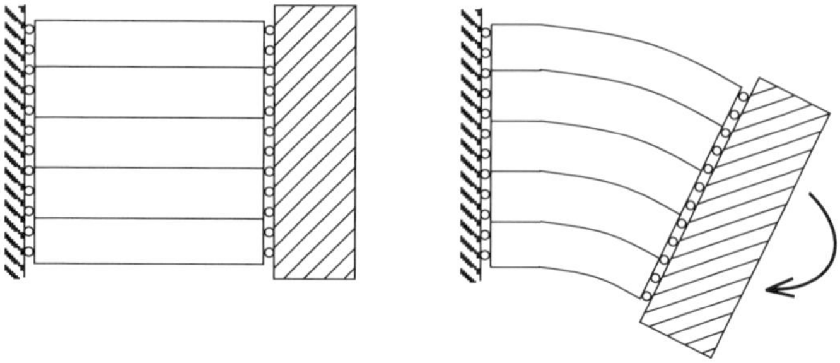

(a) (b)

Fig. 6. Schematic representation of the undeformed (a) and deformed (b) configurations of the macroscopically bended specimen.

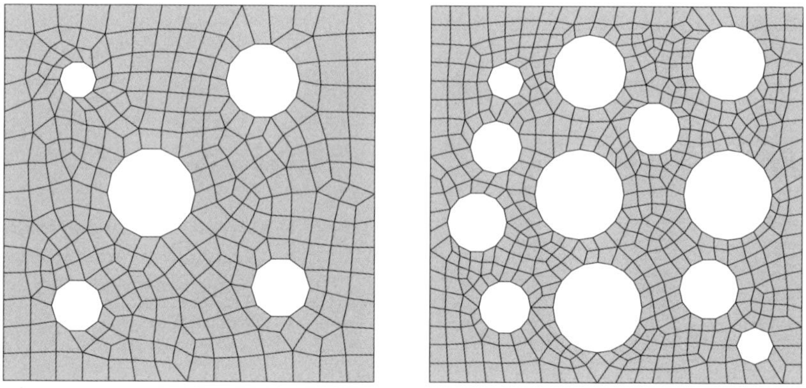

(a) (b)

Fig. 7. Microstructural cells used in the calculations with 12% voids (a) and 30% voids (b).

Micro–macro calculations for the heterogeneous structure, represented by the RVEs shown in Fig. 7, have been carried out, simulating pure bending at a prescribed moment rate equal to $5 \times 10^5\,\mathrm{N\,m\,s^{-1}}$. Figure 8 shows the distribution plots of the effective plastic strain for the case of the RVE with 12% volume fraction voids at an applied moment equal to $6.8 \times 10^5\,\mathrm{N\,m}$ in the deformed macrostructure and in three deformed, initially identical RVEs at different locations in the macrostructure. Each hole acts as a plastic strain concentrator and causes higher strains in the RVE than those occurring in the homogenised macrostructure. In the present calculations

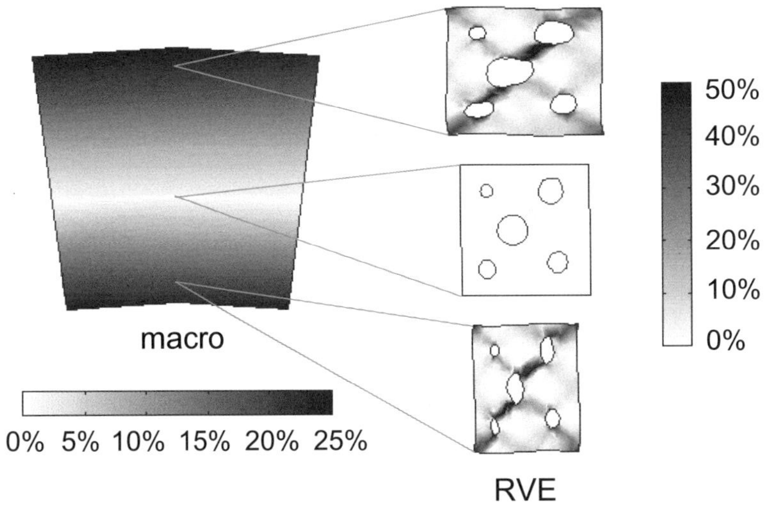

Fig. 8. Distribution of the effective plastic strain in the deformed macrostructure and in three deformed RVEs, corresponding to different points of the macrostructure.

the maximum effective plastic strain in the macrostructure is about 25%, whereas at RVE level this strain reaches 50%. It is obvious from the deformed geometry of the holes in Fig. 8 that the RVE in the upper part of the bended strip is subjected to tension and the RVE in the lower part to compression, while the RVE in the vicinity of the neutral axis is loaded considerably less than the other RVEs. This confirms the conclusion that the method realistically describes the deformation modes of the microstructure.

In Fig. 9, the moment–curvature (curvature defined for the bottom edge of the specimen) diagram resulting from the computational homogenisation approach is presented. To give an impression of the influence of the holes as well, the response of a homogeneous configuration (without cavities) is shown. It can be concluded that even the presence of 12% voids induces a reduction in the bending moment (at a certain curvature) of more than 25% in the plastic regime. This significant reduction in the bending moment may be attributed to the formation of microstructural shear bands, which are clearly observed in Fig. 8. This indicates that in order to capture such an effect a detailed microstructural analysis is required. A straightforward application of, for example, the rule of mixtures would lead to erroneous results.

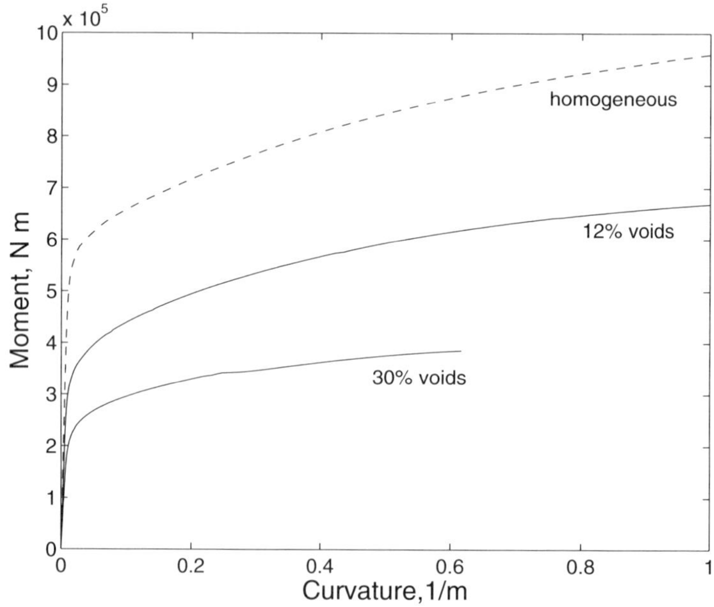

Fig. 9. Moment–curvature diagram resulting from the first-order computational homogenisation analysis.

8. Concept of an RVE within Computational Homogenisation

The computational homogenisation approach, as well as most of other homogenisation techniques, is based on the concept of a representative volume element (RVE). An RVE is a model of a material microstructure to be used to obtain the response of the corresponding homogenised macroscopic continuum in a macroscopic material point. Thus, the proper choice of the RVE largely determines the accuracy of the modelling of a heterogeneous material.

There appear to be two significantly different ways to define an RVE.[54] The first definition requires an RVE to be a statistically representative sample of the microstructure, i.e. to include virtually a sampling of all possible microstructural configurations that occur in the composite. Clearly, in the case of a non-regular and non-uniform microstructure such a definition leads to a considerably large RVE. Therefore, RVEs that rigorously satisfy this definition are rarely used in actual homogenisation analyses. This concept is usually employed when a computer model of the microstructure is being constructed based on experimentally obtained statistical information.[55,56]

Another definition characterises an RVE as the smallest microstructural volume that sufficiently accurately represents the overall macroscopic properties of interest. This usually leads to much smaller RVE sizes than the statistical definition described above. However, in this case the minimum required RVE size also depends on the type of material behaviour (e.g. for elastic behaviour usually much smaller RVEs suffice than for plastic behaviour), macroscopic loading path and contrast in properties between heterogeneities. Moreover, the minimum RVE size, which results in a good approximation of the overall material properties, does not always lead to adequate distributions of the microfields within the RVE. This may be important if, for example, microstructural damage initiation or evolving microstructures are of interest.

The latter definition of an RVE is closely related to the one established by Hill,[48] who argued that an RVE is well defined if it reflects the material microstructure and if the responses under uniform displacement and traction boundary conditions coincide. If a microstructural cell does not contain sufficient microstructural information, its overall responses under uniform displacement and traction boundary conditions will differ. The homogenised properties determined in this way are called "apparent", a notion introduced by Huet.[57] The apparent properties obtained by application of uniform displacement boundary conditions on a microstructural cell usually overestimate the real effective properties, while the uniform traction boundary conditions lead to underestimation. For a given microstructural cell size, the periodic boundary conditions provide a better estimation of the overall properties than the uniform displacement and uniform traction boundary conditions.[46,47,58,59] This conclusion also holds if the microstructure does not really possess geometrical periodicity. Increasing the size of the microstructural cell leads to a better estimation of the overall properties, and, finally, to a "convergence" of the results obtained with the different boundary conditions to the real effective properties of the composite material, as schematically illustrated in Fig. 10. The convergence of the apparent properties towards the effective ones at increasing size of the microstructural cell has been investigated in by a number of authors.[47,57–63]

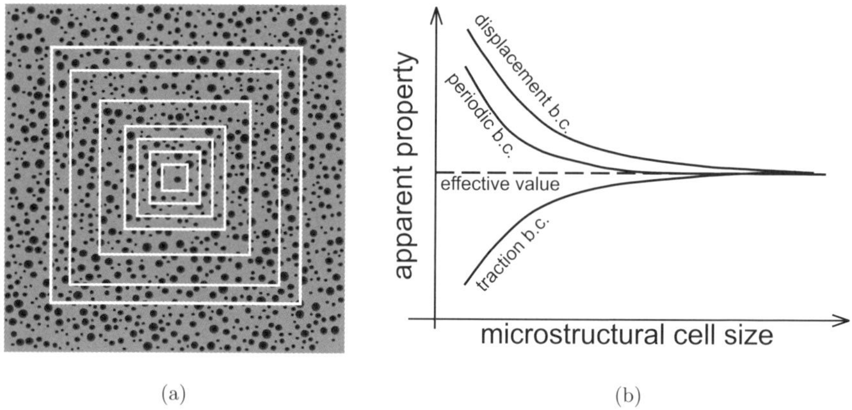

Fig. 10. (a) Several microstructural cells of different sizes. (b) Convergence of the apparent properties to the effective values with increasing microstructural cell size for different types of boundary conditions.

9. Extensions of the Classical Computational Homogenisation Scheme

9.1. *Homogenisation towards second gradient continuum*

The classical first-order computational homogenisation framework, as presented above, relies on the principle of scale separation, which restricts its applicability limits. The fundamental scale separation concept used in the first-order scheme (and also accepted in most other classical homogenisation approaches) requires that the microstructural length scale is negligible in comparison with the macrostructural characteristic length (determined by the characteristic wave length of the macroscopic load). In this case it is justified to assume macroscopic uniformity of the deformation field over the microstructural RVE. As a result, only simple first-order deformation modes (tension, compression, shear, or combinations thereof) of the microstructure are found. As can be noticed, for example, in Fig. 8, a typical bending mode, which from a physical point of view should appear for small, but finite, microstructural cells in the macroscopically bended specimen, is not retrieved. Moreover, the dimensions of the microstructural heterogeneities do not influence the averaging procedure. Increasing the scale of the entire microstructure then leads to identical results. All of this is not surprising, since the first-order approach is fully in line with standard continuum mechanics concepts, where one of the fundamental points of departure is the principle of local action. In fact this principle states that material points are local, i.e. are identified with an infinitesimal volume only. This infinitesimal character is exactly represented in the behaviour of the microstructural RVEs, which are considered as macroscopic material points. This implies that the size of the microstructure is considered as irrelevant and hence microstructural and geometrical size effects are not taken into account. Furthermore, it has been demonstrated[42,64] that if a microstructural RVE exhibits overall softening behaviour (due to geometrical softening or material softening of constituents), the macroscopic solution obtained from the first-order computational homogenisation approach fully localises according to the size of the elements used in the macromesh, i.e. the macroscopic BVP becomes ill-posed leading to a mesh-dependent macroscopic response.

In order to deal with these limitations, the classical (first-order) computational homogenisation has been extended to a so-called second-order computational homogenisation framework,[42,65,66] which aims at capturing of macroscopic localisation and microstructural size effects. A general scheme of the second-order computational homogenisation approach is shown in Fig. 11 (cf. Fig. 3).

In the second-order homogenisation approach, the macroscopic deformation gradient tensor $\mathbf{F}_\mathrm{M}$ and its gradient $\nabla_{0\mathrm{M}}\mathbf{F}_\mathrm{M}$ are used to formulate boundary conditions for a microstructural RVE. Every microstructural constituent is modelled as a classical continuum, characterised by standard first-order equilibrium and constitutive equations. Therefore, for the

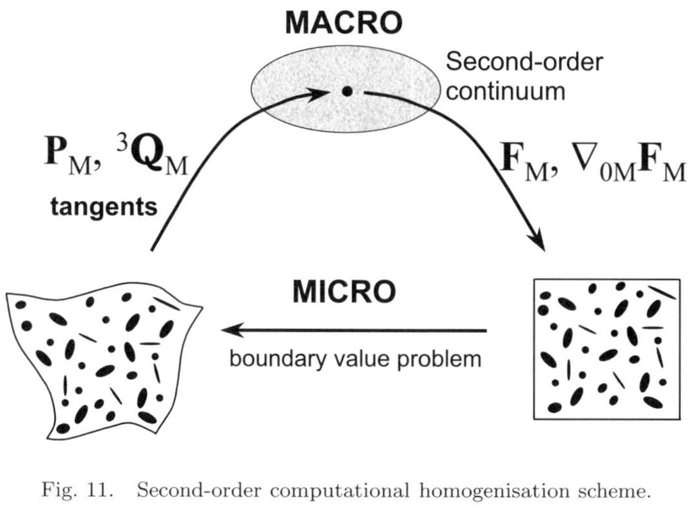

Fig. 11. Second-order computational homogenisation scheme.

description of the microstructural phenomena already available models developed for the first-order homogenisation can be directly employed. On the macrolevel, however, a full second gradient equilibrium problem (of the type originally proposed by Mindlin[67,68]) appears. From the solution of the underlying microstructural BVP, the macroscopic stress tensor $\mathbf{P}_{\mathrm{M}}$ and a higher-order stress tensor ${}^{3}\mathbf{Q}_{\mathrm{M}}$ (a third-order tensor, defined as the work conjugate of the gradient of the deformation gradient tensor) are derived based on an extension of the classical Hill–Mandel condition. This automatically delivers the microstructurally based constitutive response of the second gradient macrocontinuum. Consistent (higher-order) tangents of this second-gradient continuum are extracted from the microstructural stiffness using a procedure similar to the one presented for the classical homogenisation. Moreover, it has been shown[69] that the size of the microstructural RVE used in the second-order computational homogenisation scheme may be related to the length scale of the associated macroscopic homogenised higher-order continuum. Details of the second-order computational homogenisation, its implementation and application examples may be found in the literature.[42,65,66,69]

9.2. *Computational homogenisation for beams and shells*

Beam and shell structures have been efficiently and economically applied in various fields of engineering for centuries. Structured and layered thin sheets are used in a variety of innovative applications as well. A typical example is flexible electronics, e.g. flexible displays, where stacks of different materials with complex geometries and interconnects between layers, prohibit the use of classical layer-wise composite shell theory.[70] For these complex applications, a computational homogenisation technique for thin structured sheets has recently been proposed.[71,72] In this case the actual three-dimensional heterogeneous sheet is represented by a homogenised shell continuum for which the constitutive response is obtained from the analysis of a microstructural RVE, representing the full thickness of the sheet and an in-plane cell of the macroscopic structure (e.g. a single pixel of a flexible display). The computational homogenisation for structured thin sheets is schematically illustrated in Fig. 12.

Consider a material point of the shell continuum (in-plane integration point in an FE setting). At this macroscopic point generalised strains are assumed to be known. In the particular case of a Mindlin–Reisner shell, these generalised strains are the membrane strain tensor $\mathcal{E}_{\mathrm{M}}$, the curvature

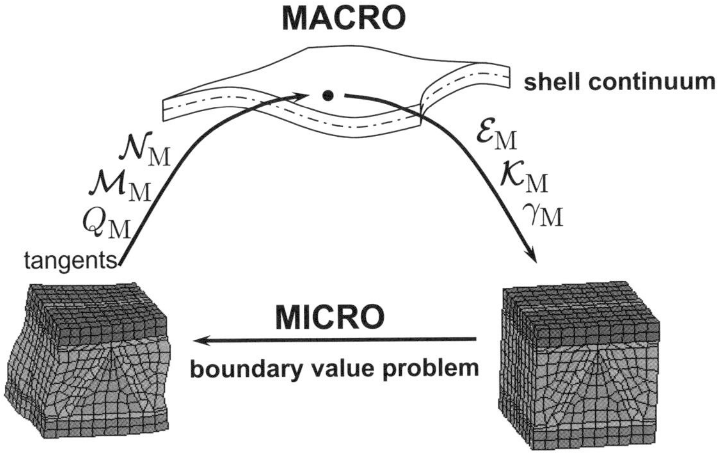

Fig. 12. Scheme of the computational homogenisation for structured thin sheets.

tensor $\mathcal{K}_{\mathrm{M}}$ and the transverse shear strain γ_{M}. The application of the scheme to other shell formulations (e.g. solid-like shells) can also be developed. The vicinity of this macroscopic point is represented by a microstructural through-thickness RVE. At the RVE scale all microstructural constituents are treated as an ordinary continuum, described by the standard first-order equilibrium and constitutive equations. The microscopic BVP is completed by essential and natural boundary conditions, whereby the macroscopic generalised strains are used to formulate the kinematical boundary conditions on the lateral faces of the RVE, while the top and bottom RVE faces (corresponding to the faces of the macroscopic shell) can be left traction-free (which is typically relevant for shells that are not loaded in the out-of-plane direction, e.g. flexible displays) or other boundary conditions consistent with the out-of-plane loading of the shell can be prescribed.

Upon the solution of the microstructural BVP, the macroscopic generalised stress resultants, i.e. the stress resultant $\mathcal{N}_{\mathrm{M}}$, the couple-stress resultant (moment) $\mathcal{M}_{\mathrm{M}}$ and the transverse shear resultant Q_{M}, are obtained by proper averaging the resulting RVE stress field. In this way, the in-plane homogenisation is directly combined with a through-thickness stress integration. Thus, from a macroscopic point of view, a (numerical) generalised stress–strain constitutive response at every macroscopic in-plane integration point is obtained. The macroscopic consistent tangent operators are also extracted through the condensation of the total microstructural stiffness in a structurally similar manner as discussed above. Additionally, the simultaneously resolved microscale RVE local deformation and stress fields provide valuable information for assessing the reliability of a particular microstructural design. More details on the computational homogenisation for shell structures can be found in the literature.[71,72]

9.3. *Computational homogenisation for heat conduction problems*

Materials and structures are often subjected to thermal loading, which may also be transient in nature, e.g. in the case of thermoshock. Typical examples of materials subjected to strong temperature changes and cycles include thermal coatings, refractories in furnaces, microelectronics components and engines. Deterioration and failure of the components at the macroscale is known to originate from non-uniformity and mismatches between microstructural constituents at the microscale resulting in thermal expansion anisotropy and internal stress gradients. A computational homogenisation approach for the coupled multiscale analysis of evolving thermal fields in heterogeneous solids with complex microstructures and including temperature- and orientation-dependent conductivities has recently been proposed by Özdemir *et al.*[73] The computational homogenisation framework for heat conduction problems is schematically illustrated in Fig. 13.

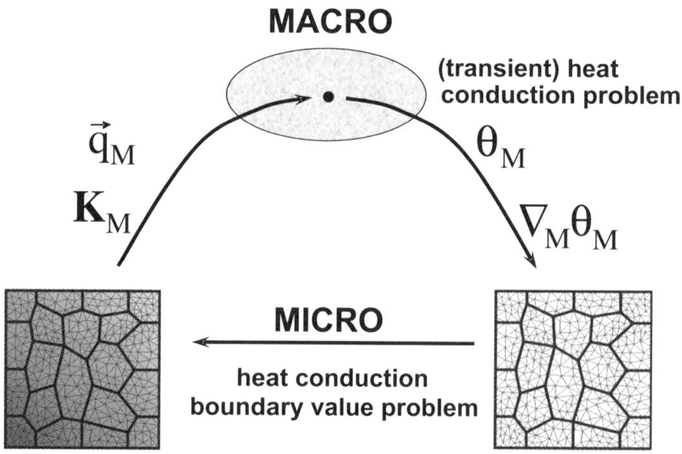

Fig. 13. Scheme of the computational homogenisation for heat conduction problems.

At the macro level, a (transient) heat conduction problem is considered, for which the thermal constitutive behaviour is not formulated explicitly, but which is numerically obtained through a multiscale analysis. At each macroscopic (integration) point, temperature $\theta_{\rm M}$ and temperature gradient $\nabla_{\rm M}\theta_{\rm M}$ are calculated and used to define the boundary conditions to be imposed on the microscopic RVE associated with this particular point. The thermal constitutive behaviour of each phase at the micro level is assumed to be known. After solving the microscopic heat conduction BVP, the macroscopic heat flux $\vec{q}_{\rm M}$ is obtained by volume averaging the resulting heat flux field over the RVE. Additionally, the macroscopic (tangent) conductivity $\mathbf{K}_{\rm M}$ is extracted from the microstructural conductivity. Although the development of the computational homogenisation framework for the heat conduction problems follows the same philosophy as its mechanical counterpart discussed above, it poses some fundamental differences. More details can be found in the literature.[73] Combining the heat conduction and the purely mechanical computational homogenisation schemes, a coupled thermo-mechanical computational homogenisation framework can be established and will be published in forthcoming works.

Acknowledgements

Parts of this research were carried out under projects No. ME97020 and No. MC2.03148 "Multi–scale computational homogenisation" in the framework of the Strategic Research Programme of the Netherlands Institute for Metals Research in the Netherlands.

References

1. J. D. Eshelby, The determination of the field of an ellipsoidal inclusion and related problems, *Proc. R. Soc. Lond A* **241**, 376–396 (1957).
2. Z. Hashin, The elastic moduli of heterogeneous materials, *J. Appl. Mech.* **29**, 143–150 (1962).
3. B. Budiansky, On the elastic moduli of some heterogeneous materials, *J. Mech. Phys. Solids* **13**, 223–227 (1965).
4. T. Mori and K. Tanaka, Average stress in the matrix and average elastic energy of materials with misfitting inclusions, *Acta Metall.* **21**, 571–574 (1973).
5. R. Hill, A self-consistent mechanics of composite materials, *J. Mech. Phys. Solids* **13**, 213–222 (1965).

6. R. M. Christensen and K. H. Lo, Solutions for effective shear properties in three phase sphere and cylinder models, *J. Mech. Phys. Solids* **27**, 315–330 (1979).
7. Z. Hashin and S. Shtrikman, A variational approach to the theory of the elastic behaviour of multiphase materials, *J. Mech. Phys. Solids* **11**, 127–140 (1963).
8. Z. Hashin, Analysis of composite materials. A survey, *J. Appl. Mechanics* **50**, 481–505 (1983).
9. J. R. Willis, Variational and related methods for the overall properties of composites, *Adv. Appl. Mech.* **21**, 1–78 (1981).
10. P. Ponte Castañeda and P. Suquet, Nonlinear composites, *Adv. Appl. Mech.* **34**, 171–302 (1998).
11. A. Bensoussan, J.-L. Lionis and G. Papanicolaou, *Asymptotic Analysis for Periodic Structures* (North-Holland, Amsterdam, 1978).
12. E. Sanchez-Palencia, *Non-homogeneous Media and Vibration Theory*, Lecture Notes in Physics, Vol. 127 (Springer-Verlag, Berlin, 1980).
13. A. Tolenado and H. Murakami, A high-order mixture model for periodic particulate composites, *Int. J. Solids Struct.* **23**, 989–1002 (1987).
14. F. Devries, H. Dumontet, G. Duvaut and F. Lene, Homogenization and damage for composite structures, *Int. J. Numer. Meth. Eng.* **27**, 285–298 (1989).
15. J. M. Guedes and N. Kikuchi, Preprocessing and postprocessing for materials based on the homogenization method with adaptive finite element methods, *Comput. Meth. Appl. Mech. Eng.* **83**, 143–198 (1990).
16. S. J. Hollister and N. Kikuchi, A comparison of homogenization and standard mechanics analysis for periodic porous composites, *Comput. Mech.* **10**, 73–95 (1992).
17. J. Fish, Q. Yu and K. Shek, Computational damage mechanics for composite materials based on mathematical homogenisation, *Int. J. Numer. Meth. Eng.* **45**, 1657–1679 (1999).
18. S. Nemat-Nasser and M. Hori, *Micromechanics: Overall Properties of Heterogeneous Materials* (Elsevier, Amsterdam, 1993).
19. T. Christman, A. Needleman and S. Suresh, An experimental and numerical study of deformation in metal-ceramic composites, *Acta Metall.* **37**, 3029–3050 (1989).
20. V. Tvergaard, Analysis of tensile properties for wisker-reinforced metal-matrix composites, *Acta Metall. Mater.* **38**, 185–194 (1990).
21. G. Bao, J. W. Hutchinson and R. M. McMeeking, Plastic reinforcement of ductile materials against plastic flow and creep, *Acta Metall. Mater.* **39**, 1871–1882 (1991).
22. J. R. Brockenbrough, S. Suresh and H. A. Wienecke, Deformation of metal-matrix composites with continuous fibers: Geometrical effect of fiber distribution and shape, *Acta Metall. Mater.* **39**, 735–752 (1991).
23. T. Nakamura and S. Suresh, Effect of thermal residual stress and fiber packing on deformation of metal-matrix composites, *Acta Metall. Mater.* **41**, 1665–1681 (1993).

24. P. E. McHugh, R. J. Asaro and C. F. Shin, Computational modeling of metal matrix composite materials — II. Isothermal stress–strain behaviour, *Acta Metall. Mater.* **41**, 1477–1488 (1993).
25. O. van der Sluis, P. J. G. Schreurs and H. E. H. Meijer, Effective properties of a viscoplastic constitutive model obtained by homogenisation, *Mech. Mater.* **31**, 743–759 (1999).
26. O. van der Sluis, Homogenisation of structured elastoviscoplastic solids, PhD thesis, Eindhoven University of Technology, Eindhoven, The Netherlands (2001).
27. S. Suresh, A. Mortensen and A. Needleman (eds.), *Fundamentals of Metal-Matrix Composites* (Butterworth-Heinemann, Boston, 1993).
28. P. E. McHugh, R. J. Asaro and C. F. Shin, Computational modeling of metal matrix composite materials — III. Comparison with phenomenological models, *Acta Metall. Mater.* **41**, 1489–1499 (1993).
29. P. M. Suquet, Local and global aspects in the mathematical theory of plasticity, in *Plasticity Today: Modelling, Methods and Applications*, eds. A. Sawczuk and G. Bianchi (Elsevier Applied Science Publishers, London, 1985), pp. 279–310
30. K. Terada and N. Kikuchi, Nonlinear homogenisation method for practical applications, in *Computational Methods in Micromechanics*, eds. S. Ghosh and M. Ostoja-Starzewski, AMD-Vol. 212/MD-Vol. 62 (ASME, 1995), pp. 1–16.
31. S. Ghosh, K. Lee and S. Moorthy, Multiple scale analysis of heterogeneous elastic structures using homogenisation theory and Voronoi cell finite element method, *Int. J. Solids Struct.* **32**, 27–62 (1995).
32. S. Ghosh, K. Lee and S. Moorthy, Two scale analysis of heterogeneous elastic–plastic materials with asymptotic homogenisation and Voronoi cell finite element model, *Comput. Meth. Appl. Mech. Eng.* **132**, 63–116 (1996).
33. R. J. M. Smit, W. A. M. Brekelmans and H. E. H. Meijer, Prediction of the mechanical behaviour of non-linear heterogeneous systems by multi-level finite element modeling, *Comput. Meth. Appl. Mech. Eng.* **155**, 181–192 (1998).
34. C. Miehe, J. Schröder and J. Schotte, Computational homogenization analysis in finite plasticity. Simulation of texture development in polycrystalline materials, *Comput. Meth. Appl. Mech. Eng.* **171**, 387–418 (1999).
35. C. Miehe, J. Schotte and J. Schröder, Computational micro-macro transitions and overall moduli in the analysis of polycrystals at large strains, *Comput. Mater. Sci.* **16**, 372–382 (1999).
36. J. C. Michel, H. Moulinec and P. Suquet, Effective properties of composite materials with periodic microstructure: a computational approach, *Comput. Meth. Appl. Mech. Eng.* **172**, 109–143 (1999).
37. F. Feyel and J.-L. Chaboche, FE^2 multiscale approach for modelling the elastoviscoplastic behaviour of long fiber SiC/Ti composite materials, *Comput. Meth. Appl. Mech. Eng.* **183**, 309–330 (2000).
38. K. Terada and N. Kikuchi, A class of general algorithms for multi-scale analysis of heterogeneous media, *Comput. Meth. Appl. Mech. Eng.* **190**, 5427–5464 (2001).

39. S. Ghosh, K. Lee and P. Raghavan, A multi-level computational model for multi-scale damage analysis in composite and porous materials, *Int. J. Solids Struct.* **38**, 2335–2385 (2001).
40. V. Kouznetsova, W. A. M. Brekelmans and F. P. T. Baaijens, An approach to micro–macro modeling of heterogeneous materials, *Comput. Mech.* **27**, 37–48 (2001).
41. C. Miehe and A. Koch, Computational micro-to-macro transition of discretized microstructures undergoing small strain, *Arch. Appl. Mech.* **72**, 300–317 (2002).
42. V. Kouznetsova, Computational homogenization for the multi-scale analysis of multi-phase materials, PhD thesis, Eindhoven University of Technology, Eindhoven, The Netherlands (2002).
43. G. K. Sfantos and M. H. Aliabadi, Multi-scale boundary element modelling of material degradation and fracture, *Comput. Meth. Appl. Mech. Eng.* **196**, 1310–1329 (2007).
44. H. Moulinec and P. Suquet, A numerical method for computing the overall response of non-linear composites with complex microstructure, *Comput. Meth. Appl. Mech. Eng.* **157**, 69–94 (1998).
45. C. Oskay and J. Fish, Eigendeformation-based reduced order homogenization for failure analysis of heterogeneous materials, *Comput. Meth. Appl. Mech. Eng.* **196**, 1216–1243 (2007).
46. O. van der Sluis, P. J. G. Schreurs, W. A. M. Brekelmans and H. E. H. Meijer, Overall behaviour of heterogeneous elastoviscoplastic materials: Effect of microstructural modelling, *Mech. Mater.* **32**, 449–462 (2000).
47. K. Terada, M. Hori, T. Kyoya and N. Kikuchi, Simulation of the multi-scale convergence in computational homogenization approach, *Int. J. Solids Struct.* **37**, 2285–2311 (2000).
48. R. Hill, Elastic properties of reinforced solids: Some theoretical principles, *J. Mech. Phys. Solids.* **11**, 357–372 (1963).
49. R. Hill, On macroscopic effects of heterogeneity in elastoplastic media at finite strain, *Math. Proc. Cam. Phil. Soc.* **95**, 481–494 (1984).
50. S. Nemat-Nasser, Averaging theorems in finite deformation plasticity, *Mech. Mater.* **31**, 493–523 (1999).
51. R. D. Cook, D. S. Malkus and M. E. Plesha, *Concepts and Applications of Finite Element Analysis* (Wiley, Chichester, 1989).
52. C. Miehe, Numerical computation of algorithmic (consistent) tangent moduli in large-strain computational inelasticity, *Comput. Meth. Appl. Mech. Eng.* **134**, 223–240 (1996).
53. H. C. E. van der Aa, M. A. H. van der Aa, P. J. G. Schreurs, F. P. T. Baaijens and W. J. van Veenen, An experimental and numerical study of the wall ironing process of polymer coated sheet metal, *Mech. Mater.* **32**, 423–443 (2000).
54. W. J. Drugan and J. R. Willis, A micromechanics-based nonlocal constitutive equation and estimates of representative volume element size for elastic composites, *J. Mech. Phys. Solids* **44**(4), 497–524 (1996).

55. J. Zeman and M. Šejnoha, Numerical evaluation of effective elastic properties of graphite fiber tow impregnated by polymer matrix, *J. Mech. Phys. Solids* **49**, 69–90 (2001).
56. Z. Shan and A. M. Gokhale, Representative volume element for non-uniform micro-structure, *Comput. Mater. Sci.* **24**, 361–379 (2002).
57. C. Huet, Application of variational concepts to size effects in elastic heterogeneous bodies, *J. Mech. Phys. Solids* **38**(6), 813–841 (1990).
58. T. Kanit, S. Forest, I. Galliet, V. Mounoury and D. Jeulin, Determination of the size of the representative volume element for random composites: Statistical and numerical approach, *Int. J. Solids Struct.* **40**, 3647–3679 (2003).
59. T. Kanit, F. N'Guyen, S. Forest, D. Jeulin, M. Reed and S. Singleton, Apparent and effective physical properties of heterogeneous materials: Representativity of samples of two materials from food industry, *Comput. Meth. Appl. Mech. Eng.* **195**, 3960–3982 (2006).
60. C. Huet, Coupled size and boundary-condition effects in viscoelastic heterogeneous and composite bodies, *Mech. Mater.* **31**, 787–829 (1999).
61. M. Ostoja-Starzewski, Random field models of heterogeneous materials, *Int. J. Solids Struct.* **35**(19), 2429–2455 (1998).
62. M. Ostoja-Starzewski, Scale effects in materials with random distributions of needles and cracks, *Mech. Mater.* **31**, 883–893 (1999).
63. S. Pecullan, L. V. Gibiansky and S. Torquato, Scale effects on the elastic behavior of periodic and hierarchical two-dimensional composites, *J. Mech. Phys. Solids* **47**, 1509–1542 (1999).
64. M. G. D. Geers, V. G. Kouznetsova and W. A. M. Brekelmans, Multi-scale first-order and second-order computational homogenisation of microstructures towards continua, *Int. J. Multiscale Comput. Eng.* **1**, 371–386 (2003).
65. V. Kouznetsova, M. G. D. Geers and W. A. M. Brekelmans, Multi-scale constitutive modelling of heterogeneous materials with a gradient-enhanced computational homogenisation scheme, *Int. J. Numer. Meth. Eng.* **54**, 1235–1260 (2002).
66. V. G. Kouznetsova, M. G. D. Geers and W. A. M. Brekelmans, Multi-scale second-order computational homogenization of multi-phase materials: A nested finite element solution strategy, *Comput. Meth. Appl. Mech. Eng.* **193**, 5525–5550 (2004).
67. R. D. Mindlin, Second gradient of strain and surface-tension in linear elasticity, *Int. J. Solids Struct.* **1**, 417–438 (1965).
68. R. D. Mindlin and N. N. Eshel, On first strain-gradient theories in linear elasticity, *Int. J. Solids Struct.* **4**, 109–124 (1968).
69. V. G. Kouznetsova, M. G. D. Geers and W. A. M. Brekelmans, Size of a representative volume element in a second-order computational homogenization framework, *Int. J. Multiscale Comput. Eng.* **2**, 575–598 (2004).
70. J. Hohe and W. Becker, Effective stress–strain relations for two dimensional cellular sandwich cores: Homogenization, material models, and properties, *Appl. Mech. Rev.* **55**, 61–87 (2002).

71. M. G. D. Geers, E. W. C. Coenen and V. G. Kouznetsova, Multi-scale computational homogenisation of structured thin sheets, *Modelling Simul. Mater. Sci. Eng.* **15**, S393–S404 (2007).
72. E. W. C. Coenen, V. G. Kouznetsova and M. G. D. Geers, Computational homogeneization for heterogeneous thin sheets, in preparation (2007).
73. I. Özdemir, W. A. M. Brekelmans and M. G. D. Geers, Computational homogenization for heat conduction in heterogeneous solids, *Int. J. Numer. Meth. Eng.* **73**(2), 185–204 (2008).

TWO-SCALE ASYMPTOTIC HOMOGENISATION-BASED FINITE ELEMENT ANALYSIS OF COMPOSITE MATERIALS

Qi-Zhi Xiao

LUSAS FEA Ltd., Forge House, 66 High Street
Kingston-upon-Thames, KT1 1HN, UK
qizhi.xiao@lusas.com

Bhushan Lal Karihaloo

School of Engineering, Cardiff University
Cardiff CF24 3AA, UK
karihaloob@cardiff.ac.uk

Numerous micro-mechanical models have been developed to estimate the equivalent moduli of composite materials. A more important but also more difficult problem is the estimation of the residual strength of a composite because it needs the evaluation of the deformation at the micro-scale, e.g. strains and stresses at the interface between different phases. The powerful finite element method (FEM) cannot realistically model all the details at the micro level, even with the help of the most powerful computers available today. The two-scale asymptotic homogenisation method, which has attracted the attention of many researchers, is most suitable for such problems because it gives not only the equivalent material properties but also detailed information of local micro fields with less computational cost. However, the widely used first-order homogenisation gives micro fields with very low accuracy. In this chapter, the higher-order homogenisation theory and corresponding consistent solution strategies are fully described. Modern high-performance FEMs, which are powerful for the solution of sub-problems from homogenisation analysis, are also discussed. Numerical results from the first-order homogenisation are provided to illustrate some features of the method.

1. Introduction

In the mechanics of composite materials, numerous analytical, semi-analytical and computational micro-mechanical models have been developed to determine the effective properties of the composite from the distribution and basic properties of constituents and the detailed distribution of fields on the scale of micro-constituents.[1,2] These models generally predict satisfactorily effective response of the composite; it is found, however, that the complexity and computational cost of each of the methods are proportional to the accuracy of the micro-stress field, as might be expected. The finite element method (FEM)[3–5] is arguably the most accurate and universal method to perform micro-mechanical analysis of composite materials but it is also the most expensive. If the volume fraction of reinforcement is very large, it is still not realistic to model the entire macro-domain with a grid size comparable to that of the micro-scale features even with the help of the most powerful computers available today.

In the linear analysis of composite materials, the concept of representative unit cell (RUC) or representative volume element (RVE) is usually used. It gives properties applicable to the whole macro-domain. When nonlinear deformation starts, RVE will become location dependent. Local analysis, which predicts effective properties for the global analysis, should be coupled with the global analysis and the boundary conditions derived from it. In this case, the asymptotic homogenisation method, which has received considerable attention of many researchers,[6–33] seems to be the most suitable one. It is a kind of singular perturbation method suitable for problems with boundary layers[34] that exist at regions near the interfaces of different phases in a heterogeneous medium. With the help of two-scale expansion, it gives not only the effective properties of the composite but also detailed distribution of fields on the scale of micro-constituents at an acceptable cost. In contrast to the most widely used methods in determining the macro properties,[1,2] i.e. the Eshelby method, the self-consistent method, the Mori–Tanaka method, the differential scheme and the bound theories, the homogenisation method takes into account the interaction between phases naturally and avoids assumptions other than the assumption of periodic distribution of constituents. On the other hand, it accounts for micro-structural effects on the macroscopic response without explicitly representing the details of the microstructure in the global analysis. The computational model at the lower scales is only needed if and when there is a necessity to do so. In recent years, the first-order homogenisation approach has been employed for the solution of complex problems in conjunction with the FEM.[9–33] Since accuracy of the widely used isoparametric compatible elements is not satisfactory, high-performance incompatible and multivariable elements are also introduced in the homogenisation method to improve the accuracy.[19–21,23]

As argued by Kouznetsova *et al.*,[35] the conventional first-order asymptotic homogenisation and other classical micro–macro computational approaches have two major disadvantages: (1) they can account for the volume fraction, distribution and morphology of the constituents, but cannot take into account the absolute size of the microstructure, thus making it impossible to treat micro-structural size effects; (2) the intrinsic assumption of the uniformity of the macroscopic (stress–strain) fields attributed to each micro-structural RUC is not appropriate in critical regions of high gradients. Furthermore, Karihaloo *et al.*[20] showed that, the accuracy of the local stress is generally quite low, although the global displacements are quite accurate.[24]

As demonstrated by several researchers,[35] the disadvantages mentioned above can be remedied, and the accuracy of local fields from various homogenisation/micro-mechanical approaches can be significantly improved by employing higher-order theories.

Kouznetsova *et al.*[35] proposed a gradient-enhanced computational homogenisation procedure that allows for the modelling of micro-structural size effects and nonuniform macroscopic deformation fields within the micro-structural RVE, within a general nonlinear framework. The macroscopic deformation gradient tensor and its gradient are used to prescribe the essential boundary conditions on a micro-structural RVE allowing for periodic micro-structural fluctuations. From the solution of the classical (all micro-structural constituents are treated as a classical continuum), micro-structural boundary value problem (BVP) of RVE, the macroscopic stress tensor and the higher-order stress tensor are derived based on an extension of the Hill–Mandel condition; the (numerical) macroscopic constitutive relations between stresses, higher-order stresses, deformation and its gradient are also obtained by integration of the micro fields. Williams[36] discussed a 2D homogenisation theory, which utilises a higher-order, elasticity-based cell model analysis. It models the material microstructure as a 2D periodic array of RUCs with each RUC being discretised into four subregions (or subcells). The displacement field within each subcell is approximated by an (truncated) eigenfunction expansion up to the fifth order. The governing equations for the theory are developed by satisfying the pointwise governing equations of geometrically linear continuum mechanics exactly up through the given order of the subcell displacement fields. The specified governing equations are valid for any type of constitutive model used to describe the behaviour of the material in a subcell. For 3D cases, each RUC is discretised into eight subcells.[37] These two approaches inherit some features of the asymptotic homogenisation. However, they are not strict asymptotic homogenisation.

Chen and Fish[38] studied the dynamic response of a 1D composite bar subjected to impact loading using the higher-order asymptotic homogenisation. However, they did not consider the difference in the time scales. Moreover, they assumed that higher-order expansions are composed of a term dependent only on the macro-scale, and terms dependent on both macro- and micro-scales whose area (2D) or volume (3D) averages over the RUC are zero. Obviously, their assumptions are in contradiction with the philosophy of asymptotic homogenisation, since their solution for the macro scale is also an expansion of the small parameter. Their assumptions are also in contradiction with the widely used first-order homogenisation, where the first-order expansion is the multiplication of a macro-scale-dependent function and a micro-scale-dependent function. However, it does not include a term dependent only on the macro scale, and its average over the RUC does not necessarily vanish.

In this chapter, the higher-order homogenisation theory and corresponding consistent solution strategies are fully developed. Modern high-performance numerical methodologies, which are powerful for the solution of subproblems from homogenisation analysis, are reviewed. Control differential equations from asymptotic homogenisation and solution strategies are discussed in Sec. 2. The corresponding variational principles are then deduced as the basis of the FEM in Sec. 3. The compatible, incompatible, hybrid and enhanced-strain elements, and the element-free Galerkin (EFG) method are discussed in Sec. 4. In Sec. 5, a penalty function method is discussed to enforce the periodicity boundary condition of the RUC and constraints from higher-order equilibrium. In Sec. 6, accurate recovery schemes for the derivatives are introduced. In Sec. 7, results for composite shafts[20] reinforced with circular fibers aligned along their axis and subjected to pure torsion are given to illustrate some common features of the method. Conclusions and discussion follow in Sec. 8.

2. Mathematical Formulation of First- and Higher-Order Two-Scale Asymptotic Homogenisation

Assume the microstructure of the domain Ω^ε occupied by the composite material to be locally periodic with a period defined by a statistically homogeneous volume element, denoted by the RUC or RVE Y, as shown

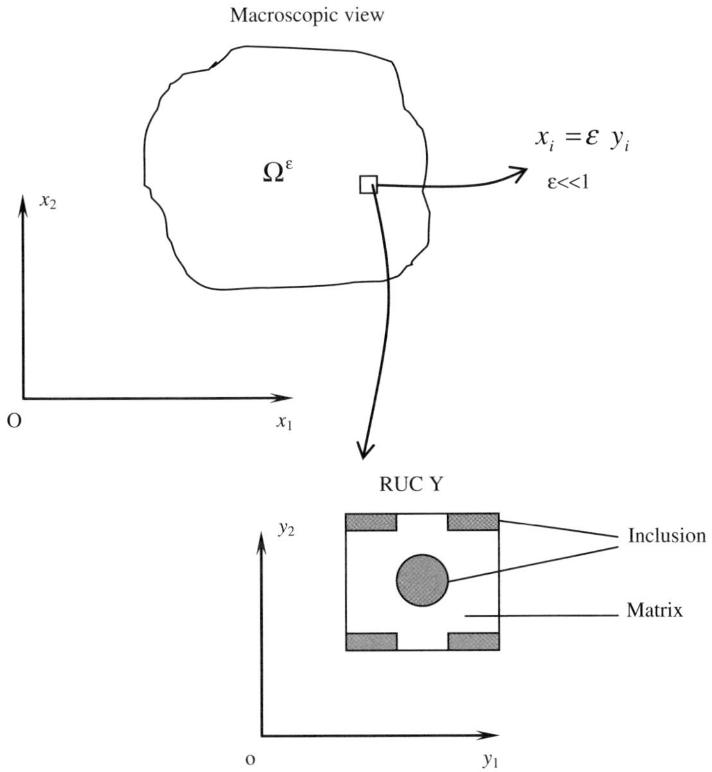

Fig. 1. Illustration of a problem with two length scales.

in Fig. 1. In other words, the composite material is formed by a spatial repetition of the RUC. The problem has two length scales: a global length scale D that is of the order of the size of domain Ω^{ε}, and a local length scale d that is of the order of the RUC and proportional to the wavelength of the variation of the micro-structure. The size of the RUC is much larger than that of the constituents but much smaller than that of the domain. The relation between the global coordinate system x_i for the domain and the local system y_i for the minimum RUC can then be written as

$$y_i = \frac{x_i}{\varepsilon}, \quad i = 1, 2, 3, \tag{1}$$

where ε is a very small positive number representing the scaling factor between the two length scales. The local coordinate vector y_i is regarded as a stretched coordinate vector in the microscopic domain. From (1) we have

$$\frac{\partial x_i}{\partial y_j} = \varepsilon \delta_{ij}. \tag{2}$$

The convention of summation over the repeated indices is used throughout the text. δ_{ij} is the Kronecker Delta.

For an actual heterogeneous body subjected to external forces, field quantities such as displacements u_i, strains e_{ij} and stresses σ_{ij} are assumed to have slow variations from point to point with macroscopic (global) coordinate $\mathbf{x}$ as well as fast variations with local microscopic coordinate $\mathbf{y}$ within a small neighbourhood of size ε of a given point $\mathbf{x}$. That is, displacements u_i, strains e_{ij} and stresses σ_{ij} have two explicit dependencies: one on the macroscopic level with coordinates x_i, and the other on the level of micro-constituents with coordinates y_i:

$$\begin{aligned} u_i^\varepsilon &= u_i^\varepsilon(\mathbf{x}, \mathbf{y}), \\ e_{ij}^\varepsilon &= e_{ij}^\varepsilon(\mathbf{x}, \mathbf{y}), \\ \sigma_{ij}^\varepsilon &= \sigma_{ij}^\varepsilon(\mathbf{x}, \mathbf{y}), \end{aligned} \tag{3}$$

where $i, j = 1, 2$ for a 2D problem; and 1, 2, 3 for a 3D problem. The superscript ε denotes Y-periodicity of the corresponding function. Owing to the periodicity of the microstructure, the functions u_i^ε, e_{ij}^ε and σ_{ij}^ε are assumed to be Y-periodic, i.e.

$$\begin{aligned} u_i^\varepsilon(\mathbf{x}, \mathbf{y}) &= u_i^\varepsilon(\mathbf{x}, \mathbf{y} + kY), \\ e_{ij}^\varepsilon(\mathbf{x}, \mathbf{y}) &= e_{ij}^\varepsilon(\mathbf{x}, \mathbf{y} + kY), \\ \sigma_{ij}^\varepsilon(\mathbf{x}, \mathbf{y}) &= \sigma_{ij}^\varepsilon(\mathbf{x}, \mathbf{y} + kY), \end{aligned} \tag{4}$$

where $Y(y_i)$ is the size of the RUC, or the basic period of the stretched coordinate system $\mathbf{y}$ and k is a nonzero integer.

The unknown displacements u_i^ε, strains e_{ij}^ε and stresses σ_{ij}^ε can be solved from the following equations[39]:

Equilibrium:

$$\frac{\partial \sigma_{ij}^\varepsilon}{\partial x_j} + f_i = 0 \quad \text{in } \Omega^\varepsilon, \tag{5}$$

Kinematical:

$$e_{ij}^{\varepsilon} = \frac{1}{2}\left(\frac{\partial u_i^{\varepsilon}}{\partial x_j} + \frac{\partial u_j^{\varepsilon}}{\partial x_i}\right) \quad \text{in } \Omega^{\varepsilon}, \tag{6}$$

Constitutive:

$$\sigma_{ij}^{\varepsilon} = D_{ijkl}^{\varepsilon}\left(e_{kl}^{\varepsilon} - e_{kl}^{0}\right) + \sigma_{ij}^{0} \quad \text{in } \Omega^{\varepsilon}, \tag{7}$$

Prescribed boundary displacements:

$$u_i^{(0)} = \bar{u}_i \quad \text{on } \mathrm{S_u}, \tag{8}$$

Prescribed boundary tractions:

$$\sigma_{ij}^{(0)} n_j = \bar{t}_i \quad \text{on } \mathrm{S}_{\sigma}, \tag{9}$$

together with the traction and displacement conditions at the interfaces between the micro-constituents. For the sake of simplicity and clarity, we assume that the fields are continuous across the interfaces. The material property tensor D_{ijkl}^{ε} is symmetric with respect to indices (i, j, k, l). f_i represent body forces. σ_{ij}^{0} and e_{ij}^{0} are initial stresses and strains, respectively. Superscript (0) in parenthesis represents the zeroth-order solution, which will be clarified in the following. n_j are the direction cosines of the unit outward normal to $\partial\Omega$, the boundary of the domain Ω. $\partial\Omega$ is composed of the segment $\mathrm{S_u}$ on which the displacements $\bar{u}_i$ are prescribed and the segment S_{σ} on which the tractions $\bar{t}_i$ are prescribed.

2.1. *Two-scale expansion*

The displacement $u_i^{\varepsilon}(\mathbf{x}, \mathbf{y})$ is expanded in powers of the small number ε as[6–33]

$$u_i^{\varepsilon}(\mathbf{x}, \mathbf{y}) = u_i^{(0)}(\mathbf{x}, \mathbf{y}) + \varepsilon u_i^{(1)}(\mathbf{x}, \mathbf{y}) + \varepsilon^2 u_i^{(2)}(\mathbf{x}, \mathbf{y}) + \varepsilon^3 u_i^{(3)}(\mathbf{x}, \mathbf{y}) + \cdots, \tag{10}$$

where $u_i^{(0)}, u_i^{(1)}, u_i^{(2)}, \ldots$ are Y-periodic functions with respect to $\mathbf{y}$. Note that the partial derivatives in (5) and (6) with respect to coordinate $\mathbf{x}$ must also include the two-scale dependence. To show this explicitly, we will denote it here as d/dx_j

$$\frac{d}{dx_j} = \frac{\partial}{\partial x_j} + \frac{\partial}{\partial y_k}\frac{\partial y_k}{\partial x_j},$$

so that the derivatives of displacements in (6) should be understood as

$$\frac{du_i^\varepsilon(\mathbf{x},\mathbf{y})}{dx_j} = \frac{\partial u_i^\varepsilon(\mathbf{x},\mathbf{y})}{\partial x_j} + \frac{\partial u_i^\varepsilon(\mathbf{x},\mathbf{y})}{\partial y_k}\frac{\partial y_k}{\partial x_j} = \frac{\partial u_i^\varepsilon(\mathbf{x},\mathbf{y})}{\partial x_j} + \varepsilon^{-1}\frac{\partial u_i^\varepsilon(\mathbf{x},\mathbf{y})}{\partial y_j}, \tag{11}$$

where

$$\begin{aligned}\frac{\partial u_i^\varepsilon}{\partial x_j} &= \frac{\partial u_i^{(0)}}{\partial x_j} + \varepsilon\frac{\partial u_i^{(1)}}{\partial x_j} + \varepsilon^2\frac{\partial u_i^{(2)}}{\partial x_j} + \varepsilon^3\frac{\partial u_i^{(3)}}{\partial x_j} + \cdots,\\ \frac{\partial u_i^\varepsilon}{\partial y_j} &= \frac{\partial u_i^{(0)}}{\partial y_j} + \varepsilon\frac{\partial u_i^{(1)}}{\partial y_j} + \varepsilon^2\frac{\partial u_i^{(2)}}{\partial y_j} + \varepsilon^3\frac{\partial u_i^{(3)}}{\partial y_j} + \cdots.\end{aligned} \tag{12}$$

Substituting (10) into (6) gives the expansion of the strain e_{ij}^ε:

$$\begin{aligned}e_{ij}^\varepsilon = {}& \varepsilon^{-1}\frac{\partial u_i^{(0)}}{\partial y_j} + \frac{\partial u_i^{(0)}}{\partial x_j} + \frac{\partial u_i^{(1)}}{\partial y_j} + \varepsilon\left(\frac{\partial u_i^{(1)}}{\partial x_j} + \frac{\partial u_i^{(2)}}{\partial y_j}\right)\\ &+ \varepsilon^2\left(\frac{\partial u_i^{(2)}}{\partial x_j} + \frac{\partial u_i^{(3)}}{\partial y_j}\right) + \varepsilon^3\left(\frac{\partial u_i^{(3)}}{\partial x_j} + \frac{\partial u_i^{(4)}}{\partial y_j}\right) + \cdots\end{aligned}$$

or

$$e_{ij}^\varepsilon = \varepsilon^{-1}e_{ij}^{(-1)} + e_{ij}^{(0)} + \varepsilon^1 e_{ij}^{(1)} + \varepsilon^2 e_{ij}^{(2)} + \cdots \tag{13}$$

where

$$\begin{aligned}2e_{ij}^{(-1)} &= \frac{\partial u_i^{(0)}}{\partial y_j} + \frac{\partial u_j^{(0)}}{\partial y_i}\\ 2e_{ij}^{(k)} &= \frac{\partial u_i^{(k)}}{\partial x_j} + \frac{\partial u_i^{(k+1)}}{\partial y_j} + \frac{\partial u_j^{(k)}}{\partial x_i} + \frac{\partial u_j^{(k+1)}}{\partial y_i}, \quad k \geq 0.\end{aligned} \tag{14}$$

Substituting (13) into the constitutive relation (7) gives the expansion of the stress σ_{ij}^ε:

$$\begin{aligned}\sigma_{ij}^\varepsilon &= D_{ijkl}\left(\varepsilon^{-1}e_{kl}^{(-1)} + e_{kl}^{(0)} + \varepsilon^1 e_{kl}^{(1)} + \varepsilon^2 e_{kl}^{(2)} - e_{kl}^0\right) + \sigma_{ij}^0\\ &= \varepsilon^{-1}\sigma_{ij}^{(-1)} + \sigma_{ij}^{(0)} + \varepsilon\sigma_{ij}^{(1)} + \varepsilon^2\sigma_{ij}^{(2)},\end{aligned} \tag{15}$$

where

$$\sigma_{ij}^{(-1)} = D_{ijkl} e_{kl}^{(-1)},$$
$$\sigma_{ij}^{(0)} = D_{ijkl} \left(e_{kl}^{(0)} - e_{kl}^{0} \right) + \sigma_{ij}^{0}, \tag{16}$$
$$\sigma_{ij}^{(k)} = D_{ijkl} e_{kl}^{(k)}.$$

Note that (we have again used the full derivative to emphasise the two-scale dependence)

$$\frac{d\sigma_{ij}^{\varepsilon}}{dx_j} = \varepsilon^{-2} \frac{\partial \sigma_{ij}^{(-1)}}{\partial y_j} + \varepsilon^{-1} \left(\frac{\partial \sigma_{ij}^{(-1)}}{\partial x_j} + \frac{\partial \sigma_{ij}^{(0)}}{\partial y_j} \right) + \frac{\partial \sigma_{ij}^{(0)}}{\partial x_j} + \frac{\partial \sigma_{ij}^{(1)}}{\partial y_j} + \varepsilon \left(\frac{\partial \sigma_{ij}^{(1)}}{\partial x_j} + \frac{\partial \sigma_{ij}^{(2)}}{\partial y_j} \right). \tag{17}$$

Inserting the asymptotic expansion for the stress field (15) into the equilibrium equation (5) and collecting the terms of like powers in ε give the following sequence of equilibrium equations:

$O(\varepsilon^{-2})$:

$$\frac{\partial \sigma_{ij}^{(-1)}}{\partial y_j} = 0, \tag{18}$$

$O(\varepsilon^{-1})$:

$$\frac{\partial \sigma_{ij}^{(-1)}}{\partial x_j} + \frac{\partial \sigma_{ij}^{(0)}}{\partial y_j} = 0, \tag{19}$$

$O(\varepsilon^{0})$:

$$\frac{\partial \sigma_{ij}^{(0)}}{\partial x_j} + \frac{\partial \sigma_{ij}^{(1)}}{\partial y_j} + f_i = 0, \tag{20}$$

$O(\varepsilon^{k})$:

$$\frac{\partial \sigma_{ij}^{(k)}}{\partial x_j} + \frac{\partial \sigma_{ij}^{(k+1)}}{\partial y_j} = 0 \quad (k \geq 1). \tag{21}$$

In the following will discuss the method for solving these equations.

2.2. $O(\varepsilon^{-2})$ equilibrium: Solution structure of $u_i^{(0)}$

We first consider the $O(\varepsilon^{-2})$ equilibrium equation (18) in Y. Premultiplying it by $u_i^{(0)}$, integrating over Y, followed by integration by parts, yields

$$\begin{aligned}\int_Y u_i^{(0)} \frac{\partial \sigma_{ij}^{(-1)}}{\partial y_j} dY &= \oint_{\partial Y} u_i^{(0)} \sigma_{ij}^{(-1)} n_j \, dS - \int_Y \frac{\partial u_i^{(0)}}{\partial y_j} \sigma_{ij}^{(-1)} \, dY \\ &= - \int_Y \frac{\partial u_i^{(0)}}{\partial y_j} \sigma_{ij}^{(-1)} \, dY = - \int_Y \frac{\partial u_i^{(0)}}{\partial y_j} D_{ijkl} \frac{\partial u_k^{(0)}}{\partial y_l} \, dY \\ &= 0, \end{aligned} \tag{22}$$

where ∂Y denotes the boundary of Y. The boundary integral term in (22) vanishes due to the periodicity of the boundary conditions in Y, because $u_i^{(0)}$ and $\sigma_{ij}^{(-1)}$ are identical on the opposite sides of the unit cell, while the corresponding normals n_j are in opposite directions. Taking into account the positive definiteness of the symmetric constitutive tensor D_{ijkl} (its superscript ε has been omitted for clarity), we have

$$\frac{\partial u_i^{(0)}}{\partial y_j} = 0 \quad \text{or} \quad u_i^{(0)}(\mathbf{x}, \mathbf{y}) = u_i^{(0)}(\mathbf{x}), \tag{23}$$

and

$$\varepsilon_{ij}^{(-1)}(\mathbf{x}, \mathbf{y}) = 0, \quad \sigma_{ij}^{(-1)}(\mathbf{x}, \mathbf{y}) = 0. \tag{24}$$

2.3. $O(\varepsilon^{-1})$ equilibrium: First-order homogenisation and solution structure of $u_m^{(1)}$

Next, we proceed to the $O(\varepsilon^{-1})$ equilibrium equation (19). From (14) and (16) and taking into account (24), it follows that

$$\frac{\partial \sigma_{ij}^{(0)}}{\partial y_j} = 0,$$

or

$$\frac{\partial}{\partial y_j}\left(D_{ijkl} \frac{\partial u_k^{(1)}}{\partial y_l}\right) + \frac{\partial}{\partial y_j}\left(D_{ijkl} \frac{\partial u_k^{(0)}}{\partial x_l}\right) + \frac{\partial}{\partial y_j}\left(\sigma_{ij}^0 - D_{ijkl} e_{kl}^0\right) = 0. \tag{25}$$

Without loss of generality, assume that

$$\sigma_{ij}^{0} = \sigma_{ij}^{0}(\mathbf{x}), \quad e_{ij}^{0} = e_{ij}^{0}(\mathbf{x}). \tag{26}$$

Equations (25) can be rewritten as

$$\frac{\partial}{\partial y_j}\left(D_{ijkl}\frac{\partial u_k^{(1)}}{\partial y_l}\right) = -\frac{\partial D_{ijkl}}{\partial y_j}\left(\frac{\partial u_k^{(0)}}{\partial x_l} - e_{kl}^{0}\right). \tag{27}$$

Based on the form of the right-hand side of (27) that permits a separation of variables, $u_k^{(1)}$ may be expressed as

$$u_m^{(1)}(\mathbf{x},\mathbf{y}) = \chi_m^{kl}(\mathbf{y})\left(\frac{\partial u_k^{(0)}(\mathbf{x})}{\partial x_l} - e_{kl}^{0}(\mathbf{x})\right), \tag{28}$$

where $\chi_m^{kl}(\mathbf{y})$ is a Y-periodic function defined in the unit cell Y. Substituting (28) into (27) and taking into account the arbitrariness of the macroscopic strain field,

$$\frac{\partial u_k^{(0)}(\mathbf{x})}{\partial x_l} - e_{kl}^{0}(\mathbf{x})$$

within an RUC, we have

$$\frac{\partial}{\partial y_j}\left(D_{ijmn}\frac{\partial \chi_m^{kl}}{\partial y_n}\right) = -\frac{\partial D_{ijkl}}{\partial y_j}. \tag{29}$$

We can also write

$$\begin{aligned}\sigma_{ij}^{(0)} &= D_{ijkl}\left(\frac{\partial u_k^{(0)}}{\partial x_l} + \frac{\partial u_k^{(1)}}{\partial y_l} - e_{kl}^{0}\right) + \sigma_{ij}^{0} \\ &= D_{ijkl}\frac{\partial u_k^{(0)}}{\partial x_l} + D_{ijkl}\frac{\partial \chi_k^{mn}}{\partial y_l}\left(\frac{\partial u_m^{(0)}}{\partial x_n} - e_{mn}^{0}\right) + \sigma_{ij}^{0} - D_{ijkl}e_{kl}^{0}.\end{aligned} \tag{30}$$

2.4. *$O(\varepsilon^0)$ equilibrium: Second-order homogenisation*

We now consider the $O(\varepsilon^0)$ equilibrium equation (20).

2.4.1. *Solution structure of* $u_k^{(2)}$

Differentiating equation (20) with respect to y_i,

$$\frac{\partial^2 \sigma_{ij}^{(0)}}{\partial y_i \partial x_j} + \frac{\partial^2 \sigma_{ij}^{(1)}}{\partial y_i \partial y_j} + \frac{\partial f_i}{\partial y_i} = 0. \tag{31}$$

Without loss of generality, assume

$$\frac{\partial f_i}{\partial y_i} = 0, \tag{32}$$

and make use of (25), so that (31) becomes

$$\frac{\partial^2 \sigma_{ij}^{(1)}}{\partial y_i \partial y_j} = 0. \tag{33}$$

From (14) and (16) and making use of (28)

$$\begin{aligned}\sigma_{ij}^{(1)} &= D_{ijkl} e_{kl}^{(1)} = D_{ijkl} \left(\frac{\partial u_k^{(1)}}{\partial x_l} + \frac{\partial u_k^{(2)}}{\partial y_l} \right) \\ &= D_{ijkl} \left[\chi_k^{mn}(\mathbf{y}) \frac{\partial}{\partial x_l} \left(\frac{\partial u_m^{(0)}(\mathbf{x})}{\partial x_n} - e_{mn}^0(\mathbf{x}) \right) + \frac{\partial u_k^{(2)}}{\partial y_l} \right].\end{aligned} \tag{34}$$

We thus have

$$u_k^{(2)}(\mathbf{x}, \mathbf{y}) = \psi_k^{mno}(\mathbf{y}) \frac{\partial}{\partial x_o} \left(\frac{\partial u_m^{(0)}(\mathbf{x})}{\partial x_n} - e_{mn}^0(\mathbf{x}) \right) \tag{35}$$

from (33).

2.4.2. *Solution of* $u_m^{(0)}$

Integrating (20) over the unit cell domain Y yields

$$\frac{\partial}{\partial x_j} \int_Y \sigma_{ij}^{(0)} \, dY + \int_Y \frac{\partial \sigma_{ij}^{(1)}}{\partial y_j} \, dY + \int_Y f_i \, dY = 0. \tag{36}$$

Taking into account the periodicity of $\sigma_{ij}^{(1)}$ on Y, the second term vanishes

$$\int_Y \frac{\partial \sigma_{ij}^{(1)}}{\partial y_j} \, dY = \oint_{\partial Y} \sigma_{ij}^{(1)} n_j \, dS = 0.$$

Substituting (30) into (36) yields

$$\frac{\partial}{\partial x_j}\left(\bar{D}_{ijkl}\frac{\partial u_k^{(0)}}{\partial x_l}\right)+\frac{1}{Y}\frac{\partial}{\partial x_j}\int_Y \sigma_{ij}^0\,dY-\frac{\partial}{\partial x_j}(\bar{D}_{ijkl}e_{kl}^0)$$
$$+\frac{1}{Y}\int_Y f_i\,dY=0. \tag{37}$$

This is an equilibrium equation for a homogeneous medium (cf. (5)) with constant material properties $\bar{D}_{ijkl}$, which are usually termed as the homogenised or effective material properties and are given by

$$\begin{aligned}\bar{D}_{ijmn}&=\frac{1}{Y}\int_Y \tilde{D}_{ijmn}\,dY,\\ \tilde{D}_{ijmn}&=D_{ijkl}\left(\delta_{km}\delta_{ln}+\frac{\partial\chi_k^{mn}(\mathbf{y})}{\partial y_l}\right),\end{aligned} \tag{38}$$

where the integration is over the area or volume Y of the RUC.

In the widely used first-order homogenisation, displacements to order $u_m^{(1)}$ are solved; in a like manner the equations to order $O(\varepsilon^{-1})$ are considered. Equation (37) results from constraints from higher-order equilibrium and is used directly to solve for $u_m^{(0)}$. Hence no more constraints are required.

2.4.3. *Solution of* $\psi_k^{mno}(\mathbf{y})$

$\psi_k^{mno}(\mathbf{y})$ can be solved out from (33) on Y with the use of (35). In order to avoid higher-order derivatives, we can solve them from (20) instead. With the use of (30), (34), (35) and (38), (20) becomes

$$\frac{\partial}{\partial x_j}\left(\tilde{D}_{ijkl}\frac{\partial u_k^{(0)}}{\partial x_l}\right)+\frac{\partial}{\partial x_j}(\sigma_{ij}^0-\tilde{D}_{ijkl}e_{kl}^0)+\frac{\partial}{\partial y_j}$$
$$\times\left\{D_{ijkl}\left[\chi_k^{mn}(\mathbf{y})\delta_{lo}+\frac{\partial\psi_k^{mno}(\mathbf{y})}{\partial y_l}\right]\frac{\partial}{\partial x_o}\left(\frac{\partial u_m^{(0)}}{\partial x_n}-e_{mn}^0\right)\right\}+f_i=0,$$

or

$$\frac{\partial}{\partial y_j}\left[D_{ijkl}\frac{\partial\psi_k^{mno}(\mathbf{y})}{\partial y_l}\right]\frac{\partial}{\partial x_o}\left(\frac{\partial u_m^{(0)}}{\partial x_n}-e_{mn}^0\right)+\hat{f}_i=0, \tag{39}$$

where

$$\hat{f}_i = f_i + \frac{\partial}{\partial x_j}\left(\tilde{D}_{ijkl}\frac{\partial u_k^{(0)}}{\partial x_l}\right) + \frac{\partial}{\partial x_j}(\sigma_{ij}^0 - \tilde{D}_{ijkl}e_{kl}^0)$$
$$+ \frac{\partial}{\partial y_j}\left[D_{ijkl}\chi_k^{mn}(\mathbf{y})\right]\frac{\partial}{\partial x_l}\left(\frac{\partial u_m^{(0)}}{\partial x_n} - e_{mn}^0\right). \tag{40}$$

Although $u_k^{(2)}$ can also be separated into **x**- and **y**-functions as in (35), this variable separation does not benefit the solution process. Hence, we can solve u_k^2 directly from

$$\frac{\partial}{\partial y_j}\left[D_{ijkl}\frac{\partial u_k^{(2)}(\mathbf{x},\mathbf{y})}{\partial y_l}\right] + \hat{f}_i = 0 \tag{41}$$

instead of (39).

2.4.4. *Constraints from higher-order solutions*

If the expansion is truncated to the second-order term $u_k^{(2)}$, its contribution to the $O(\varepsilon^1)$ order equilibrium equation also needs to be considered. The unwanted higher-order term $u_k^{(3)}$ in the equation can be eliminated by integrating the complete $O(\varepsilon^1)$ order equilibrium equation over Y. We thus have

$$\frac{\partial}{\partial x_j}\int_Y \left\{D_{ijkl}\left[\chi_k^{mn}(\mathbf{y})\delta_{lo} + \frac{\partial \psi_k^{mno}(\mathbf{y})}{\partial y_l}\right]\right\}dY \cdot \frac{\partial}{\partial x_o}\left(\frac{\partial u_m^{(0)}}{\partial x_n} - e_{mn}^0\right) = 0. \tag{42}$$

For the convenience of solution but without loss of generality, we can assume

$$\int_Y D_{ijkl}\chi_k^{mn}(\mathbf{y})\,dY \cdot \frac{\partial}{\partial x_l}\left(\frac{\partial u_m^{(0)}}{\partial x_n} - e_{mn}^0\right) = 0,$$
$$\int_Y D_{ijkl}\frac{\partial \psi_k^{mno}(\mathbf{y})}{\partial y_l}\,dY \cdot \frac{\partial}{\partial x_o}\left(\frac{\partial u_m^{(0)}}{\partial x_n} - e_{mn}^0\right) = 0, \tag{43}$$

or

$$\int_Y D_{ijkl}\frac{\partial u_k^{(2)}}{\partial y_l}dY = 0. \tag{44}$$

Note that it is not necessary to take (43) into account in the first-order homogenisation.

2.5. $O(\varepsilon^1)$ *equilibrium: Third-order homogenisation*

2.5.1. *Solution of* $u_k^{(3)}$

With the use of (14), (16) and (34), Eq. (21) becomes

$$\frac{\partial \sigma_{ij}^{(1)}}{\partial x_j} + \frac{\partial \sigma_{ij}^{(2)}}{\partial y_j} = \frac{\partial}{\partial x_j}\left\{D_{ijkl}\left[\chi_k^{mn}(\mathbf{y})\frac{\partial}{\partial x_l}\left(\frac{\partial u_m^{(0)}}{\partial x_n} - e_{mn}^0\right) + \frac{\partial u_k^{(2)}}{\partial y_l}\right]\right\}$$
$$+ \frac{\partial}{\partial y_j}\left\{D_{ijkl}\left[\frac{\partial u_k^{(2)}}{\partial x_l} + \frac{\partial u_k^{(3)}}{\partial y_l}\right]\right\} = 0. \tag{45}$$

In the solution of $u_m^{(1)}$, it is advantageous to separate it into **x**- and **y**-dependent terms, i.e. **x**-dependent terms disappear in the solution of **y**-dependent terms. In the solution of $u_k^{(2)}$, this advantage disappears, although it can also be theoretically expressed in separable **x**- and **y**-dependent terms. From (45), if we assume that D_{ijkl} is not explicitly dependent on **x**, $u_k^{(3)}$ can also be separated into **x**- and **y**-dependent terms with the **x**-dependent terms being

$$\frac{\partial^2}{\partial x_l \partial x_o}\left(\frac{\partial u_m^{(0)}}{\partial x_n} - e_{mn}^0\right).$$

As this variable separation does not benefit the solution process, $u_k^{(3)}$ is solved directly from (45) on Y. Equation (45) can be rewritten into the following equations:

$$\frac{\partial}{\partial y_j}\left(D_{ijkl}\frac{\partial u_k^{(3)}}{\partial y_l}\right) - f_i^{(3)} = 0, \tag{46}$$

where

$$f_i^{(3)} = -\frac{\partial}{\partial x_j}\left\{D_{ijkl}\left[\chi_k^{mn}(\mathbf{y})\frac{\partial}{\partial x_l}\left(\frac{\partial u_m^{(0)}}{\partial x_n} - e_{mn}^0\right) + \frac{\partial u_k^{(2)}}{\partial y_l}\right]\right\}$$
$$- \frac{\partial}{\partial y_j}\left\{D_{ijkl}\frac{\partial u_k^{(2)}}{\partial x_l}\right\}. \tag{47}$$

2.5.2. *Constraints from higher-order terms*

For the same reason as in Sec. 2.4.4, we need also to consider the $O(\varepsilon^2)$ order equilibrium equation. Again, integration of the complete $O(\varepsilon^2)$ order equilibrium equation over Y gives

$$\frac{\partial}{\partial x_j}\int_Y \left\{ D_{ijkl}\left[\psi_k^{mno}(\mathbf{y})\frac{\partial^2}{\partial x_l \partial x_o}\left(\frac{\partial u_m^{(0)}}{\partial x_n} - e_{mn}^o\right) + \frac{\partial u_k^{(3)}}{\partial y_l}\right]\right\} dY = 0. \tag{48}$$

As in Sec. 2.2.4, we can now assume

$$\int_Y D_{ijkl}\psi_k^{mno}(\mathbf{y})\frac{\partial^2}{\partial x_l \partial x_o}\left(\frac{\partial u_m^{(0)}}{\partial x_n} - e_{mn}^0\right) dY$$
$$= \int_Y D_{ijkl}\frac{\partial u_k^{(2)}}{\partial x_l} dY = 0, \tag{49}$$

$$\int_Y D_{ijkl}\frac{\partial u_k^{(3)}}{\partial y_l} dY = 0. \tag{50}$$

Obviously, terms higher than the third-order can be solved in a similar way. The controlling equations for the pth order ($p \geq 3$) displacements are

$$\frac{\partial}{\partial y_j}\left(D_{ijkl}\frac{\partial u_k^{(p)}}{\partial y_l}\right) - f_i^{(p)} = 0, \tag{51}$$

$$f_i^{(p)} = -\frac{\partial}{\partial x_j}\left\{D_{ijkl}\left[\frac{\partial u_k^{(p-2)}}{\partial x_l} + \frac{\partial u_k^{(p-1)}}{\partial y_l}\right]\right\}$$
$$- \frac{\partial}{\partial y_j}\left\{D_{ijkl}\frac{\partial u_k^{(p-1)}}{\partial x_l}\right\}. \tag{52}$$

However, in the numerical implementation, although it is only required to solve a second-order equilibrium equation on the RUC (cf. (41) and (46)), it is actually limited by the requirement of the higher-order derivatives of the solution $u_i^{(0)}$ at the macro scale (cf. (40) and (47)).

3. Variational Formulation of Problem (29)

To solve the deformation of composite materials or structures by the first- or higher-order homogenisation method, together with the numerical methods, e.g. the FEM, we will first solve for $\chi_i^{kl}(\mathbf{y})$ from Eq. (29) assuming it to be a Y-periodic function defined in Y. The effective material properties $\bar{D}_{ijkl}$ are given by (38). We then solve the homogeneous macro problem (37) and obtain the macroscopic displacement fields $u_i^{(0)}$. $u_i^{(1)}$ (28) can then be obtained from $\chi_i^{kl}(\mathbf{y})$ and $u_i^{(0)}$. $u_i^{(2)}$ can next be solved from (41) with constraints (43) and (44); $u_i^{(3)}$can be solved from (46) with constraints (49) and (50). Higher-order displacement terms can be solved in a similar way. The strains e_{ij} and stresses σ_{ij} can be calculated from (14) and (16), respectively. Equations (37), (41), (46) and (51) are standard second-order partial differential equations in solid mechanics. They can be solved in a similar way. However, for a problem defined on the RUC Y the periodic boundary conditions and constraints from higher-order equilibrium should be enforced appropriately. Equation (29) is slightly different. However, it is also a second-order partial differential equation. In the remainder of this section, we will derive the corresponding variational formulation following Karihaloo *et al.*[20]

Corresponding to the equilibrium equation (29), the virtual work principle states that

$$\int_Y \delta\chi_i^{kl} \frac{\partial}{\partial y_j}\left(D_{ijmn}\frac{\partial \chi_m^{kl}}{\partial y_n}\right) dY + \int_Y \delta\chi_i^{kl} \frac{\partial D_{ijkl}}{\partial y_j}\, dY = 0,$$

where $\delta\chi_i^{kl}$ are arbitrary Y-periodic functions defined in the RUC Y. Integration of the above equation by parts yields

$$\oint_{\partial Y} \delta\chi_i^{kl} D_{ijmn}\frac{\partial \chi_m^{kl}}{\partial y_n} n_j\, ds + \oint_{\partial Y} \delta\chi_i^{kl} D_{ijkl} n_j\, ds$$
$$- \int_Y \frac{\partial \delta\chi_i^{kl}}{\partial y_j} D_{ijmn}\frac{\partial \chi_m^{kl}}{\partial y_n}\, dY - \int_Y \frac{\partial \delta\chi_i^{kl}}{\partial y_j} D_{ijkl}\, dY = 0.$$

The boundary integral terms in the above equation vanish due to the Y-periodicity of χ_i^{kl} and $\delta\chi_i^{kl}$. Thus, we have

$$\int_Y \frac{\partial \delta\chi_i^{kl}}{\partial y_j} D_{ijmn}\frac{\partial \chi_m^{kl}}{\partial y_n} dY + \int_Y \frac{\partial \delta\chi_i^{kl}}{\partial y_j} D_{ijkl}\, dY = 0. \tag{53}$$

Based on Eq. (53), displacement elements can be constructed in a standard manner.

It is easy to prove that (53) is the first-order variation of the following potential functional:

$$\Pi_P(\chi_i^{kl}) = \int_Y \frac{1}{2}\frac{\partial \chi_i^{kl}}{\partial y_j} D_{ijmn} \frac{\partial \chi_m^{kl}}{\partial y_n}\, dY + \int_Y \frac{\partial \chi_i^{kl}}{\partial y_j} D_{ijkl}\, dY,$$

or in matrix form,

$$\Pi_P(\boldsymbol{\chi}^{kl}) = \int_Y \frac{1}{2}(\tilde{\mathbf{e}}^{kl})^{\mathrm{T}}\mathbf{D}\tilde{\mathbf{e}}^{kl}\, dY + \int_Y (\tilde{\mathbf{e}}^{kl})^{\mathrm{T}}\mathbf{D}\, dY. \tag{54}$$

If we define the strain

$$\tilde{e}_{ij}^{kl} = \frac{\partial \chi_i^{kl}}{\partial y_j} \tag{55}$$

and the stress

$$\tilde{\sigma}_{ij}^{kl} = D_{ijmn}\tilde{e}_{mn}^{kl}, \tag{56}$$

so that

$$\tilde{e}_{ij}^{kl} = D_{ijmn}^{-1}\tilde{\sigma}_{mn}^{kl}, \tag{57}$$

which are Y-periodic functions in the RUC, we have a 2-field Hellinger–Reissner functional

$$\Pi_{\mathrm{HR}}\left(\chi_i^{kl}, \tilde{\sigma}_{ij}^{kl}\right) = \int_Y \left[-\frac{1}{2}\tilde{\sigma}_{ij}^{kl} D_{ijmn}^{-1}\tilde{\sigma}_{mn}^{kl} + \tilde{\sigma}_{ij}^{kl}\frac{\partial \chi_i^{kl}}{\partial y_j} - \frac{\partial D_{ijkl}}{\partial y_j}\chi_i^{kl}\right] dY,$$

or equivalently

$$\Pi_{\mathrm{HR}}\left(\chi_i^{kl}, \tilde{\sigma}_{ij}^{kl}\right) = \int_Y \left[-\frac{1}{2}\tilde{\sigma}_{ij}^{kl} D_{ijmn}^{-1}\tilde{\sigma}_{mn}^{kl} + \tilde{\sigma}_{ij}^{kl}\frac{\partial \chi_i^{kl}}{\partial y_j} + D_{ijkl}\frac{\partial \chi_i^{kl}}{\partial y_j}\right] dY,$$

and

$$\Pi_{\mathrm{HR}}(\boldsymbol{\chi}^{kl}, \tilde{\boldsymbol{\sigma}}^{kl}) = \int_Y \left[-\frac{1}{2}(\tilde{\boldsymbol{\sigma}}^{kl})^{\mathrm{T}}\mathbf{D}^{-1}\tilde{\boldsymbol{\sigma}}^{kl} + (\tilde{\boldsymbol{\sigma}}^{kl})^{\mathrm{T}}\partial(\boldsymbol{\chi}^{kl}) + \mathbf{D}\partial(\boldsymbol{\chi}^{kl})\right] dY \tag{58}$$

in matrix form.

By making use of the Lagrange multiplier method and relaxing the compatibility condition in the potential principle (54), or by employing Legendre transformation,

$$\tilde{\sigma}_{ij}^{kl}\tilde{e}_{ij}^{kl} = \frac{1}{2}\tilde{e}_{ij}^{kl}D_{ijmn}\tilde{e}_{mn}^{kl} + \frac{1}{2}\tilde{\sigma}_{ij}^{kl}D_{ijmn}^{-1}\tilde{\sigma}_{mn}^{kl} \tag{59}$$

on the Hellinger–Reissner functional (58), one arrives at the 3-field Hu–Washizu functional

$$\begin{aligned}&\Pi_{\mathrm{HW}}(\chi_i^{kl}, \tilde{e}_{ij}^{kl}, \tilde{\sigma}_{ij}^{kl})\\ &\quad = \int_Y \left[\frac{1}{2}\tilde{e}_{ij}^{kl}D_{ijmn}\tilde{e}_{mn}^{kl} - \tilde{\sigma}_{ij}^{kl}\left(\tilde{e}_{ij}^{kl} - \frac{\partial \chi_i^{kl}}{\partial y_j}\right) + D_{ijkl}\frac{\partial \chi_i^{kl}}{\partial y_j}\right] dY\end{aligned}$$

or

$$\begin{aligned}&\Pi_{\mathrm{HW}}(\boldsymbol{\chi}^{kl}, \tilde{\mathbf{e}}^{kl}, \tilde{\boldsymbol{\sigma}}^{kl})\\ &\quad = \int_Y \left\{\frac{1}{2}(\tilde{\mathbf{e}}^{kl})^{\mathrm{T}}\mathbf{D}\tilde{\mathbf{e}}^{kl} - (\tilde{\boldsymbol{\sigma}}^{kl})^{\mathrm{T}}[\tilde{\mathbf{e}}^{kl} - \partial(\boldsymbol{\chi}^{kl})] + \mathbf{D}\partial(\boldsymbol{\chi}^{kl})\right\} dY\end{aligned} \tag{60}$$

in matrix form.

Based on the functionals (58) and (60), multivariable finite elements (FEs) can be established.

In the following, we will give the differential operator ∂ and material modulus matrix $\mathbf{D}$.[39]

For plane stress, $\sigma_3 = \sigma_{13} = \sigma_{23} = 0$, which is suitable for analysing structures that are thin in the out of plane direction, e.g. thin plates subject to in-plane loading, the differential operator ∂ for infinitesimal strain–displacement relationship is defined as

$$\begin{Bmatrix} e_{y_1} \\ e_{y_2} \\ e_{y_1y_2} \end{Bmatrix} = \partial\mathbf{u} = \begin{bmatrix} \dfrac{\partial}{\partial y_1} & 0 \\ 0 & \dfrac{\partial}{\partial y_2} \\ \dfrac{\partial}{\partial y_2} & \dfrac{\partial}{\partial y_1} \end{bmatrix} \begin{Bmatrix} u_1 \\ v_1 \end{Bmatrix}. \tag{61}$$

The isotropic elastic modulus matrix is

$$D = \frac{E}{1-\nu^2}\begin{bmatrix} 1 & \nu & 0 \\ \nu & 1 & 0 \\ 0 & 0 & \dfrac{1-\nu}{2} \end{bmatrix}, \tag{62}$$

where E and ν are Young's modulus and Poisson's ratio, respectively. The orthotropic elastic modulus matrix in the principal coordinates of orthotropy is

$$D = \begin{bmatrix} \dfrac{1}{E_1} & -\dfrac{\nu_{12}}{E_1} & 0 \\ -\dfrac{\nu_{12}}{E_1} & \dfrac{1}{E_2} & 0 \\ 0 & 0 & \dfrac{1}{G_{12}} \end{bmatrix}^{-1}, \tag{63}$$

where ν_{21} has been set to $\nu_{12}E_2/E_1$ to maintain symmetry. For a valid material $\nu_{12} < (E_1/E_2)^{1/2}$. The out of plane strain component is

$$e_3 = -\frac{\nu}{E}(\sigma_1 + \sigma_2) \quad \text{for isotropic materials,}$$
$$e_3 = -\frac{\nu_{13}}{E_1}\sigma_1 - \frac{\nu_{23}}{E_2}\sigma_2 \quad \text{for orthotropic materials.}$$

For plane strain, $e_3 = e_{23} = e_{13} = 0$, which is suitable for analysing structures that are thick in the out of plane direction, e.g. dams or thick cylinders, the differential operator ∂ for infinitesimal strain–displacement relationship is the same as in plane stress, i.e. (61). The isotropic elastic modulus matrix is

$$D = \frac{E}{(1+\nu)(1-2\nu)} \begin{bmatrix} 1-\nu & \nu & 0 \\ \nu & 1-\nu & 0 \\ 0 & 0 & \dfrac{1-2\nu}{2} \end{bmatrix}, \tag{64}$$

and the orthotropic elastic modulus matrix in the principal coordinates of orthotropy is

$$D = \begin{bmatrix} \dfrac{E_3 - \nu_{13}^2 E_1}{E_1 E_3} & \dfrac{-\nu_{12}E_3 - \nu_{13}\nu_{23}E_2}{E_2 E_3} & 0 \\ \dfrac{-\nu_{12}E_3 - \nu_{23}\nu_{13}E_1}{E_1 E_3} & \dfrac{E_3 - \nu_{23}^2 E_2}{E_2 E_3} & 0 \\ 0 & 0 & \dfrac{1}{G_{12}} \end{bmatrix}^{-1}, \tag{65}$$

where for symmetry

$$E_2(\nu_{12}E_3 + \nu_{23}\nu_{13}E_1) = E_1(\nu_{12}E_3 + \nu_{13}\nu_{23}E_2).$$

To obtain a valid material

$$\nu_{12} < (E_1/E_2)^{1/2}, \quad \nu_{13} < (E_1/E_3)^{1/2}, \quad \nu_{23} < (E_2/E_3)^{1/2}.$$

The out of plane stress component is

$$\sigma_3 = \nu(\sigma_1 + \sigma_2) \quad \text{for isotropic materials,}$$
$$\sigma_3 = \nu_{13}\frac{E_3}{E_1}\sigma_1 + \nu_{23}\frac{E_3}{E_2}\sigma_2 \quad \text{for orthotropic materials.}$$

The differential operator ∂ for 3D infinitesimal strain–displacement relationship is defined as

$$\begin{Bmatrix} e_1 \\ e_2 \\ e_3 \\ 2e_{12} \\ 2e_{23} \\ 2e_{13} \end{Bmatrix} = \partial\mathbf{u} = \begin{bmatrix} \frac{\partial}{\partial y_1} & 0 & 0 \\ 0 & \frac{\partial}{\partial y_2} & 0 \\ 0 & 0 & \frac{\partial}{\partial y_3} \\ \frac{\partial}{\partial y_2} & \frac{\partial}{\partial y_1} & 0 \\ 0 & \frac{\partial}{\partial y_3} & \frac{\partial}{\partial y_2} \\ \frac{\partial}{\partial y_3} & 0 & \frac{\partial}{\partial y_1} \end{bmatrix} \begin{Bmatrix} u_1 \\ u_2 \\ u_3 \end{Bmatrix}. \tag{66}$$

The isotropic elastic modulus matrix is

$$D = \frac{E}{(1+\nu)(1-2\nu)} \begin{bmatrix} 1-\nu & \nu & \nu & 0 & 0 & 0 \\ \nu & 1-\nu & \nu & 0 & 0 & 0 \\ \nu & \nu & 1-\nu & 0 & 0 & 0 \\ 0 & 0 & 0 & \frac{1-2\nu}{2} & 0 & 0 \\ 0 & 0 & 0 & 0 & \frac{1-2\nu}{2} & 0 \\ 0 & 0 & 0 & 0 & 0 & \frac{1-2\nu}{2} \end{bmatrix}, \tag{67}$$

and the orthotropic elastic modulus matrix in the principal coordinates of orthotropy is

$$D = \begin{bmatrix} \frac{1}{E_1} & -\frac{\nu_{21}}{E_2} & -\frac{\nu_{31}}{E_3} & 0 & 0 & 0 \\ -\frac{\nu_{12}}{E_1} & \frac{1}{E_2} & -\frac{\nu_{32}}{E_3} & 0 & 0 & 0 \\ -\frac{\nu_{13}}{E_1} & -\frac{\nu_{23}}{E_2} & \frac{1}{E_3} & 0 & 0 & 0 \\ 0 & 0 & 0 & \frac{1}{G_{12}} & 0 & 0 \\ 0 & 0 & 0 & 0 & \frac{1}{G_{23}} & 0 \\ 0 & 0 & 0 & 0 & 0 & \frac{1}{G_{13}} \end{bmatrix}^{-1}, \tag{68}$$

where ν_{21}, ν_{31} and ν_{32} are defined by

$$\nu_{21} = \nu_{12}\frac{E_2}{E_1}, \quad \nu_{31} = \nu_{13}\frac{E_3}{E_1}, \quad \nu_{32} = \nu_{23}\frac{E_3}{E_2}$$

to maintain symmetry. To obtain a valid material, ν_{12}, ν_{13} and ν_{23} need to meet the same constraints as in plane strain.

4. Finite Element Methods

From Sec. 2, all subproblems derived from the first as well as higher-order homogenisation are second-order elliptic partial differential equations. Therefore, in this section we will give an overview of all high-performance FEMs applicable to the solution of these subproblems. We will start our discussion from the standard solid mechanics problems defined on the RUC Y. Subdivide the RUC domain Y into FE subdomains Y^e, such that $\cup Y^e = Y$, $Y^a \cap Y^b = \emptyset$ and $\partial Y^a \cap \partial Y^b = S_{ab}$ (a, b are arbitrary elements). The elemental potential functional is

$$\Pi_p^{(e)}(u_i) = \int_{Y^e} \left(\frac{1}{2}e_{ij}D_{ijkl}e_{kl} - f_i u_i\right) dY - \int_{S_\sigma^e} \bar{t}_i u_i \, ds$$

or

$$\Pi_p^{(e)}(\mathbf{u}) = \int_{Y^e} \left(\frac{1}{2}\mathbf{e}^{\mathrm{T}}\mathbf{D}\mathbf{e} - \mathbf{u}^{\mathrm{T}}\mathbf{f}\right) dY - \int_{S_\sigma^e} \mathbf{u}^{\mathrm{T}}\bar{\mathbf{t}} \, ds. \tag{69}$$

The 2-field Hellinger–Reissner elemental functional is

$$\Pi_{\mathrm{HR}}^{(e)}(u_i, \sigma_{ij}) = \int_{Y^e} \left(-\frac{1}{2}\sigma_{ij} D_{ijkl}^{-1} \sigma_{kl} + \sigma_{ij} \frac{\partial u_k}{\partial y_l} - f_i u_i \right) dY - \int_{S_\sigma^e} \bar{t}_i u_i \, ds$$

or

$$\Pi_{\mathrm{HR}}^{(e)}(\mathbf{u}, \boldsymbol{\sigma}) = \int_{Y^e} \left[-\frac{1}{2}\boldsymbol{\sigma}^{\mathrm{T}} \mathbf{S} \sigma + \boldsymbol{\sigma}^{\mathrm{T}} (\partial \mathbf{u}) - \mathbf{u}^{\mathrm{T}} \mathbf{f} \right] dY - \int_{S_\sigma^e} \mathbf{u}^{\mathrm{T}} \bar{\mathbf{t}} \, ds. \tag{70}$$

The 3-field Hu–Washizu elemental functional is

$$\Pi_{\mathrm{HW}}^{(e)}(u_i, e_{ij}, \sigma_{ij}) = \int_{Y^e} \left[\frac{1}{2} e_{ij} D_{ijkl} e_{kl} - \sigma_{ij} \left(e_{ij} - \frac{\partial u_i}{\partial y_j} \right) - f_i u_i \right] dY - \int_{S_\sigma^e} \bar{t}_i u_i \, ds,$$

or

$$\Pi_{\mathrm{HW}}^{(e)}(\mathbf{u}, \mathbf{e}, \boldsymbol{\sigma}) = \int_{Y^e} \left[\frac{1}{2} \mathbf{e}^{\mathrm{T}} \mathbf{D} \mathbf{e} - \boldsymbol{\sigma}^{\mathrm{T}} (\mathbf{e} - \partial \mathbf{u}) - \mathbf{u}^{\mathrm{T}} \mathbf{f} \right] dY - \int_{S_\sigma^e} \mathbf{u}^{\mathrm{T}} \bar{\mathbf{t}} \, ds \tag{71}$$

where Y^e is the area or volume of element 'e', S_σ^e is the part of the element boundary on which traction is prescribed.

Solution of the macro counterpart problems is similar. The only difference is that in the micro level, periodic boundary conditions and constraints from higher-order equilibrium need to be properly enforced. However, they will be considered in a later section.

4.1. *Displacement compatible elements from the potential principle*

Isoparametric compatible FEs utilise the same shape functions to interpolate both the displacements and geometry.[3–5] The approximate displacement field $\mathbf{u}$ in element 'e' is given as

$$\mathbf{u} = \mathbf{N}\mathbf{q}^e, \tag{72}$$

where $\mathbf{N}$ is the element shape function matrix and $\mathbf{q}^e$ is the vector of nodal displacements. For the 4-node quadrilateral isoparametric element

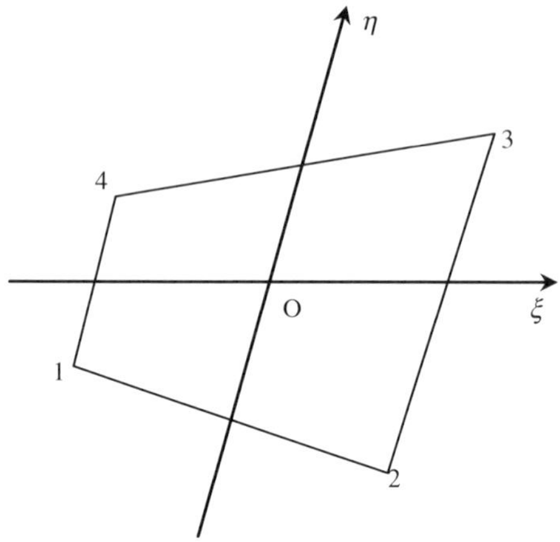

Fig. 2. A plane 4-node quadrilateral element.

shown in Fig. 2,

$$\mathbf{N} = \begin{bmatrix} N_1 & 0 & N_2 & 0 & N_3 & 0 & N_4 & 0 \\ 0 & N_1 & 0 & N_2 & 0 & N_3 & 0 & N_4 \end{bmatrix}, \tag{73}$$

where the bilinear interpolation function for node i

$$N_i = \frac{1}{4}(1 + \xi_i \xi)(1 + \eta_i \eta) \tag{74}$$

with (ξ, η) being the isoparametric coordinates, (ξ_i, η_i) being the isoparametric coordinates of point i with the global coordinates $(y_1^{(i)}, y_2^{(i)})$, $i = 1, 2, 3, 4$.

For the 8-node 3D hexahedral isoparametric element shown in Fig. 3,

$$\mathbf{N} = \left[\begin{matrix} N_1 & 0 & 0 & N_2 & 0 & 0 & N_3 & 0 & 0 & N_4 & 0 & 0 \\ 0 & N_1 & 0 & 0 & N_2 & 0 & 0 & N_3 & 0 & 0 & N_4 & 0 \\ 0 & 0 & N_1 & 0 & 0 & N_2 & 0 & 0 & N_3 & 0 & 0 & N_4 \end{matrix} \right. \\ \left. \begin{matrix} N_5 & 0 & 0 & N_6 & 0 & 0 & N_7 & 0 & 0 & N_8 & 0 & 0 \\ 0 & N_5 & 0 & 0 & N_6 & 0 & 0 & N_7 & 0 & 0 & N_8 & 0 \\ 0 & 0 & N_5 & 0 & 0 & N_6 & 0 & 0 & N_7 & 0 & 0 & N_8 \end{matrix} \right]. \tag{75}$$

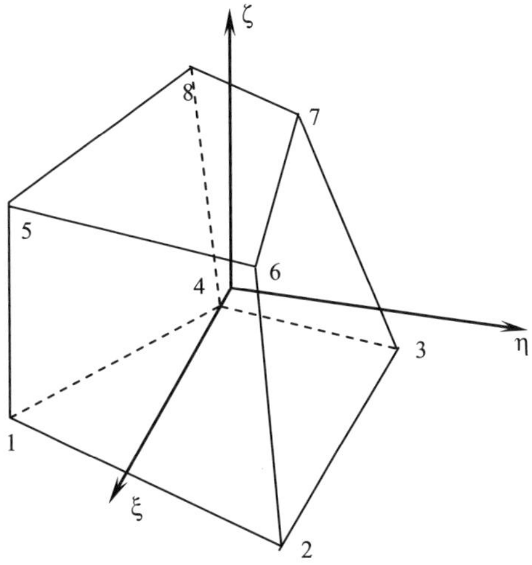

Fig. 3. A 3D 8-node hexahedral element.

The tri-linear interpolation function for node i

$$N_i = \frac{1}{8}(1+\xi_i\xi)(1+\eta_i\eta)(1+\varsigma_i\varsigma), \tag{76}$$

where (ξ, η, ζ) represents the isoparametric coordinates, (ξ_i, η_i, ζ_i) are the isoparametric coordinates of node i with the global coordinates $(y_1^{(i)}, y_2^{(i)}, y_3^{(i)})$, $i = 1, \ldots, 8$.

The corresponding strains are

$$\mathbf{e} = \partial\mathbf{u} = \partial\mathbf{N}\mathbf{q}^e = \mathbf{B}\mathbf{q}^e, \tag{77}$$

where $\mathbf{B}$ is the strain–displacement relation matrix. Substituting (72) and (77) into (69) and denoting the element stiffness matrix

$$\mathbf{K}^e = \int_{Y^e} \mathbf{B}^{\mathrm{T}}\mathbf{D}\mathbf{B}\, dY \tag{78}$$

and element nodal force vector

$$\mathbf{F}^e = \int_{Y^e} \mathbf{N}^{\mathrm{T}}\mathbf{f}\, dY + \int_{S_\sigma^e} \mathbf{N}^{\mathrm{T}}\bar{\mathbf{t}}\, ds, \tag{79}$$

we have

$$\Pi_p^{(e)}(\mathbf{q}^e) = \frac{1}{2}(\mathbf{q}^e)^{\mathrm{T}}\mathbf{K}^e\mathbf{q}^e - (\mathbf{q}^e)^{\mathbf{T}}\mathbf{F}^e. \tag{80}$$

Vanishing of the first-order variation of $\Pi_p^{(e)}(\mathbf{q}^e)$ in (80) with respect to $\mathbf{q}^e$ gives

$$\mathbf{K}^e\mathbf{q}^e = \mathbf{F}^e. \tag{81}$$

4.2. *Element-free Galerkin method from the potential principle*

The element-free Galerkin (EFG) method[40] is a meshless compatible method based on the potential principle. It uses a moving least squares (MLS) method to interpolate the approximate displacement field. In this section, we will briefly discuss the MLS interpolant, treatment of the essential boundary conditions and the handling of discontinuities in EFG.

4.2.1. *MLS interpolant*

The MLS interpolant $u^h(x)$ of the function $u(x)$ is defined in the domain Ω by[40]

$$\mathbf{u}^h(\mathbf{x}) = \sum_j^m p_j(\mathbf{x})a_j(\tilde{\mathbf{x}}) = \mathbf{p}^{\mathrm{T}}(\mathbf{x})\mathbf{a}(\tilde{\mathbf{x}}) \tag{82}$$

where $p_j(\mathbf{x}), j = 1, 2, \ldots, m$ are complete basis functions in the spatial coordinates $\mathbf{x}$. The coefficients $a_j(\tilde{\mathbf{x}})$ are also functions of $\mathbf{x}$ and obtained at any point $\tilde{\mathbf{x}}$ by minimising a weighted L_2 norm as follows:

$$J = \sum_I^n w(\tilde{\mathbf{x}} - \mathbf{x}_I)[\mathbf{p}^{\mathrm{T}}(\mathbf{x}_I)\mathbf{a}(\tilde{\mathbf{x}}) - \hat{u}_I]^2 \tag{83}$$

where n is the number of points in the neighbourhood, or the domain of influence (DOI) of $\tilde{\mathbf{x}}$ for which the weight function

$$w(\tilde{\mathbf{x}} - \mathbf{x}_I) \neq 0 \tag{84}$$

and $\hat{u}_I$ is the virtual nodal value of $u(\mathbf{x})$ at $\mathbf{x} = \mathbf{x}_I$.

The stationarity of J in (83) with respect to $\mathbf{a}(\tilde{\mathbf{x}})$ leads to the following linear relation between $\mathbf{a}(\tilde{\mathbf{x}})$ and $\hat{u}_I$:

$$\mathbf{A}(\tilde{\mathbf{x}})\mathbf{a}(\tilde{\mathbf{x}}) = \mathbf{B}(\tilde{\mathbf{x}})\hat{\mathbf{u}} \quad \text{or} \quad \mathbf{a}(\tilde{\mathbf{x}}) = \mathbf{A}^{-1}(\tilde{\mathbf{x}})\mathbf{B}(\tilde{\mathbf{x}})\hat{\mathbf{u}}, \tag{85}$$

where the matrices $\mathbf{A}(\tilde{\mathbf{x}})$ and $\mathbf{B}(\tilde{\mathbf{x}})$ are defined by

$$\mathbf{A}(\tilde{\mathbf{x}}) = \sum_I^n w(\tilde{\mathbf{x}} - \mathbf{x}_I)\mathbf{p}^{\mathrm{T}}(\mathbf{x}_I)\mathbf{p}(\mathbf{x}_I), \tag{86}$$

$$\mathbf{B}(\tilde{\mathbf{x}}) = \begin{bmatrix} w(\tilde{\mathbf{x}} - \mathbf{x}_1)\mathbf{p}(\mathbf{x}_1) & w(\tilde{\mathbf{x}} - \mathbf{x}_2)\mathbf{p}(\mathbf{x}_2) & \cdots & w(\tilde{\mathbf{x}} - \mathbf{x}_n)\mathbf{p}(\mathbf{x}_n) \end{bmatrix}, \tag{87}$$

$$\hat{\mathbf{u}}^{\mathrm{T}} = \begin{bmatrix} \hat{u}_1 & \hat{u}_2 & \cdots & \hat{u}_n \end{bmatrix}. \tag{88}$$

Hence, we have

$$u^h(\mathbf{x}) = \sum_I^n \sum_j^m p_j(\mathbf{x})(\mathbf{A}^{-1}(\mathbf{x})\mathbf{B}(\mathbf{x}))_{jI}\hat{u}_I = \sum_I^n \phi_I(\mathbf{x})\hat{u}_I, \tag{89}$$

where the shape function $\phi_I(\mathbf{x})$ is defined by

$$\phi_I(\mathbf{x}) = \sum_j^m p_j(\mathbf{x})(\mathbf{A}^{-1}(\mathbf{x})\mathbf{B}(\mathbf{x}))_{jI}. \tag{90}$$

Its derivatives are given by

$$\begin{aligned}\phi_{I,k}(\mathbf{x}) = \sum_{j=1}^m [&p_{j,k}(\mathbf{x})(\mathbf{A}^{-1}(\mathbf{x})\mathbf{B}(\mathbf{x}))_{jI} + p_j(\mathbf{x})(\mathbf{A}^{-1}_{,k}(\mathbf{x})\mathbf{B}(\mathbf{x}))_{jI} \\ &+ p_j(\mathbf{x})(\mathbf{A}^{-1}(\mathbf{x})\mathbf{B}_{,k}(\mathbf{x}))].\end{aligned} \tag{91}$$

Note that

$$\mathbf{A}^{-1}(\mathbf{x})\mathbf{A}(\mathbf{x}) = \mathbf{I},$$

where $\mathbf{I}$ is a unit matrix. We have

$$\frac{\partial \mathbf{A}^{-1}(\mathbf{x})}{\partial x_k} = -\mathbf{A}^{-1}(\mathbf{x})\frac{\partial \mathbf{A}(\mathbf{x})}{\partial x_k}\mathbf{A}^{-1}(\mathbf{x}). \tag{92}$$

4.2.2. *Imposition of the essential boundary conditions*

EFG uses MLS interpolants to construct the trial and test functions for the variational principle or weak form of the BVP. MLS interpolants do not pass through all the data points because the interpolation functions are not equal to unity at the nodes unless the weighting functions are singular.[41] Therefore, it complicates the imposition of the essential boundary conditions (including the application of point loads).

Several methods have been introduced for the imposition of essential boundary conditions, such as the Lagrange multiplier method,[40] modified variational principle approach,[42] FEM,[43,44] the collocation method[45] and the penalty function method.[46]

4.2.3. *Discontinuity in the displacement field*

MLS interpolants are highly smooth. However, we need to handle discontinuity in the displacement for particular cases, i.e. accounting for cracks. Two widely used criteria, the visibility criterion and the diffraction method, have been introduced to introduce discontinuity in the displacement in the MLS interpolant.

In the visibility criterion,[47] any surface with discontinuous displacements is considered opaque, which cannot be crossed by DOIs, and the approximation at a point $\mathbf{x}$ is not affected by node J if node J is not visible from point $\mathbf{x}$. Quadrature point q includes a surrounding node in its neighbour list (i.e. the nodal weight function is nonzero at point q) only if a straight line connecting point q to the node will not intersect any discontinuity surface like a crack. The visibility criterion has been applied with success in many static and dynamic fracture simulations. However, it results in discontinuous displacements within the domain in the vicinity of the crack-tip, in addition to the required discontinuities. Krysl and Belytschko[47] have shown by theoretical arguments that solutions generated with these discontinuous approximations are convergent, and the numerical simulations in the literature support this finding.

The diffraction method[48] increases the distance $|\mathbf{x} - \mathbf{x}_I|$ from the evaluation point $\mathbf{x}$ to node I for points on opposite sides of a crack (the line connecting $\mathbf{x}_I$ and $\mathbf{x}$ intersects the crack line) by bending the ray connecting the two points around the crack-tip, similar to the way light diffracts. The DOI effectively wraps around the crack-tip, so that the weight function is continuous in the material but remains discontinuous across the crack. Organ *et al.*[48] have shown that for high-order bases (e.g. a basis including singular crack-tip terms) and large DOIs, there are significant improvements in solutions generated by the diffraction method over the visibility criterion.

4.2.4. *Interfaces with discontinuous first-order derivatives*

For a 2D model adjacent to the Jth line of discontinuity, the displacement approximation is given by[49]

$$u^h(\mathbf{x}) = u^{EFG}(\mathbf{x}) + q^J(s)\psi_J(r), \tag{93}$$

where u^{EFG} is the standard EFG approximation (89). r denotes the signed distance from point $\mathbf{x}$ to the closest point on the line of discontinuity; s provides a parametric representation of the line of discontinuity. ψ_J are the jump shape functions. The amplitude of the jump $q^J(s)$ can be discretised as follows:

$$q^J(s) = \sum_I N_I(s) q_I^J, \tag{94}$$

where N_I are 1D shape functions that need to be C^1 so that they do not introduce any discontinuities in the derivatives other than across the discontinuity line. N_I can thus be constructed by MLS along the line of discontinuity.

Cubic-spline jump:

$$\psi_J(\bar{r}) = \begin{cases} -\dfrac{1}{6}\bar{r}^3 + \dfrac{1}{2}\bar{r}^2 - \dfrac{1}{2}\bar{r} + \dfrac{1}{6}, & \bar{r} \leq 1, \\ 0, & \bar{r} > 1, \end{cases} \qquad \bar{r} = \frac{|r|}{r_{cJ}} \tag{95}$$

where r_{cJ} is the characteristic length over which the jump function $\psi_J(r)$ for Jth line of discontinuity is non zero.

Ramp jump:

$$\psi_J(r) = \left(\langle r \rangle - \sum_I \phi_I(x) \langle r_I \rangle \right) w(\bar{r}), \tag{96}$$

where $w(\bar{r})$ is a weight function to make the jump function to have compact support, and the ramp function $\langle x \rangle$ is defined by

$$\langle x \rangle = \begin{cases} 0, & x < 0 \\ x, & x \geq 0. \end{cases} \tag{97}$$

From the results of Krongauz and Belytschko,[49] both jump functions work well. However, the ramp jump function (93) is even better since the oscillations are weaker adjacent to the interface.

For 1D problems, $q^J(s)$ in (93) reduces to a constant; and $w(\bar{r})$ is not necessary for the ramp jump function (96).

4.3. Displacement incompatible element from the potential principle

In each element, u_i is divided into a compatible part u_{qi} (72) and an incompatible part $u_{\lambda i}$, so that the functional (69) can be rewritten as

$$\Pi_p(u_i = u_{qi} + u_{\lambda i}) = \sum_e \int_{Y^e} \left(\frac{1}{2} \frac{\partial u_i}{\partial y_j} D_{ijkl} \frac{\partial u_k}{\partial y_l} - f_i u_{qi} \right) dY - \int_{S_\sigma^e} \bar{t}_i u_{qi} \, ds \tag{98}$$

for the whole domain. Taking the variation of the above functional and integrating by parts yield

$$\begin{aligned} \delta\Pi_p(u_i) = & \sum_e - \int_{Y^e} \delta u_{qi} \left[\frac{\partial}{\partial y_j} \left(D_{ijkl} \frac{\partial u_k}{\partial y_l} \right) - f_i \right] dY \\ & + \sum_{a,b} \int_{S_{ab}} \delta u_{qi} \left\{ \left[\left(D_{ijkl} \frac{\partial u_k}{\partial y_l} \right) n_j \right]^{(a)} + \left[\left(D_{ijkl} \frac{\partial u_k}{\partial y_l} \right) n_j \right]^{(b)} \right\} ds \\ & + \int_{S_\sigma^e} \delta u_{qi} \left[\left(D_{ijkl} \frac{\partial u_k}{\partial y_l} \right) n_j - \bar{t}_i \right] ds \\ & + \sum_e \int_{Y^e} \frac{\partial \delta u_{\lambda i}}{\partial y_j} D_{ijkl} \frac{\partial u_k}{\partial y_l} dY. \end{aligned}$$

Hence, the stationary condition of the functional (98) gives the equilibrium equation, the equilibrium of traction between the elements and prescribed boundary conditions on traction, if the following condition is met *a priori*

$$\sum_e \int_{Y^e} \frac{\partial \delta u_{\lambda i}}{\partial y_j} D_{ijkl} \frac{\partial u_k}{\partial y_l} dY = 0.$$

A convenient way to meet this condition (i.e. the sufficient but not the necessary condition) is to satisfy the following strong form in each element:

$$\int_{Y^e} \frac{\partial \delta u_{\lambda i}}{\partial y_j} D_{ijkl} \frac{\partial u_k}{\partial y_l} \, dY = 0.$$

Since a constant stress state is recovered in each element as its size is reduced to zero and since $\delta u_{\lambda i}$ is arbitrary, the above constraint reduces to the general constant stress patch test condition (PTC)[5]

$$\int_{Y^e} \frac{\partial u_{\lambda i}}{\partial y_j} \, dY = 0 \quad \text{or equivalently} \quad \oint_{\partial Y^e} u_{\lambda i} n_j \, ds = 0. \tag{99}$$

The incompatible functions meeting the PTC can now be easily formulated.

If the compatible displacement $\mathbf{u}_q$ (72) is related to the nodal values $\mathbf{q}^e$ via the bilinear interpolation functions (73) and (74), then the incompatible term $\mathbf{u}_\lambda$ is related to the element inner parameters $\boldsymbol{\lambda}^e$ via the shape functions $\mathbf{N}_\lambda$

$$\mathbf{u}_\lambda = \mathbf{N}_\lambda \boldsymbol{\lambda}^e. \tag{100}$$

With the above assumed displacements (72) and (100), we have the strains

$$e_{ij} = \partial \mathbf{u} = \partial(\mathbf{u}_q + \mathbf{u}_\lambda) = \begin{bmatrix} \mathbf{B} & \mathbf{B}_\lambda \end{bmatrix} \begin{Bmatrix} \mathbf{q}^e \\ \boldsymbol{\lambda}^e \end{Bmatrix}, \tag{101}$$

where

$$\mathbf{B} = \partial \mathbf{N}, \quad \mathbf{B}_\lambda = \partial \mathbf{N}_\lambda. \tag{102}$$

Substituting (72) and (101) into the elemental functional in (98) and denoting

$$\begin{bmatrix} \mathbf{K}_{qq} & \mathbf{K}_{q\lambda} \\ \mathbf{K}_{q\lambda}^{\mathrm{T}} & \mathbf{K}_{\lambda\lambda} \end{bmatrix} = \int_{Y^e} \begin{bmatrix} \mathbf{B}^{\mathrm{T}} \\ \mathbf{B}_\lambda^{\mathrm{T}} \end{bmatrix} \mathbf{D} \begin{bmatrix} \mathbf{B} & \mathbf{B}_\lambda \end{bmatrix} dY \tag{103}$$

yield

$$\begin{aligned} \Pi_p^{(e)}(\mathbf{q}^e, \boldsymbol{\lambda}^e) &= \frac{1}{2}(\mathbf{q}^e)^{\mathrm{T}} \mathbf{K}_{qq} \mathbf{q}^e + (\mathbf{q}^e)^{\mathrm{T}} \mathbf{K}_{q\lambda} \boldsymbol{\lambda}^e \\ &\quad + \frac{1}{2}(\boldsymbol{\lambda}^e)^{\mathrm{T}} \mathbf{K}_{\lambda\lambda} \boldsymbol{\lambda}^e - (\mathbf{q}^e)^{\mathrm{T}} \mathbf{F}^e \end{aligned}$$

where the element nodal force vector $\mathbf{F}^e$ is still (79). Making use of the stationary condition of $\Pi_p^{(e)}(\mathbf{q}^e, \boldsymbol{\lambda}^e)$ with respect to $\boldsymbol{\lambda}^e$ yields

$$\mathbf{K}_{q\lambda}^{\mathrm{T}} \mathbf{q}^e + \mathbf{K}_{\lambda\lambda} \boldsymbol{\lambda}^e = 0;$$

hence, the element inner parameters $\boldsymbol{\lambda}^e$ are recovered as follows:

$$\boldsymbol{\lambda}^e = -\mathbf{K}_{\lambda\lambda}^{-1} \mathbf{K}_{q\lambda}^{\mathrm{T}} \mathbf{q}^e. \tag{104}$$

Vanishing of the first-order variation of $\Pi_p^{(e)}(\mathbf{q}^e, \boldsymbol{\lambda}^e)$ with respect to $\mathbf{q}^e$ gives

$$\mathbf{K}_{qq} \mathbf{q}^e + \mathbf{K}_{q\lambda} \boldsymbol{\lambda}^e = \mathbf{F}^e.$$

With the use of (104), we have

$$\tilde{\mathbf{K}}^e \mathbf{q}^e = \mathbf{F}^e, \tag{105}$$

where the element stiffness matrix is

$$\tilde{\mathbf{K}}^e = \mathbf{K}_{qq} - \mathbf{K}_{q\lambda}\mathbf{K}_{\lambda\lambda}^{-1}\mathbf{K}_{q\lambda}^{\mathrm{T}}. \tag{106}$$

4.3.1. *2D 4-node incompatible element*

Refer to the 4-node element shown in Fig. 2, the element compatible displacements are (72)–(74); and the interpolation matrix $\mathbf{N}_\lambda$ for $\mathbf{u}_\lambda$ is defined as follows:

$$\mathbf{N}_\lambda = \begin{bmatrix} N_{\lambda 1} & 0 & N_{\lambda 2} & 0 \\ 0 & N_{\lambda 1} & 0 & N_{\lambda 2} \end{bmatrix}. \tag{107}$$

Here, the two incompatible terms are[5]

$$\begin{aligned} N_{\lambda 1} &= \xi^2 - \Delta, \quad N_{\lambda 2} = \eta^2 + \Delta, \\ \Delta &= \frac{2}{3}\left(\frac{J_1}{J_0}\xi - \frac{J_2}{J_0}\eta\right), \end{aligned} \tag{108}$$

where J_0, J_1 and J_2 are related to the element Jacobian as follows:

$$\begin{aligned} |J| &= J_0 + J_1\xi + J_2\eta \\ &= (a_1b_3 - a_3b_1) + (a_1b_2 - a_2b_1)\xi + (a_2b_3 - a_3b_2)\eta, \end{aligned} \tag{109}$$

and coefficients a_i and b_i $(i = 1, 2, 3)$ are dependent on the element nodal coordinates

$$\begin{bmatrix} a_1 & b_1 \\ a_2 & b_2 \\ a_3 & b_3 \end{bmatrix} = \frac{1}{4}\begin{bmatrix} -1 & 1 & 1 & -1 \\ 1 & -1 & 1 & -1 \\ -1 & -1 & 1 & 1 \end{bmatrix}\begin{bmatrix} y_1^{(1)} & y_2^{(1)} \\ y_1^{(2)} & y_2^{(2)} \\ y_1^{(3)} & y_2^{(3)} \\ y_1^{(4)} & y_2^{(4)} \end{bmatrix}. \tag{110}$$

A 2×2 Gauss quadrature is employed for the element formulation.

4.3.2. *3D 8-node incompatible element*

If we refer to the 8-node 3D isoparametric element shown in Fig. 3, the compatible displacement field $\mathbf{u}_q$ is related to the nodal values via the tri-linear interpolation functions (75) and (76). The incompatible interpolation functions $\mathbf{N}_\lambda$ are

$$\mathbf{N}_\lambda = \begin{bmatrix} \xi^2 & \eta^2 & \varsigma^2 \end{bmatrix} - \begin{bmatrix} \xi & \eta & \varsigma \end{bmatrix}\mathbf{P}^*\mathbf{P}_\lambda, \tag{111}$$

$$\mathbf{P}^* = \oint_{\partial Y} \begin{Bmatrix} l \\ m \\ n \end{Bmatrix} [\xi \quad \eta \quad \varsigma]\, ds = \int_Y \begin{Bmatrix} \dfrac{\partial}{\partial y_1} \\ \dfrac{\partial}{\partial y_2} \\ \dfrac{\partial}{\partial y_3} \end{Bmatrix} [\xi \quad \eta \quad \varsigma]\, dY = \int_Y \mathbf{J}^{-1}\, dY, \tag{112}$$

$$\mathbf{P}_\lambda = \oint_{\partial Y} \begin{Bmatrix} l \\ m \\ n \end{Bmatrix} [\xi^2 \quad \eta^2 \quad \varsigma^2]\, ds = \int_Y \begin{Bmatrix} \dfrac{\partial}{\partial y_1} \\ \dfrac{\partial}{\partial y_2} \\ \dfrac{\partial}{\partial y_3} \end{Bmatrix} [\xi^2 \quad \eta^2 \quad \varsigma^2]\, dY$$

$$= 2\int_Y \mathbf{J}^{-1} \begin{bmatrix} \xi & 0 & 0 \\ 0 & \eta & 0 \\ 0 & 0 & \varsigma \end{bmatrix} dY, \tag{113}$$

where (l, m, n) are components of the unit outward normal $\mathbf{n}$. $\mathbf{J}$ is related to the element Jacobian

$$\mathbf{J} = \begin{bmatrix} a_1 + a_4\eta + a_5\varsigma + a_7\eta\varsigma & b_1 + b_4\eta + b_5\varsigma + b_7\eta\varsigma & c_1 + c_4\eta + c_5\varsigma + c_7\eta\varsigma \\ a_2 + a_4\xi + a_6\varsigma + a_7\xi\varsigma & b_2 + b_4\xi + b_6\varsigma + b_7\xi\varsigma & c_2 + c_4\xi + c_6\varsigma + c_7\xi\varsigma \\ a_3 + a_5\xi + a_6\eta + a_7\xi\eta & b_3 + b_5\xi + b_6\eta + b_7\xi\eta & c_3 + c_5\xi + c_6\eta + c_7\xi\eta \end{bmatrix} \tag{114}$$

and coefficients a_k, b_k and c_k $(k = 1, \ldots, 7)$ are dependent on the element nodal coordinates $y_i^{(j)}$ $(i = 1, 2, 3;\ j = 1, \ldots, 8)$

$$\begin{bmatrix} a_1 & b_1 & c_1 \\ \vdots & \vdots & \vdots \\ a_7 & b_7 & c_7 \end{bmatrix} = \begin{bmatrix} -1 & 1 & 1 & -1 & -1 & 1 & 1 & -1 \\ -1 & -1 & 1 & 1 & -1 & -1 & 1 & 1 \\ 1 & 1 & 1 & 1 & -1 & -1 & -1 & -1 \\ 1 & -1 & 1 & -1 & 1 & -1 & 1 & -1 \\ -1 & 1 & 1 & -1 & 1 & -1 & -1 & 1 \\ -1 & -1 & 1 & 1 & 1 & 1 & -1 & -1 \\ 1 & -1 & 1 & -1 & -1 & 1 & -1 & 1 \end{bmatrix} \times \begin{bmatrix} y_1^{(1)} & y_2^{(1)} & y_3^{(1)} \\ \vdots & \vdots & \vdots \\ y_1^{(8)} & y_2^{(8)} & y_3^{(8)} \end{bmatrix}. \tag{115}$$

4.4. *Hybrid stress elements from the Hellinger–Reissner principle*

Consider an element whose displacement $\mathbf{u}$ is related to nodal values $\mathbf{q}^e$ via the shape functions $\mathbf{N}$ as in (72). The relevant strain array is (77). The stress is related to stress parameters $\boldsymbol{\beta}$ via the stress interpolation function $\boldsymbol{\varphi}$

$$\boldsymbol{\sigma} = \boldsymbol{\varphi}\boldsymbol{\beta}. \tag{116}$$

If the displacements are compatible along the common boundary of elements, the stresses from adjacent elements are not required to be in equilibrium. Substitution of (72), (77) and (116) into (70) and denoting the characteristic matrices of the element

$$\mathbf{H} = \int_{Y^e} \boldsymbol{\varphi}^{\mathrm{T}}\mathbf{D}^{-1}\boldsymbol{\varphi}\, dY, \quad \mathbf{G} = \int_{Y^e} \boldsymbol{\varphi}^{\mathrm{T}}\mathbf{B}\, dY \tag{117}$$

give[5]

$$\Pi_{\mathrm{HR}}^{(e)}(\mathbf{q}^e, \boldsymbol{\beta}) = \boldsymbol{\beta}^{\mathrm{T}}\mathbf{G}\mathbf{q}^e - \frac{1}{2}\boldsymbol{\beta}^{\mathrm{T}}\mathbf{H}\boldsymbol{\beta} - (\mathbf{q}^e)^{\mathrm{T}}\mathbf{F}^e,$$

where the element nodal force vector $\mathbf{F}^e$ is still (79). Vanishing of the first-order variation of $\Pi_{\mathrm{HR}}^{(e)}(\mathbf{q}^e, \boldsymbol{\beta})$ with respect to $\boldsymbol{\beta}$ gives

$$\mathbf{G}\mathbf{q}^e - \mathbf{H}\boldsymbol{\beta} = 0,$$

which determines the stress parameters β via nodal displacement parameters as

$$\boldsymbol{\beta} = \mathbf{H}^{-1}\mathbf{G}\mathbf{q}^e. \tag{118}$$

Vanishing of the first-order variation of $\Pi_{\mathrm{HR}}^{(e)}(\mathbf{q}^e, \boldsymbol{\beta})$ with respect to $\mathbf{q}^e$ gives

$$\mathbf{G}^{\mathrm{T}}\boldsymbol{\beta} = \mathbf{F}^e$$

and the discretised equations of equilibrium of element "e"

$$\mathbf{K}^e\mathbf{q}^e = \mathbf{F}^e \tag{119}$$

with the substitution of (118) and denoting the stiffness matrix of element "e"

$$\mathbf{K}^e = \mathbf{G}^{\mathrm{T}}\mathbf{H}^{-1}\mathbf{G}. \tag{120}$$

Equation (119) cannot be solved uniquely unless the displacement and stress parameters are selected appropriately so that they satisfy the condition given in Eq. (121)[5]

$$n_{\boldsymbol{\beta}} \geq n_q - n_r, \tag{121}$$

where $n_{\boldsymbol{\beta}}$ and n_q represent the number of element stress parameters $\boldsymbol{\beta}$ and nodal displacement parameters $\mathbf{q}^e$, respectively, and n_r is the number of independent rigid body motions.

In the above hybrid formulation, the stresses are condensed in the element level. If we do not condense them in the element level, we will have a mixed formulation.

In the formulation of the hybrid stress element, the performance or the capability of the element in predicting stresses can be improved through the introduction of incompatible displacements.[5] Let us append an incompatible term (100) to the compatible displacement (72) and let us substitute the resulting displacement field into Eq. (70). The first-order variation of the Hellinger–Reissner functional for the whole domain becomes

$$\begin{aligned}
&\delta\Pi_{\mathrm{HR}}(u_i = u_{qi} + u_{\lambda i}, \sigma_{ij}) \\
&\quad + \sum_e \int_{Y^e} \left[\delta\sigma_{ij}\left(-D^{-1}_{ijkl}\sigma_{kl} + \frac{\partial u_i}{\partial y_j}\right) - \delta u_{qi}\left(\frac{\partial \sigma_{ij}}{\partial y_j} + f_i\right)\right] dY \\
&\quad + \sum_{a,b} \int_{S_{ab}} \delta u_{qi}[(\sigma_{ij}n_j)^{(a)} + (\sigma_{ij}n_j)^{(b)}] ds + \int_{S^e_\sigma} (\sigma_{ij}n_j - \bar{t}_i)\, ds\delta u_{qi} \\
&\quad + \sum_e \int_{Y^e} \frac{\partial \delta u_{\lambda i}}{\partial y_j}\sigma_{ij}\, dY.
\end{aligned} \tag{122}$$

The stationary condition of the functional (70) provides equilibrium, compatibility, equilibrium of traction between elements and the prescribed traction constraints if and only if the following integral vanishes

$$\sum_e \int_{Y^e} \frac{\partial \delta u_{\lambda i}}{\partial y_j}\sigma_{ij}\, dY = 0.$$

As argued in Sec. 4.3, a convenient way to meet this condition is to satisfy the following strong form (i.e. the sufficient but not the necessary condition) in each element:

$$\int_{Y^e} \frac{\partial \delta u_{\lambda i}}{\partial y_j} \sigma_{ij}\, dY = 0. \tag{123}$$

If the body forces are absent, or handled by using the equivalent element nodal loads, the term relevant to derivatives of the stresses from (123) can be combined with the corresponding terms in (122). Hence the condition (123) can be written in the widely used form

$$\oint_{\partial Y^e} \delta u_{\lambda i} \sigma_{ij} n_j\, ds = 0, \quad \text{or equivalently} \quad \oint_{\partial Y^e} \boldsymbol{\sigma}^{\mathrm{T}} \mathbf{n}^{\mathrm{T}} \delta \mathbf{u}_\lambda\, ds = 0. \tag{124}$$

Since inner incompatible displacements $\mathbf{u}_\lambda$ can be selected arbitrarily, Eq. (124) can be rewritten as

$$\oint_{\partial Y^e} \boldsymbol{\sigma}^{\mathrm{T}} \mathbf{n}^{\mathrm{T}} \mathbf{u}_\lambda\, ds = 0. \tag{125}$$

Physically, (125) means that along the element boundary stresses do not work on the incompatible part of the displacement.

If the assumed stress is divided into lower constant and higher-order parts

$$\sigma = \boldsymbol{\varphi\beta} = [\boldsymbol{\varphi}_c \quad \boldsymbol{\varphi}_h] \begin{Bmatrix} \boldsymbol{\beta}_c \\ \boldsymbol{\beta}_h \end{Bmatrix} = [\boldsymbol{\varphi}_c \quad \boldsymbol{\varphi}_I \quad \boldsymbol{\varphi}_{II}] \begin{Bmatrix} \boldsymbol{\beta}_c \\ \boldsymbol{\beta}_I \\ \boldsymbol{\beta}_{II} \end{Bmatrix}, \tag{126}$$

we then obtain from Eq. (125) the PTC as shown in Eq. (127) for evaluating the incompatible displacement fields that pass the PTC,

$$\oint_{\partial Y^e} \boldsymbol{\sigma}_c^{\mathrm{T}} \mathbf{n}^{\mathrm{T}} \mathbf{u}_\lambda\, ds = 0 \tag{127}$$

and the stress optimization condition (OPC) for optimizing the trial stresses

$$\oint_{\partial Y^e} \boldsymbol{\sigma}_h^{\mathrm{T}} \mathbf{n}^{\mathrm{T}} \mathbf{u}_\lambda\, ds = 0. \tag{128}$$

For the hybrid stress element formulated from the Hellinger–Reissner principle, it has been proved that the PTC (127) is equivalent to the stability condition popularly known as the Babuska–Brezzi (BB) condition.[5]

4.4.1. *Plane 4-node Pian and Sumihara (PS) 5β element*

With reference to the 4-node isoparametric element shown in Fig. 2, the employed displacement field is defined in Eq. (72) with the widely used bilinear interpolation functions (73) and (74). The stress interpolation function φ in (116) is defined as

$$\begin{bmatrix} \varphi_1 \\ \varphi_2 \\ \varphi_{12} \end{bmatrix} = \begin{bmatrix} 1 & 0 & 0 & a_1^2\eta & a_3^2\xi \\ 0 & 1 & 0 & b_1^2\eta & b_3^2\xi \\ 0 & 0 & 1 & a_1b_1\eta & a_3b_3\xi \end{bmatrix}, \tag{129}$$

where coefficients a_i and b_i $(i = 1, 2, 3)$ are dependent on the element nodal coordinates as (110).

4.4.2. *3D 8-node 18β hybrid stress element*

The approximate compatible displacement field is the same as an 8-node hexahedral isoparametric element defined in (72), (75) and (76). The stress interpolation function φ in (116) is defined as

$$\varphi = \left[\begin{matrix} 1 & 0 & 0 & 0 & 0 & 0 & d_{22}^2\eta & d_{33}^2\varsigma & \eta\varsigma & 0 \\ 0 & 1 & 0 & 0 & 0 & 0 & d_{12}^2\eta & 0 & 0 & d_{33}^2\varsigma \\ 0 & 0 & 1 & 0 & 0 & 0 & 0 & d_{13}^2\varsigma & 0 & d_{23}^2\varsigma \\ 0 & 0 & 0 & 1 & 0 & 0 & -d_{12}d_{22}\eta & 0 & 0 & 0 \\ 0 & 0 & 0 & 0 & 1 & 0 & 0 & 0 & 0 & -d_{23}d_{33}\varsigma \\ 0 & 0 & 0 & 0 & 0 & 1 & 0 & -d_{13}d_{33}\varsigma & 0 & 0 \end{matrix}\right.$$

$$\begin{matrix} d_{21}^2\xi & 0 & d_{31}^2\xi & 0 & 0 & 0 \\ d_{11}^2\xi & \xi\varsigma & 0 & d_{32}^2\eta & 0 & 0 \\ 0 & 0 & d_{11}^2\xi & d_{22}^2\eta & \xi\eta & 2d_{13}d_{23}\varsigma \\ -d_{21}d_{11}\xi & 0 & 0 & 0 & 0 & d_{33}^2\varsigma \\ 0 & 0 & 0 & -d_{32}d_{22}\eta & 0 & -d_{13}d_{33}\varsigma \\ 0 & 0 & -d_{31}d_{11}\xi & 0 & 0 & -d_{23}d_{33}\varsigma \end{matrix}$$

$$\left.\begin{matrix} 2d_{21}d_{31}\xi & 0 \\ 0 & 2d_{12}d_{32}\eta \\ 0 & 0 \\ -d_{31}d_{11}\xi & -d_{32}d_{22}\eta \\ d_{11}^2\xi & -d_{12}d_{22}\eta \\ -d_{21}d_{11}\xi & d_{22}^2\eta \end{matrix}\right], \tag{130}$$

where

$$\mathbf{d} = \begin{bmatrix} b_2c_3 - b_3c_2 & b_3c_1 - b_1c_3 & b_1c_2 - b_2c_1 \\ c_2a_3 - c_3a_2 & c_3a_1 - c_1a_3 & c_1a_2 - c_2a_1 \\ a_2b_3 - a_3b_2 & a_3b_1 - a_1b_3 & a_1b_2 - a_2b_1 \end{bmatrix}, \tag{131}$$

and coefficients a_i, b_i and c_i, $i = 1, 2, 3$, are defined in (115).

4.5. *Enhanced-strain element based on the Hu–Washizu principle*

One can formulate hybrid or mixed elements from the Hu–Washizu principle (71) with u_i, σ_{ij} and e_{ij} being interpolated independently from one another. However, a more efficient way is to formulate the so-called enhanced-strain elements as follows.

Only the compatible displacement field is used here, and the strain field is enhanced by appending to the strain corresponding to the compatible displacement an enhanced incompatible strain field $\mathbf{e}_\lambda$ as follows[50,51]:

$$e_{ij} = \frac{\partial u_i}{\partial y_j} + e_{\lambda ij}. \tag{132}$$

The Hu–Washizu functional (71) can now be rewritten for the whole domain as

$$\begin{aligned} &\Pi_{\mathrm{HW}}(u_i, e_{\lambda ij}, \sigma_{ij}) \\ &\quad = \sum_e \int_{Y^e} \left[\frac{1}{2} \left(\frac{\partial u_i}{\partial y_j} + e_{\lambda ij} \right) D_{ijkl} \left(\frac{\partial u_k}{\partial y_l} + e_{\lambda kl} \right) - \sigma_{ij} e_{\lambda ij} - f_i u_i \right] dY \\ &\qquad - \int_{S_\sigma^e} \bar{t}_i u_i \, ds, \end{aligned}$$

or the elemental functional in matrix form

$$\begin{aligned} \Pi_{\mathrm{HW}}^{(e)}(\mathbf{u}, \mathbf{e}_\lambda, \boldsymbol{\sigma}) &= \int_{Y^e} \left[\frac{1}{2} (\partial \mathbf{u} + \mathbf{e}_\lambda)^{\mathrm{T}} \mathbf{D} (\partial \mathbf{u} + \mathbf{e}_\lambda) - \boldsymbol{\sigma}^{\mathrm{T}} \mathbf{e}_\lambda - \mathbf{u}^{\mathrm{T}} \mathbf{f} \right] dY \\ &\quad - \int_{S_\sigma^e} \mathbf{u}^{\mathrm{T}} \bar{\mathbf{t}} \, ds. \end{aligned} \tag{133}$$

Taking the variation of the above functional and integrating by parts yield

$$\begin{aligned} \delta \Pi_{\mathrm{HW}}(u_i, e_{\lambda ij}, \sigma_{ij}) = \sum_e \int_{Y^e} \Bigg\{ &-\delta u_i \left[\frac{\partial}{\partial y_j} \left\langle D_{ijkl} \left(\frac{\partial u_k}{\partial y_l} + e_{\lambda kl} \right) \right\rangle - f_i \right] \\ &+ \delta e_{\lambda ij} \left[D_{ijkl} \left(\frac{\partial u_k}{\partial y_l} + e_{\lambda kl} \right) - \sigma_{ij} \right] \Bigg\} dY \end{aligned}$$

$$+\sum_{a,b}\oint_{S_{\rm ab}}\delta u_i\left\{\left[D_{ijkl}\left(\frac{\partial u_k}{\partial y_l}+e_{\lambda kl}\right)n_j\right]^{(a)}\right.$$

$$\left.+\left[D_{ijkl}\left(\frac{\partial u_k}{\partial y_l}+e_{\lambda kl}\right)n_j\right]^{(b)}\right\}ds$$

$$+\int_{S_\sigma^e}\delta u_i\left[D_{ijkl}\left(\frac{\partial u_k}{\partial y_l}+e_{\lambda kl}\right)n_j-\bar{t}_i\right]ds$$

$$-\sum_e\int_{Y^e}\delta\sigma_{ij}e_{\lambda ij}dY.$$

The stationary condition of the functional (133) gives the equilibrium equation, the stress–strain relations, the equilibrium of traction between the elements, and the prescribed boundary conditions on tractions, if the following condition is met *a priori*:

$$\sum_e\int_{Y^e}\delta\sigma_{ij}e_{\lambda ij}\,dY=0.$$

Following the procedure employed in Sec. 4.3, the above constraint can be simplified to the PTC[5]

$$\int_{Y^e}e_{\lambda ij}\,dY=0. \tag{134}$$

It is evident that (134) is an alternative formulation of the PTC (99), if the enhanced-strain $\mathbf{e}_\lambda$ corresponds to the incompatible displacement $\mathbf{u}_\lambda$.

The FE based on functional (133) requires an independent approximation of three fields: u_i, $e_{\lambda ij}$ and σ_{ij}. In the enhanced-strain element, however, the independent stress field σ_{ij} is eliminated by selecting it to be orthogonal to the enhanced-strain field $e_{\lambda ij}$, i.e.

$$\int_{Y^e}\sigma_{ij}e_{\lambda ij}\,dY=0. \tag{135}$$

Thus, the two independent fields for the enhanced-strain formulation are the displacement u_i and the enhanced assumed strains $e_{\lambda ij}$. The formulation here is the same as (103)–(106) in Sec. 4.3, provided $e_{\lambda ij}$ are interpolated from the element inner parameters as follows:

$$\mathbf{e}_\lambda=\mathbf{B}_\lambda\boldsymbol{\lambda}^e, \tag{136}$$

where λ^e are internal parameters for the enhanced strains.

Moreover, if the assumed strains $\mathbf{e}_\lambda$ in (136) correspond to the incompatible displacement $\mathbf{u}_\lambda$ in (100), the enhanced-strain formulation will be equivalent to the displacement incompatible formulation discussed in Sec. 4.3. Note, however, that the stress in the enhanced-strain formulation can be recovered with the help of the orthogonalisation condition (135).[50]

4.5.1. *Plane 4-node enhanced-strain element*

The approximate compatible displacement field is the same as the 4-node quadrilateral plane isoparametric element (72)–(74). The interpolation function matrix for the covariant enhanced-strain field is given by[51]

$$\mathbf{B}_\lambda = \begin{bmatrix} \xi & 0 & 0 & 0 & \xi\eta \\ 0 & \eta & 0 & 0 & -\xi\eta \\ 0 & 0 & \xi & \eta & \xi^2 - \eta^2 \end{bmatrix}. \tag{137}$$

$\mathbf{B}_\lambda$ passes the patch test and satisfies the L_2 orthogonality condition with the following contravariant stress field with five β parameters:

$$\begin{Bmatrix} \sigma_\xi \\ \sigma_\eta \\ \sigma_{\xi\eta} \end{Bmatrix} = \begin{bmatrix} 1 & 0 & 0 & \eta & 0 \\ 0 & 1 & 0 & 0 & \xi \\ 0 & 0 & 1 & 0 & 0 \end{bmatrix} \boldsymbol{\beta}. \tag{138}$$

4.5.2. *3D 8-node enhanced-strain element*

The approximate compatible displacement field is the same as an 8-node hexahedral isoparametric element (72), (75) and (76). The enhanced strain interpolation matrix

$$\mathbf{B}_\lambda = \left[\begin{matrix} \xi & 0 & 0 & 0 & 0 & 0 & 0 & 0 & 0 & \xi\eta \\ 0 & \eta & 0 & 0 & 0 & 0 & 0 & 0 & 0 & -\xi\eta \\ 0 & 0 & \varsigma & 0 & 0 & 0 & 0 & 0 & 0 & 0 \\ 0 & 0 & 0 & \xi & 0 & 0 & \eta & 0 & 0 & \xi^2-\eta^2 \\ 0 & 0 & 0 & 0 & \eta & 0 & 0 & \varsigma & 0 & 0 \\ 0 & 0 & 0 & 0 & 0 & \varsigma & 0 & 0 & \xi & 0 \end{matrix}\right.$$
$$\left.\begin{matrix} 0 & -\xi\varsigma & 0 & 0 & 0 & \xi\eta\varsigma & 0 & 0 \\ \eta\varsigma & 0 & 0 & 0 & 0 & 0 & \xi\eta\varsigma & 0 \\ -\eta\varsigma & \xi\varsigma & 0 & 0 & 0 & 0 & 0 & \xi\eta\varsigma \\ 0 & 0 & \xi\varsigma & \eta\varsigma & 0 & \xi^2\varsigma & \eta^2\varsigma & 0 \\ \eta^2-\varsigma^2 & 0 & 0 & \eta\xi & \varsigma\xi & 0 & \eta^2\xi & \varsigma^2\xi \\ 0 & \varsigma^2-\xi^2 & \xi\eta & 0 & \varsigma\eta & \xi^2\eta & 0 & \varsigma^2\eta \end{matrix}\right]. \tag{139}$$

Its first nine columns can also be used as an enhanced-strain interpolation matrix. Both matrices pass the PTC and satisfy the L_2 orthogonality condition with the following contravariant stresses with 12-β parameters:

$$\begin{Bmatrix} \sigma_\xi \\ \sigma_\eta \\ \sigma_\varsigma \\ \sigma_{\xi\eta} \\ \sigma_{\eta\varsigma} \\ \sigma_{\varsigma\xi} \end{Bmatrix} = \begin{bmatrix} 1 & 0 & 0 & \eta & \varsigma & 0 & 0 & 0 & 0 & 0 & 0 & 0 \\ 0 & 1 & 0 & 0 & 0 & \xi & \varsigma & 0 & 0 & 0 & 0 & 0 \\ 0 & 0 & 1 & 0 & 0 & 0 & 0 & \xi & \eta & 0 & 0 & 0 \\ 0 & 0 & 0 & 0 & 0 & 0 & 0 & 0 & 0 & 1 & 0 & 0 \\ 0 & 0 & 0 & 0 & 0 & 0 & 0 & 0 & 0 & 0 & 1 & 0 \\ 0 & 0 & 0 & 0 & 0 & 0 & 0 & 0 & 0 & 0 & 0 & 1 \end{bmatrix} \boldsymbol{\beta}. \qquad (140)$$

4.6. *Comments on the various methods*

Displacement compatible elements are simple. However, they are notorious for shear locking in slender structures and incompressible locking for nearly incompressible materials. Moreover, the stresses are obtained by differentiation of the displacements and hence their accuracy reduces.

Hybrid stress elements obtain the stresses directly without differentiating the displacements. Therefore, they predict stresses more accurately than the displacement compatible elements. They are not sensitive to the Poisson ratio and they are less sensitive to the slenderness of structures. For beams, plates and shells, hybrid elements are also powerful tools for avoiding C^1 continuity of the displacements.

Generally, incompatible and enhanced-strain elements perform as well as the hybrid element although they need also to calculate the stress via the strain. These lower-order elements exhibit improved accuracy in coarse meshes when compared with their parent compatible elements, particularly if bending predominates. In addition, these elements do not suffer from locking in the nearly incompressible limit.

EFG is efficient for modelling moving discontinuities, and when higher-order derivatives are required.

Although hybrid elements based on the Hellinger–Reissner principle or the Hu–Washizu principle can in general improve the accuracy of the approximate displacement and stress solutions, they are not suitable for the analysis of RUC, as it is difficult to meet the Y-periodicity condition of the stress on the boundary of the RUC. The general isoparametric elements are also not satisfactory because of the gradients of χ_i^{kl} that appear in (38) in the evaluation of the homogenised material properties.

5. Enforcing the Periodicity Boundary Condition and Constraints from Higher-Order Equilibrium in the Analysis of the RUC

Assembling the discretised equations of equilibrium of all elements, yields the following system of equilibrium equations:

$$\mathbf{K}\mathbf{q} = \mathbf{F}. \tag{141}$$

The periodicity condition of the boundary displacement and constraints from higher-order equilibrium can conveniently be enforced by a penalty function technique.[3] Equation (141) is the Euler–Lagrange equation of the following functional:

$$\Pi(\mathbf{q}) = \frac{1}{2}\mathbf{q}^{\mathrm{T}}\mathbf{K}\mathbf{q} - \mathbf{q}^{\mathrm{T}}\mathbf{F}. \tag{142}$$

The periodicity condition and constraints from higher-order equilibrium yield the following constraint:

$$\mathbf{R}\mathbf{q} = 0. \tag{143}$$

For the periodicity condition, if a couple of nodes, i and j, on the boundary of a 2D RUC have the same displacement because of the periodicity condition, i.e.

$$\mathbf{q}_i = \mathbf{q}_j,$$

the above condition is equivalent to

$$R(2i-1,2i-1) = R(2i,2i) = 1, \quad R(2i-1, l \neq 2i-1, 2j-1) = 0,$$
$$R(2i-1,2j-1) = R(2i,2j) = -1, \quad R(2i, l \neq 2i, 2j) = 0.$$

In order to satisfy the constraint (143) by a penalty function technique, the functional (142) is modified as

$$\tilde{\Pi}(\mathbf{q}) = \frac{1}{2}\mathbf{q}^{\mathrm{T}}\mathbf{K}\mathbf{q} - \mathbf{q}^{\mathrm{T}}\mathbf{F} + \frac{\alpha}{2}\mathbf{q}^{\mathrm{T}}\mathbf{R}^{\mathrm{T}}\mathbf{R}\mathbf{q}, \tag{144}$$

where α is a large positive number taken to be 10^4. Thus, instead of (141), we solve the following equations:

$$(\mathbf{K} + \alpha\mathbf{R}^{\mathrm{T}}\mathbf{R})\mathbf{q} = \mathbf{F}. \tag{145}$$

6. *A Posteriori* Recovery of the Gradients

Various schemes have been introduced to recover the derivatives with higher accuracy than the numerical results.

6.1. *Superconvergent patch recovery (SPR)*

The stresses sampled at certain points in an element may possess the superconvergent property, i.e. converge at the same rate as displacement at these points; at all other points the convergence will be slower. Based on this observation, Zienkiewicz *et al.* introduced a superconvergent patch recovery (SPR) technique.[4] SPR first approximates the stress field by a polynomial of appropriate order within each small patch of elements, typically the group of elements that share the node. The coefficients of the polynomial are then determined from a least square (LS) fit of the polynomial to the raw FE stress values at these superconvergent points within the elements in the patch for which the number of sampling points can be taken as greater than the number of parameters in the polynomial. SPR then uses this approximation to obtain nodal values by averaging the fitted results from those patches that include this node, and finally interpolates these nodal values by standard shape functions.

In the SPR procedure, nodal patches are established for interior nodes only, as nodes on the exterior boundary are rarely connected to enough elements. If the element node a is an interior node, $\sigma^*(x_a)$ is evaluated on the patch of elements surrounding this node. For nodes lying on the exterior boundary, $\sigma^*(x_a)$ is instead evaluated on the patch (or patches) associated with the other node(s) that are connected to node a through an interior element boundary. If in this manner more than one patch is connected to a boundary node, the corresponding values for $\sigma^*(x_a)$ computed on each patch are averaged.

Numerous studies on optimal stress points have been carried out. However, various researchers demonstrated that superconvergent points in the classical sense generally do not exist or have no fixed location; hence, applicability of SPR seems doubtful. Therefore, a more universal method is to fit the FE nodal displacements by a polynomial whose order is one higher than the employed FE shape function. Then the derivatives are obtained by differentiating the fitted polynomial. The accuracy of the derivatives so obtained is always one order higher than the direct differentiation of the shape functions.[52]

6.2. *Moving Least Squares (MLS)*

An alternative recovery procedure is based on local interpolation of nodal displacements using an MLS method.[53] A continuous stress field can be obtained directly. In most cases, the extracted derivative quantities exhibit superconvergence, i.e. a rate of convergence one order higher than the rate of convergence of the standard FE solutions. Superconvergence points are not necessary. It is useful for extracting detailed strain fields near the crack tip by adding a square root function to the monomials.

MLS can also be used to fit the derivatives at particular points in the element, e.g. Gauss points, to obtain continuous derivatives.[54,55] In fitting the EFG stresses, the MLS shape functions for recovery can be constructed by using reduced supports on the same nodal points of the original EFG analysis.[44,54]

7. Numerical Examples

To illustrate the first-order homogenisation method described above, we solve the torsion of a composite shaft with square cross-section (length of side $= 80$), as shown in Fig. 4(a). Assume that the microstructure of the cross-section is locally periodic with a period defined by an RUC shown in Fig. 4(b), i.e. it consists of an isotropic circular fibre of diameter $2a$ embedded in an isotropic square matrix with side $4a$. $a = 5$ is adopted in this study. The problem is solved in two stages. First, we solve the RUC by using the incompatible element introduced in Sec. 4.3, with the periodicity boundary condition enforced by the penalty function approach discussed in Sec. 5. We obtain the field χ_3^{3k} and its derivatives $\partial\chi_3^{3k}/\partial y_j$ and calculate the homogenised moduli from (38). Second, we solve the torsion of the square shaft shown in Fig. 4(a) with the homogenised moduli obtained at step one above, by using the hybrid stress element.[56] In this way, we calculate the warping displacement, torsional rigidity and the angle of twist per unit length, as well as the shear stresses and strains. With the results so obtained, we can calculate the first-order warping displacement from (28) and the local strain and stress fields from (14) and (16), respectively. For the present illustrative purpose, we choose $\varepsilon = 0.25$. The complete shaft section from which the RUC has been extracted is shown in Fig. 4(c). In the figures to follow, filled triangles represent computed data. In all the figures that illustrate the stress distribution, a line segment represents the distribution within an element. In Figs. 6(b), 7(b) and 7(c), the solid line

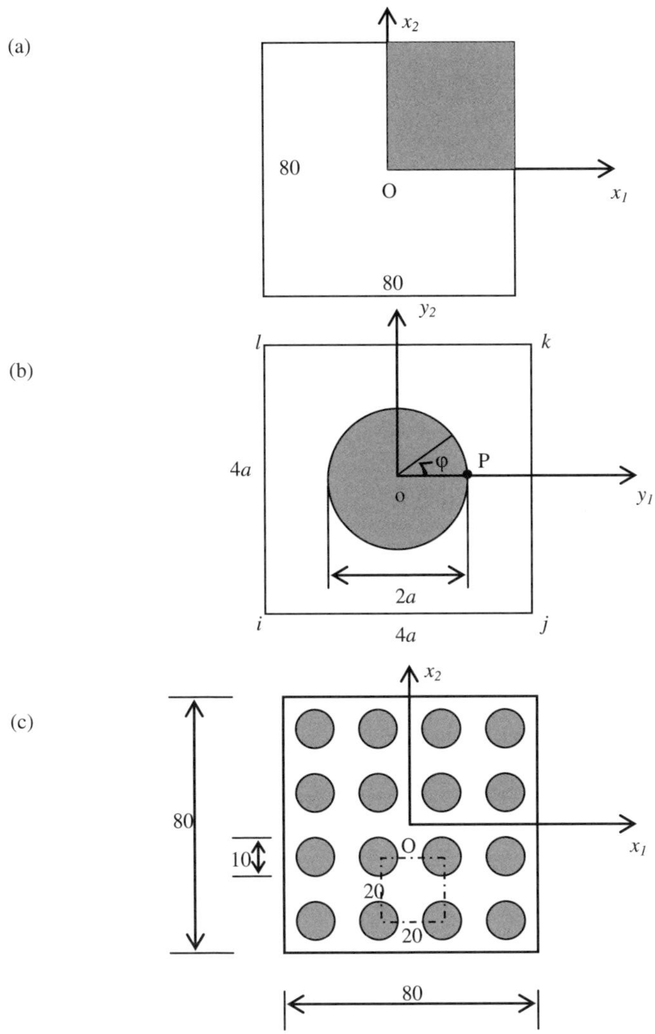

Fig. 4. Geometry of a composite shaft of square profile: (a) square profile; (b) RUC; (c) square shaft with 16 fibres.

represents the polynomial fit of the corresponding computed data that is not satisfactorily smooth.

The RUC shown in Fig. 4(b) is discretised into 896 quadrilateral elements and 929 nodes, as shown in Fig. 5(a). According to the definition of the RUC, its size should be enlarged four times as $\varepsilon = 0.25$. However,

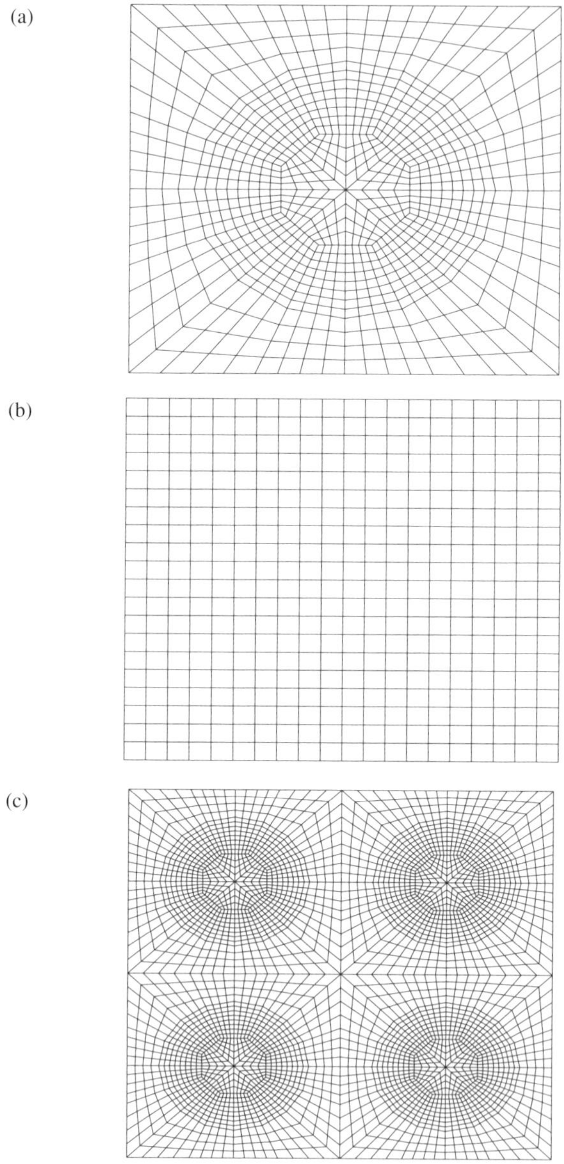

Fig. 5. Meshes used in the computation: (a) mesh of the RUC shown in Fig. 4(b); (b) mesh of a quarter of the cross-section shown in Fig. 4(a); (c) mesh of a quarter of the cross-section shown in Fig. 4(c).

numerical results show that the results are unaffected by whether or not the RUC size is enlarged, allowing us to use the original RUC size. Care must be taken in enforcing the periodicity boundary condition at corner nodes. For the four corner nodes, i, j, k and l, shown in Fig. 4(b), the periodicity condition yields

$$q_i^k = q_j^k = q_k^k = q_l^k.$$

The above condition can be rewritten as

$$\begin{aligned} q_i^k &= q_j^k, \\ q_j^k &= q_k^k, \\ q_k^k &= q_l^k, \end{aligned}$$

and treated conveniently by the procedure discussed in Sec. 5. The fibre and the matrix are considered to be isotropic with the shear moduli, G_f =10 and $G_m = 1$, respectively. The computed homogenised shear moduli are

$$\begin{bmatrix} C_{11} & C_{12} \\ Sym & C_{22} \end{bmatrix} = \begin{bmatrix} 1.38271 & -0.00138 \\ Sym & 1.38467 \end{bmatrix}. \tag{146}$$

Thus the macroscopic behaviour of the composite shaft is also isotropic. The numerical results for the characteristic displacements χ_3^{3k} and their derivatives $\partial\chi_3^{3k}/\partial y_j$ are saved for later use.

The isotropic shaft of square cross-section shown in Fig. 4(a) is now analysed with the homogenised shear moduli (146) obtained above. Only a quarter of the cross-section, the shaded part shown in Fig. 4(a), is discretised because of symmetry. The warping displacements are fixed on the axes of symmetry. The employed FE mesh with 400 quadrilateral elements and 441 nodes is shown in Fig. 5(b). One unit of torque is applied on the quarter section with its units being consistent with those of the shear moduli. The computed result for the torsional rigidity $4 \times 1.9927 \times 10^6$ is very close to the accurate value 7.9856×10^6 obtained from the formula

$$\text{Torsional rigidity} = 0.141G(2b)^4, \tag{147}$$

where the shear modulus $G = 1.38271$, and the length of side of the square cross-section $2b = 80$ in the present example. The numerical results for the local fields near or along the interface between the fibre and the matrix adjacent to the point with global co-ordinates $(x_1 = 30, x_2 = 30)$ are shown in Figs. 6 and 7. Figures 6(a)–6(c) show the results along the line

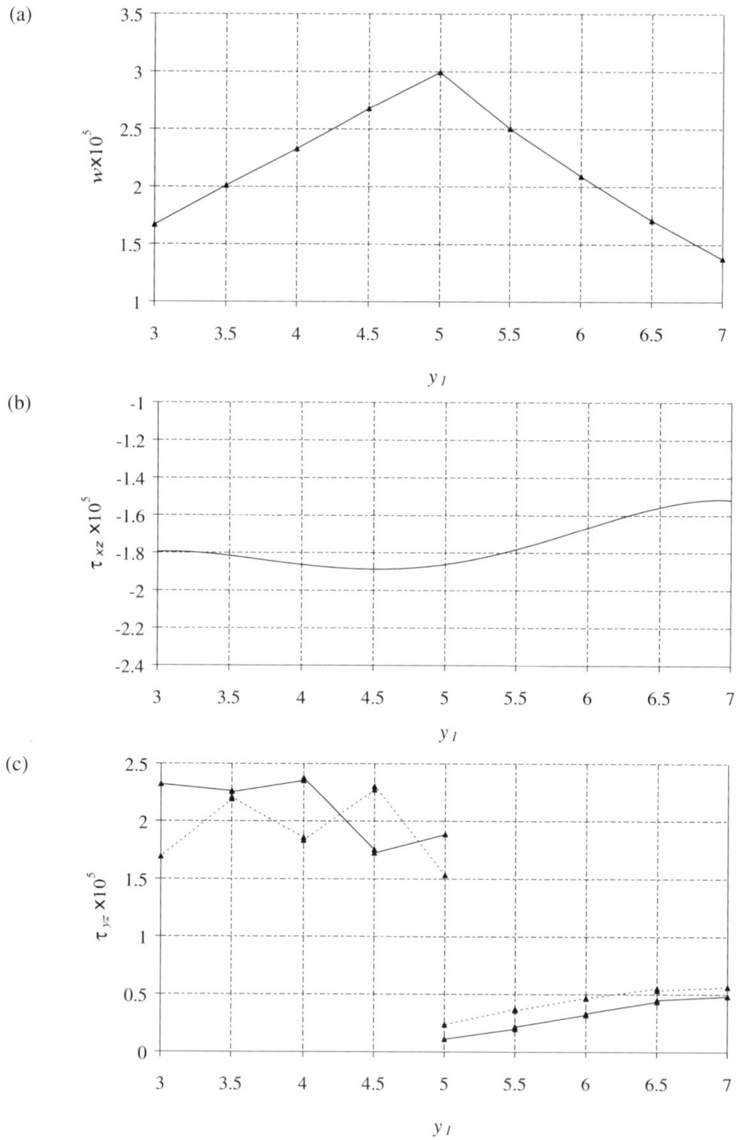

Fig. 6. Numerical results on the line $3 \leq y_1 \leq 7$, $y_2 = 0$, from the homogenisation method: (a) distribution of warping displacement; (b) distribution of τ_{xz}; and (c) distribution of τ_{yz}.

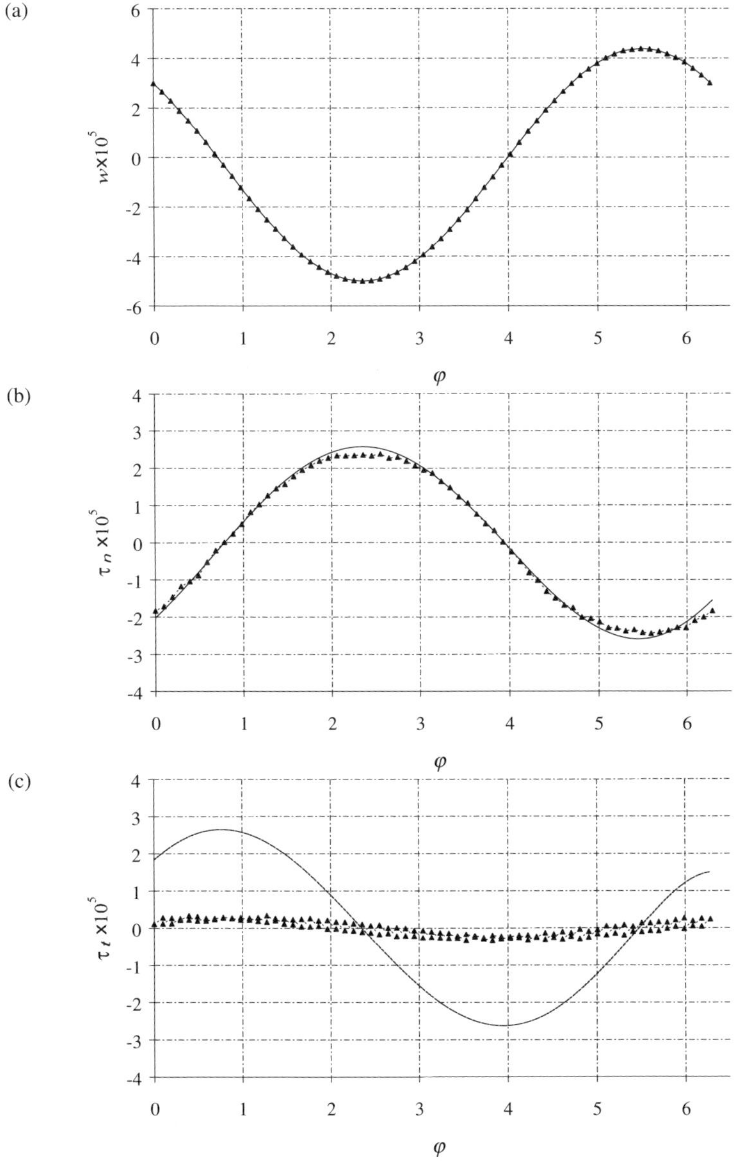

Fig. 7. Numerical results along the interface from the homogenisation method: (a) distribution of warping displacement; (b) distribution of the normal shear stress τ_n; and (c) distribution of the tangential shear stress τ_t.

$3 \leq y_1 \leq 7, y_2 = 0$ near the point P in Fig. 4(b). Figure 6(a) shows the distribution of warping displacement. Figure 6(b) shows that the polynomial fitting of the computed shear stress, τ_{xz}, on the scale of the figure results given by the upper and lower elements adjacent to the line cannot be distinguished. Figure 6(c) shows that the computed shear stress τ_{yz} data linked by solid and broken lines represent, respectively, the results obtained from the upper and lower elements adjacent to the line in question. From the results it is seen that the gradient of the warping displacement changes rapidly across the interface ($y_1 = 5$) and that the distribution of τ_{xz} but not of τ_{yz} is continuous across the interface. The distribution of warping displacement, and of normal and tangential shear stresses along the interface, which are given by

$$\begin{aligned} \tau_n &= \tau_{xz} \cos\varphi + \tau_{yz} \sin\varphi, \\ \tau_t &= -\tau_{xz} \sin\varphi + \tau_{yz} \cos\varphi, \end{aligned} \tag{148}$$

where φ is the angle from the axis y_1 as shown in Fig. 4(b), is plotted in Figs. 7(a)–7(c). In Figs. 7(b) and 7(c), data linked by broken lines represent the results obtained from the matrix side, the continuous solid line represents the polynomial fit of the results obtained from the fibre side of the interface. These results show that the warping displacement and normal shear stress τ_n vary continuously across the interface, whereas the tangential shear stress τ_t has a significant discontinuity.

Now we solve directly the torsion of the composite shaft shown in Fig. 4(c) by the hybrid stress element[56] to illustrate some typical features of local fields adjacent to the interface. Again, only a quarter of the cross-section is needed to be discretised because of symmetry. The warping displacements are fixed on the axes of symmetry. The FE mesh with 3584 quadrilateral elements and 3649 nodes is shown in Fig. 5(c). One unit of torque is applied on the quarter section with its units being consistent with those of the shear modulus. The computed result for torsional rigidity is $4 \times 1.9456356 \times 10^6$, which according to the formula (147) corresponds to an isotropic shaft with shear modulus 1.34754. The result is reasonably close to that obtained by the homogenisation method (146). The latter predicts larger values of moduli because the employment of the periodic boundary condition makes the system stiffer. The result given by the homogenisation method is also within the lower bound 1.215 and the upper bound 2.767 as per the Voigt–Reuss theory.[2]

Zhao and Weng[57] have derived the nine effective elastic constants of an orthotropic composite reinforced with monotonically aligned, uniformly dispersed elliptic cylinders using the Eshelby–Mori–Tanaka method. The problem studied above is the special case that the reinforcements are fibres with circular cross-section. The two shear moduli relevant to torsion given by Zhao and Weng[57] are

$$\frac{C_{11}}{G_m} = 1 + \frac{c_f}{\dfrac{c_m\alpha}{1+\alpha} + \dfrac{G_m}{G_f - G_m}}, \quad \frac{C_{22}}{G_m} = 1 + \frac{c_f}{\dfrac{c_m}{1+\alpha} + \dfrac{G_m}{G_f - G_m}}, \tag{149}$$

where c_f and c_m are volume fractions of fibre and matrix, respectively, and α is the cross-sectional aspect ratio of the reinforced fibre. In our case, $c_f = \pi/4$, $c_m = 1 - \pi/4$ and $\alpha = b/a = 1$, and hence the effective shear moduli $C_{11} = 4.595947 = C_{22}$ given by (149) are unreasonably higher than the results by the direct FE analysis, as well as the results (146) by the homogenisation method mentioned above. They are also above the upper bound of the Voigt–Reuss theory. The Eshelby–Mori–Tanaka method cannot give good results, especially for high volume fraction of reinforcements, because Eshelby's tensor is based on the inclusion in an infinite matrix, which takes into account the interaction between reinforcements in a very weak sense. On the other hand, it is evident that the homogenisation method has the advantage of taking the interaction between phases into account naturally and of not having to make assumptions such as isotropy of material.

The distribution of warping displacement and shear stresses along the line corresponding to Fig. 6 and the interface corresponding to Fig. 7 are plotted in Figs. 8 and 9 respectively. Equation (148) has been used to obtain the normal and tangential shear stresses in Figs. 9(b) and 9(c). A comparison of Figs. 6 and 7 with Figs. 8 and 9, respectively, shows the obvious differences of the results obtained by the homogenisation method and the direct hybrid stress element. The differences are to be expected in view of the limited number of fibres that can be economically handled by the hybrid stress element. The homogenisation method is suitable for problems involving a large number of periodically distributed reinforcements so that the RUC occupies only a "point" in the physical domain. The computed stress fields by the hybrid stress element are smoother than those obtained by the homogenisation method and smoothing techniques are unnecessary for the former since differentiations are avoided in the computations. Notwithstanding these differences, the results by the two methods reveal the common features of the local fields: a significant discontinuity exists in the tangential shear stress, while other fields are continuous adjacent to the interface.

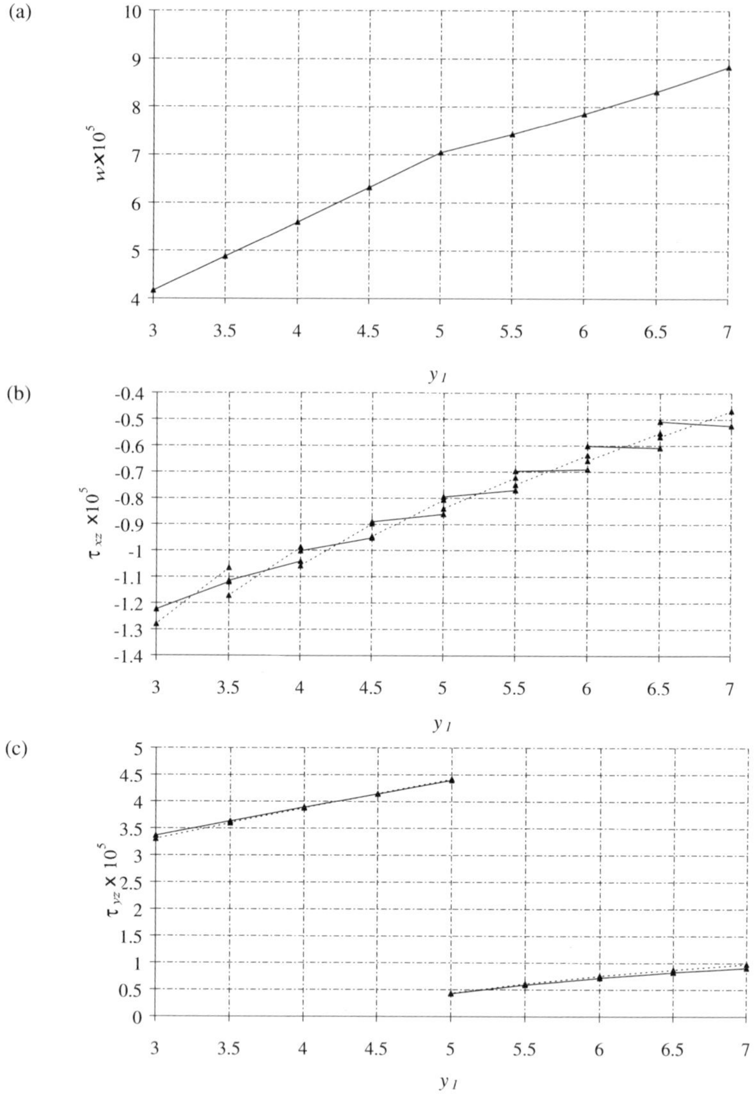

Fig. 8. Numerical results on the line $3 \leq y_1 \leq 7$, $y_2 = 0$, from the hybrid stress element method: (a) distribution of warping displacement; (b) distribution of τ_{xz}; and (c) distribution of τ_{yz}. In (b) and (c) data linked by solid and broken lines represent, respectively, the results obtained from the upper and lower elements adjacent to the line in question.

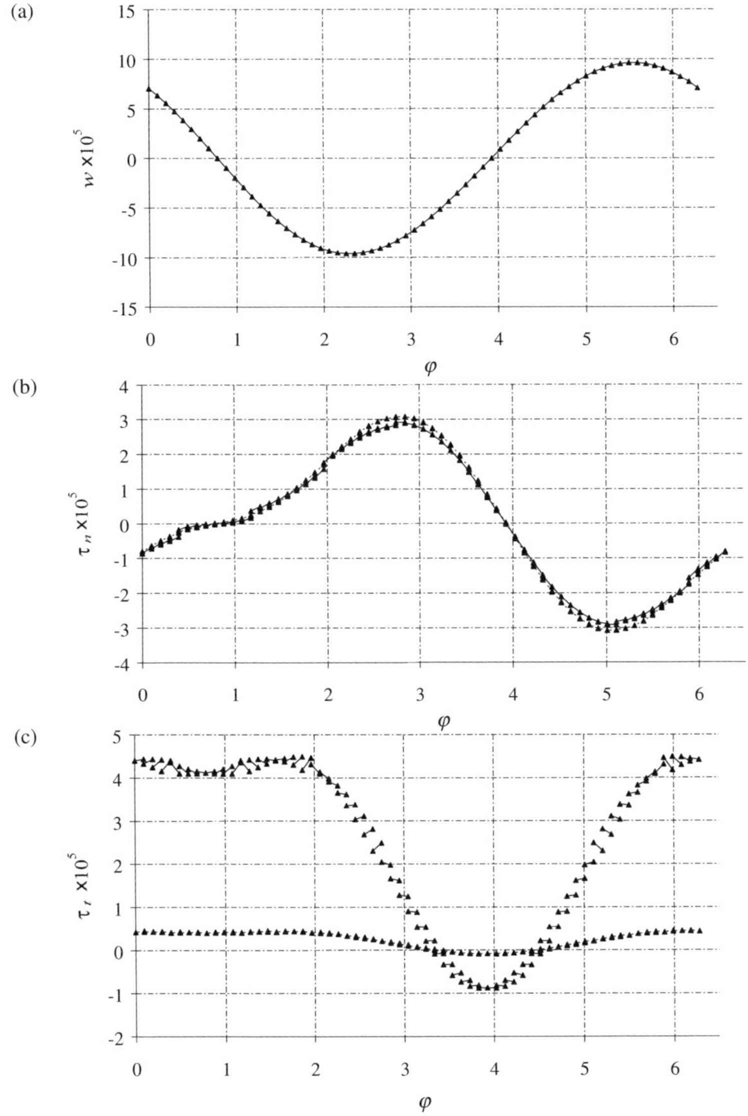

Fig. 9. Numerical results along the interface from the hybrid stress element method: (a) distribution of warping displacement; (b) distribution of the normal shear stress τ_n; and (c) distribution of the tangential shear stress τ_t. In (b) and (c), data linked by solid and broken lines represent, respectively, the results obtained from the fibre and matrix side of the interface.

8. Discussion and Conclusions

The two-scale asymptotic homogenisation method is most suitable for problems involving a large number of periodically distributed reinforcements so that the RUC can be regarded as a "point" in the physical domain. It gives not only the equivalent material properties but also detailed information of local fields with much lower computational cost. Such detailed information of the fields on the scale of micro-constituents is almost impossible to obtain by using the FEM, because of the enormous number of degrees of freedom needed to model the entire macro-domain with a grid size comparable to that of the microscale features. When the number of the reinforcements is not very large, numerical results by the homogenisation method without the terms of order higher than one are usually quantitatively different from those obtained by the direct FEM. The inclusion of higher-order terms with the methodology developed in this study should improve numerical accuracy, but it inevitably complicates the procedure.

In the homogenisation analysis, the solution of the zeroth- and first-order expansions is coupled, and equilibrium equations of orders $O(\varepsilon^{-2})$, $O(\varepsilon^{-1})$ and $O(\varepsilon^{0})$ need to be considered. Then higher-order expansions can be solved in sequence, e.g. the pth ($p \geq 2$) order expansion can be solved from the $O(\varepsilon^{p-1})$ equilibrium together with the constraints from $O(\varepsilon^{p})$ equilibrium. In the solution of χ_i^{kl} and other micro displacements, the isoparametric element and the more accurate incompatible and enhanced-strain elements, and the EFG methods can be used together with the SPR or MLS recovery strategies. The high-performance hybrid stress elements are limited, because of the difficulty in enforcing the periodic conditions of the stress. In the solution of the macro problem, all discussed methods can be used together with the SPR or MLS recovery strategies, if the displacements and/or their first-order derivatives only are required to solve the higher-order expansions. If the higher-order derivatives of the macro displacement are required, the EFG and/or the MLS recovery scheme are better.

References

1. R. M. Christensen, A critical evaluation for a class of micromechanics models, *J. Mech. Phys. Solids* **38**, 379 (1990).
2. S. Nemat-Nasser and M. Hori, *Micromechanics: Overall Properties of Heterogeneous Materials*, 2nd edn. (North-Holland, Amsterdam, 1999).

3. K. J. Bathe, *Finite Element Procedures* (Prentice-Hall, Englewood Cliffs, NJ, 1995).
4. O. C. Zienkiewicz, R. L. Taylor and J. Z. Zhu, *The Finite Element Method: Its Basis and Fundamentals* (Butterworth Heinemann, Oxford, 2005).
5. T. H. H. Pian and C. C. Wu, *Hybrid and Incompatible Finite Element Methods* (CRC Press, 2005).
6. A. Bensoussan, J. L. Lions and G. Pananicolaou, *Asymptotic Analysis for Periodic Structures* (North-Holland Publishing Company, New York, 1978).
7. A. Bakhvalov and G. P. Panassenko, *Homogenization: Averaging Process in Periodic Media* (Kluwer Academic Publisher, Dordrecht, 1989).
8. A. L. Kalamkarov, *Composite and Reinforced Elements of Construction* (John Wiley & Sons, New York, 1992).
9. J. M. Guedes and N. Kikuchi, Preprocessing and postprocessing for materials based on the homogenization method with adaptive finite element methods, *Comput. Meth. Appl. Mech. Eng.* **83**, 143 (1990).
10. S. Jansson, Homogenized nonlinear constitutive properties and local stress concentrations for composites with periodic internal structure, *Int. J. Solids Struct.* **29**, 2181 (1992).
11. J. Fish, M. Nayak and M. H. Holmes, Microscale reduction error indicators and estimators for a periodic heterogeneous medium, *Comput. Mech.* **14**, 323 (1994).
12. D. Lukkassen, L. E. Persson and P. Wall, Some engineering and mathematical aspects on the homogenization method, *Compos. Eng.* **5**, 519 (1995).
13. B. Hassani and E. Hinton, Review of homogenization and topology optimization I — homogenization theory for media with periodic structure, *Comput. Struct.* **69**, 707 (1998).
14. B. Hassani and E. Hinton, Review of homogenization and topology optimization II — analytical and numerical solution of homogenization equations, *Comput. Struct.* **69**, 719 (1998).
15. B. Hassani and E. Hinton, Review of homogenization and topology optimization III — topology optimization using optimality criteria, *Comput. Struct.* **69**, 739 (1998).
16. K. Terada, M. Hori, T. Kyoya and N. Kikuchi, Simulation of the multi-scale convergence in computational homogenization approaches, *Int. J. Solids Struct.* **37**, 2285 (2000).
17. P. W. Chung, K. K. Tamma and R. R. Namburu, Asymptotic expansion homogenization for heterogeneous media: Computational issues and applications, *Composites A* **32A**, 1291 (2001).
18. K. Terada and N. Kikuchi, A class of general algorithms for multi-scale analyses of heterogeneous media, *Comput. Meth. Appl. Mech. Eng.* **190**, 5427 (2001).
19. H. Y. Sun, S. L. Di, N. Zhang and C. C. Wu, Micromechanics of composite materials using multivariable finite element method and homogenization theory, *Int. J. Solids Struct.* **38**, 3007 (2001).

20. B. L. Karihaloo, Q. Z. Xiao and C. C. Wu, Homogenization-based multivariable element method for pure torsion of composite shafts, *Comput. Struct.* **79**, 1645 (2001).
21. H. Y. Sun, S. L. Di, N. Zhang, N. Pan and C. C. Wu, Micromechanics of braided composites via multivariable FEM, *Comput. Struct.* **81**, 2021 (2003).
22. J. B. Castillero, J. A. Otero, R. R. Ramos and A. Bourgeat, Asymptotic homogenization of laminated piezocomposite materials, *Int. J. Solids Struct.* **35**, 527 (1998).
23. M. L. Feng and C. C. Wu, Study on 3-dimensional 4-step braided piezoceramic composites by homogenization method, *Compos. Sci. Techol.* **61**, 1889 (2001).
24. H. Berger, U. Gabbert, H. Köppe, R. Rodriguez-Ramos, J. Bravo-Castillero, R. Guinovart-Diaz, J. A. Otero and G. A. Maugin, Finite element and asymptotic homogenization methods applied to smart composite materials, *Comput. Mech.* **33**, 61 (2003).
25. S. Ghosh, K. Lee and S. Moorthy, Two scale analysis of heterogeneous elastic–plastic materials with asymptotic homogenization and Voronoi cell finite element model, *Comput. Meth. Appl. Mech. Eng.* **132**, 63 (1996).
26. J. Fish, K. Shek, M. Pandheeradi and M. S. Shephard, Computational plasticity for composite structures based on mathematical homogenization: Theory and practice, *Comput. Meth. Appl. Mech. Eng.* **148**, 53 (1997).
27. J. Fish and K. Shek, Finite deformation plasticity for composite structures: Computational models and adaptive strategies, *Comput. Meth. Appl. Mech. Eng.* **172**, 145 (1999).
28. K. Lee, S. Moorthy and S. Ghosh, Multiple scale computational model for damage in composite materials, *Comput. Meth. Appl. Mech. Eng.* **172**, 175 (1999).
29. J. Fish, Q. Yu and K. Shek, Computational damage mechanics for composite materials based on mathematical homogenisation, *Int. J. Numer. Meth. Eng.* **45**, 1657 (1999).
30. Y.-M. Yi, S.-M. Park and S.-K. Youn, Asymptotic homogenization of viscoelastic composites with periodic microstructures, *Int. J. Solids Struct.* **35**, 2039 (1998).
31. Q. Yu and J. Fish, Multiscale asymptotic homogenization for multiphysics problems with multiple spatial and temporal scales: A coupled thermo-viscoelastic example problem, *Int. J. Solids Struct.* **39**, 6429 (2002).
32. C. Oskay and J. Fish, Fatigue life prediction using 2-scale temporal asymptotic homogenisation, *Int. J. Numer. Meth. Eng.* **61**, 329 (2004).
33. T. Iwamoto, Multiscale computational simulation of deformation behavior of TRIP steel with growth of martensitic particles in unit cell by asymptotic homogenization method, *Int. J. Plasticity* **20**, 841 (2004).
34. C. C. Lin and L. A. Segel, *Mathematics Applied to Deterministic Problems in the Natural Sciences* (Macmillan Publishing Co., Inc., New York, 1974).
35. V. Kouznetsova, M. G. D. Geers and W. A. M. Brekelmans, Multi-scale constitutive modelling of heterogeneous materials with a gradient-enhanced

computational homogenization scheme, *Int. J. Numer. Meth. Eng.* **54**, 1235 (2002).
36. T. O. Williams, A two-dimensional, higher-order, elasticity-based micromechanics model, *Int. J. Solids Struct.* **42**, 1009 (2005).
37. T. O. Williams, A three-dimensional, higher-order, elasticity-based micromechanics model, *Int. J. Solids Struct.* **42**, 971 (2005).
38. J. Fish and W. Chen, Higher-order homogenization of initial/boundary-value problem, *J. Eng. Mech.* **127**, 1223 (2001).
39. S. P. Timoshenko and J. N. Goodier, *Theory of Elasticity*, 3rd edn. (McGraw-Hill, New York, 1970).
40. T. Belytschko, Y. Y. Lu and L. Gu, Element-free Galerkin methods, *Int. J. Numer. Meth. Eng.* **37**, 229 (1994).
41. T. Most and C. Bucher, A moving least squares weighting function for the element-free Galerkin method which almost fulfills essential boundary conditions, *Struct. Eng. Mech.* **21**, 315 (2005).
42. Y. Y. Lu, T. Belytschko and L. Gu, A new implementation of the element-free Galerkin method, *Comput. Meth. Appl. Mech. Eng.* **113**, 397 (1994).
43. Y. Krongauz and T. Belytschko, Enforcement of essential boundary conditions in meshless approximations using finite elements, *Comput. Meth. Appl. Mech. Eng.* **131**, 133 (1996).
44. Q. Z. Xiao and M. Dhanasekar, Coupling of FE and EFG using collocation approach, *Adv. Eng. Software* **33**, 507 (2002).
45. T. Zhu and S. N. Atluri, A modified collocation and a penalty formulation for enforcing the essential conditions in the element free Galerkin method, *Comput. Mech.* **21**, 211 (1998).
46. L. Gavete, J. J. Benito, S. Falcon and A. Ruiz, Implementation of essential boundary conditions in a mesthless method, *Commun. Numer. Meth. Eng.* **16**, 409 (2000).
47. P. Krysl and T. Belytschko, Element-free Galerkin method: Convergence of the continuous and discontinuous shape functions, *Comput. Meth. Appl. Mech. Eng.* **148**, 257 (1997).
48. D. Organ, M. Fleming, T. Terry and T. Belytschko, Continuous meshless approximations for nonconvex bodies by diffraction and transparency, *Comput. Mech.* **18**, 1 (1996).
49. Y. Krongauz and T. Belytschko, EFG approximation with discontinuous derivatives, *Int. J. Numer. Meth. Eng.* **41**, 1215 (1998).
50. J. C. Simo and T. R. J. Hughes, On the variational foundations of assumed strain methods, *J. Appl. Mech.* **53**, 51 (1986).
51. J. C. Simo and M. S. Rifai, A class of mixed assumed strain methods and the method of incompatible modes, *Int. J. Numer. Meth. Eng.* **29**, 1595 (1990).
52. O. C. Zienkiewicz, The background of error estimation and adaptivity in finite element computations, *Comput. Meth. Appl. Mech. Eng.* **195**, 207 (2006).
53. M. Tabbara, T. Blacker and T. Belytschko, Finite element derivative recovery by moving least square interpolants, *Comput. Meth. Appl. Mech. Eng.* **117**, 211 (1994).

54. H. J. Chung and T. Belytschko, An error estimate in the EFG method, *Comput. Mech.* **21**, 91 (1998).
55. Q. Z. Xiao and B. L. Karihaloo, Improving the accuracy of XFEM crack tip fields using higher order quadrature and statically admissible stress recovery, *Int. J. Numer. Meth. Eng.* **66**, 1378 (2006).
56. Q. Z. Xiao, B. L. Karihaloo, Z. R. Li and F. W. Williams, An improved hybrid-stress element approach to torsion of shafts, *Comput. Struct.* **71**, 535 (1999).
57. Y. H. Zhao and G. J. Weng, Effective elastic moduli of ribbon-reinforced composites, *J. Appl. Mech.* **57**, 158 (1990).

MULTI-SCALE BOUNDARY ELEMENT MODELLING OF MATERIAL DEGRADATION AND FRACTURE

G. K. Sfantos and M. H. Aliabadi*

Department of Aeronautics, Faculty of Engineering
Imperial College, University of London, South Kensington Campus
London SW7 2AZ, UK
**m.h.aliabadi@imperial.ac.uk*

In this chapter, a multi-scale boundary element method (BEM) for modelling material degradation and fracture is proposed. The constitutive behaviour of a polycrystalline macro-continuum is described by micromechanics simulations using averaging theorems. An integral non-local approach is employed to avoid the pathological localisation of microdamage at the macro-scale. At the micro-scale, multiple intergranular crack initiation and propagation under mixed mode failure conditions is considered. A non-linear frictional contact analysis is employed for modelling the cohesive-frictional grain boundary interfaces. Both micro- and macro-scales are being modelled with the BEM. Additionally, a scheme for coupling the micro-BEM with a macro-FEM is presented. To demonstrate the accuracy of the method, the mesh independency is investigated and comparisons with two macro-FEM models are made to validate the different modelling approaches. Finally, microstructural variability of the macro-continuum is considered to investigate possible applications to heterogeneous materials.

1. Introduction

The propagation and coalescence of microcracks and similar defects in the micro-scale eventually leads to the complete fracture failure of components. However, from a modelling perspective, the transition of a microcrack to the macro-scale is still not very "clear". Continuum damage mechanics aims to fill that gap. From the early work of Kachanov,[1] continuum damage mechanics introduces an isotropic scalar multiplier that reduces the initial elastic stiffness of the material over a specific region of the macro-continuum, in order to describe the local loss of the material integrity due to the formation and propagation of microcracks. A macro-crack is subsequently represented by the region where the damage is so extensive that the material cannot sustain more load.[2,3] Even though continuum damage mechanics can actually deal with initiation of macro-cracks, it does not provide sufficient details about the actual initiation and behaviour of cracks at micro-scale. Therefore, it is evident that there is a need for modelling materials in different scales and monitoring their behaviour simultaneously.

Multi-scale modelling is receiving much attention nowadays due to the increasing need for better modelling and understanding of materials' behaviour. Engineering materials are in general heterogeneous at a certain scale. Textile composites, concrete, ceramic composites, etc. are all naturally heterogeneous. Even classic metallic materials are heterogeneous at the micro and grain scale. Multi-scale homogenisation methods provide the advantage of modelling a specific material at different scales simultaneously.[4–7] At scales where the mechanical behaviour is unknown due to the complexity of the material structure, no constitutive law is required since this can be defined at smaller scales where the behaviour may be known.

Multi-scale methods can also provide valuable information of the damage evolution in a material throughout different scales.[8–11] The macro-continuum can be modelled as in the case of continuum damage mechanics, but without considering *a priori* any constitutive law for the mechanical behaviour of the material or any damage law for the degradation of the material's integrity. Both laws can be deduced from the micromechanics *in situ*. Hence, any heterogeneities of the material in the micro-scale will affect directly the macro-continuum response and moreover, microcracking initiation and propagation in the micro-scale and their effect on the macro-scale will be monitored simultaneously as the micro-scale will pass information to the macro-scale and *vice versa*.

To date, multi-scale modelling is mainly carried out within the context of the finite element method (FEM).[4–7,9–11] The boundary element method (BEM), an alternative method to the FEM, nowadays provides a powerful tool for solving a wide range of fracture problems.[12,13] The main advantage of BEM, the reduction in the dimensionality of a problem, becomes very attractive in cases of large-scale problems that are computationally expensive as the multi-scale modelling. In this chapter, a parallel processing multi-scale boundary element method[21,22] is presented, for modelling damage initiation and progression in the micro- and macro-scale. Both micro- and macromechanics are being formulated by the proposed method, nevertheless a link for coupling the micro-BEM with a macro-FEM solution scheme is also presented.

Multi-scale modelling of intergranular microfracture in polycrystalline brittle materials is the problem in consideration. Grain boundaries of polycrystalline materials, as appears in the majority of engineering metallic alloys (ferrous, non-ferrous) and ceramics, are often characterised by the presence of deleterious features and increased surface free energy that makes them more susceptible to aggressive environmental conditions. These conditions often lead to brittle intergranular failure[14,15] and stress-corrosion cracking,[16,17] respectively. The use of cohesive surfaces inside the FEM remains the most popular approach for modelling such micromechanics failures. Among the proposed cohesive failure models, the linear law proposed by Ortiz and Pandolfi[18] for mixed mode failure initiation and propagation, and the potential-based laws proposed by Tvergaard[19] and Xu and Needleman[20] are the most popular. In the present work, boundary cohesive grain element method by Sfantos and Aliabadi[21] is used for modelling the micro-scale. Multiple intergranular microfracture initiation, propagation, branching and arresting under mixed mode failure conditions is modelled in a polycrystalline material, by incorporating a linear cohesive law.[18] Moreover, the random grain morphology, distribution and orientation are taken into consideration.

The macro-continuum is also modelled using the BEM. To monitor the material behaviour in the micro-scale and to pass information to the macro-scale, representative volume elements (RVE) are assigned to points in the domain of the macro-continuum. These RVEs represent the microstructure, at the grain level, of the macro-continuum at the infinitesimal material neighbourhood of that point. The formation and propagation of intergranular microcracks is monitored individually to each RVE. Since this micro-damage reduces the elastic stiffness of the RVE, consequently the material integrity of the local macro-element is also reduced. Therefore, a non-linear boundary element formulation is presented for the macro-continuum.

Microcracking initiation and propagation in the micro-scale results in strain softening at the macro-scale. This strain softening causes the loss of positive definiteness of the elastic stiffness resulting in an ill-posed problem.[23,24] In the FEM, the loss of ellipticity results in mesh sensitivity, where as much as the finite element (FE) discretisation is refined, the numerical solution does not converge to a physically meaningful solution.[25,26] To overcome this pathological localisation of damage, the so-called non-local models, either in integral form[23,27,28] or in gradient form,[29–31] have been proposed. In the present study, an integral non-local approach is enforced to ensure macro-mesh independency and objectivity of the results.

The macro–micro interface is being constructed in terms of *averaging theorems*.[34,35] All quantities transferred from the micro to the macro are being volume averaged over the RVE. A brief discussion on possible RVE boundary conditions is given and the implementation of the periodic boundary conditions in the context of the proposed BEM is explained in detail. The first-order computational homogenisation is being used in the present work.[6,7]

Finally, several numerical examples are presented for simulating damage and fracture in a polycrystalline brittle material. Intergranular cracking evolution at the micro-scale and the resulting damage progression and fracture at the macro-scale are illustrated. The mesh independency of the proposed formulation is discussed and comparisons with the FEM for the different damage modelling approaches conclude the chapter.

2. Macromechanics

2.1. *Modelling the continuum*

In terms of continuum mechanics, the macroscopically observed degradation of the material stiffness due to the propagation and coalescence of various microdefects in the micro-scale suggests the reduction in the local elasticity stiffness tensor. Here, the non-linear material degradation is introduced in terms of initial decremental stresses that *soften* locally the material. For this *initial stress approach*, the boundary integral equation can be written as

$$
\begin{aligned}
C_{ij}(\mathbf{x}')\dot{u}_j(\mathbf{x}') &+ \fint_S T_{ij}(\mathbf{x}',\mathbf{x})\dot{u}_j(\mathbf{x})\,dS \\
&= \int_S U_{ij}(\mathbf{x}',\mathbf{x})\dot{t}_j(\mathbf{x})\,dS + \int_V E_{ijk}(\mathbf{x}',\mathbf{X})\dot{\sigma}^D_{jk}(\mathbf{X})\,dV, \qquad (1)
\end{aligned}
$$

where $\dot{u}_j, \dot{t}_j$ denote the displacement and tractions on boundary S, respectively, T_{ij}, U_{ij}, E_{ijk} are fundamental solutions given in the Appendix, $\dot{\sigma}^D_{jk}$ denotes the decremental component of stress that is introduced by the micro-scale solution to soften locally the material in the macro-scale and

C_{ij} is the so-called free term.[12] Even though the problem in consideration is time-independent, due to the incremental formulation and to maintain a general notation with respect to other time-dependent inelastic phenomena, it is regarded as a rate problem where the field unknowns are denoted by an upper dot. Moreover, with $\mathbf{X} \in V$ a domain point is denoted while with $\mathbf{x} \in S$ a boundary point. The source point is denoted by $\mathbf{x}'$ while the field point is without the dash.

To solve Eq. (1), the boundary S of the macro-continuum is discretised into N quadratic isoparametric boundary elements while the expected non-linear domain V is discretised into M constant subparametric quadrilateral cells. For each cell, the field unknowns are evaluated at its geometrical center and is assumed to be uniformly distributed over its area. In other words, the non-linear domain is assigned M points, in which the micromechanics response will be evaluated, and this response will be uniformly distributed over the neighbourhood of the point that is limited by the neighbourhood of the adjacent points. For each point, a representative volume element (RVE) is assigned that would give all the information about the micromechanics state in the infinitesimal material neighbourhood.

After the discretisation and using the point collocation method for solution, the final system of equations can be written in matrix form as

$$\mathbf{A}\dot{\mathbf{x}} = \dot{\mathbf{f}} + \mathbf{E}\dot{\sigma}^{D}, \tag{2}$$

where the matrices $\mathbf{A}$, $\mathbf{E}$ contain known integrals of the product of shape functions, Jacobians and the fundamental fields, the vector $\dot{\mathbf{f}}$ contains contributions of the prescribed boundary values and the vector $\dot{\mathbf{x}}$ contains the unknown boundary values.

The size of the domain that must be discretised is limited by the distribution of the microdamage during the loading process that would introduce non-linear material behaviour in the macro-scale. However, in cases of non-homogeneous materials the behaviour of which depends on the location even in the elastic regime, the whole domain must be discretised. A great advantage of the proposed boundary element formulation is that even if all the macro-continuum domain was discretised and an RVE was assigned to each domain point, as long as the material remains locally undamaged, the micromechanics simulations are linear and the contribution to the computational effort is negligible. On the other hand, for the completely damaged zones, the RVE simulations are stopped and computational storage and time is saved yet again. Therefore, it should be mentioned here that even in cases where the macro-damage pattern is unknown, discretising the whole domain would not increase the computational effort substantially.

Another advantage of the proposed formulation is that the size of the final system of equations, for the macro-continuum, remains unchanged irrespective of the number of the domain points and therefore RVEs that are considered. From the final system of equations that must be solved, Eq. (2), it can be seen that the material non-linearities due to the microdamage are acting as a right-hand side vector that does not increase the system size. Hence, at every increment this right-hand vector is evaluated and the new solution is given by forward and back substitution with the L and U decomposed matrices of the coefficient matrix $\mathbf{A}$.[36]

After solving the macro-continuum, the internal strains on every domain point must be evaluated, in order to define the boundary conditions on the corresponding RVE, in the micro-scale, for the next increment. Considering Somigliana's identity for the internal displacements[12] and the Cauchy strain tensor for small deformations $\dot{\varepsilon}_{ij} = \frac{1}{2}(\dot{u}_{i,j} + \dot{u}_{j,i})$, the boundary integral equation for the internal strains can be obtained by differentiating Eq. (1) with respect to the source point $\mathbf{X}'$ and gives

$$\dot{\varepsilon}_{ij}(\mathbf{X}') = \int_S D^{\varepsilon}_{ijk}(\mathbf{X}', \mathbf{x})\dot{t}_k(\mathbf{x})\, dS - \int_S S^{\varepsilon}_{ijk}(\mathbf{X}', \mathbf{x})\dot{u}_k(\mathbf{x})\, dS + \int_V W^{\varepsilon}_{ijkl}(\mathbf{X}', \mathbf{X})\dot{\sigma}^D_{kl}(\mathbf{X})\, dV - \dot{g}^{\varepsilon}_{ij}(\mathbf{X}') \tag{3}$$

where D^{ε}_{ijk} and S^{ε}_{ijk} are fundamental solutions produced by the derivatives of the U_{ij} and T_{ij} fundamental solutions, respectively. The fourth-order fundamental solution W^{ε}_{ijkl} has been evaluated by the derivative of the domain integral, Eq. (1), using the Leibniz formula and the free term $\dot{g}^{\varepsilon}_{ij}$ is due to the treatment of the $O(r^{-2})$ singularity in the sense of Cauchy principal value.[12] All the fundamental solutions can be found in the Appendix.

Finally, the boundary integral equation for the internal stresses in the macro-continuum is derived through the application of Hooke's law and Eq. (3), i.e.

$$\dot{\sigma}_{ij}(\mathbf{X}') = \int_S D^{\sigma}_{ijk}(\mathbf{X}', \mathbf{x})\dot{t}_k(\mathbf{x})\, dS - \int_S S^{\sigma}_{ijk}(\mathbf{X}', \mathbf{x})\dot{u}_k(\mathbf{x})\, dS + \int_V W^{\sigma}_{ijkl}(\mathbf{X}', \mathbf{X})\dot{\sigma}^D_{kl}(\mathbf{X})\, dV - \dot{g}^{\sigma}_{ij}(\mathbf{X}'). \tag{4}$$

3. Artificial Microstructure Generation

For the proposed formulation to encounter the stochastic and random effects of the grain location and morphology in a polycrystalline material, different microstructures should be artificially generated. To date, the Poisson–Voronoi tessellation method is extensively used in the literature for modelling polycrystalline materials in a random manner.[14,37] In the field of grain-level material modelling, Voronoi tessellations have been coupled with the FEM[53] to simulate polycrystalline microstructures and used for modelling fragmentation of ceramic microstructures under dynamic loading,[46,47] grain boundary sliding and separation in nanocrystalline metals,[48] creep cavitation damage,[54] microdamage and microplasticity under dynamic uniaxial strains[55,56] and simulating the effective elastic constants of polycrystalline materials.[57]

In the present study, a Voronoi tessellation method is utilised for generating artificial microstructures with randomly distributed and orientated grains. Let the generator points $P = \{p_1, p_2, \ldots, p_n\} \subset \mathbb{R}^2$ bounded by a prescribed region $\mathfrak{S}$ and created by a random point generator of a uniform distribution; n denotes a finite number of points in the Euclidean plane, where $2 < n < \infty$ and $\mathbf{x}^i \neq \mathbf{x}^j$ for $i \neq j$, $i, j \in I_n : I_n = \{1, \ldots, n\}$ a set of integers. A Voronoi diagram bounded by $\mathfrak{S}$ is given by[37]

$$\mathcal{V}_{\cap\mathfrak{S}} = \{V_{(p_1)} \cap \mathfrak{S}, V_{(p_2)} \cap \mathfrak{S}, \ldots, V_{(p_n)} \cap \mathfrak{S}\}, \tag{5}$$

where $V_{(p_i)}$ denotes each Voronoi convex polygon that represents one grain. Each Voronoi polygon contains exactly one generating point and every point in a given polygon is closer to its generating point than to any other; hence,

$$V_{(p_i)} = \{\mathbf{x} : \|\mathbf{x} - \mathbf{x}^i\| \leq \|\mathbf{x} - \mathbf{x}^j\| \text{ for } j \neq i, j \in I_n\}. \tag{6}$$

In this chapter, a two-dimensional quasi-random generator using the Sobol sequence[36] was employed as a uniform random point generator. The reason was that after trials and comparisons with a pseudo-random point generator,[36] the former provided a better grain morphology than the latter in all random simulations. In the case of the quasi-random generator, the standard deviation of the resulting grain area was always less than in the case of the pseudo-random. Moreover the quasi-random generator created more equixed grains than the pseudo-random, as Fig. 1 illustrates. Figure 2 presents the resulting grain area distribution for a case of 700 grains with average grain area of $\bar{A}^{\mathrm{gr}} = 1.43 \cdot 10^{-3}\,\mathrm{mm}^2$ (ASTM $G \simeq 6.5$[58]) and a case

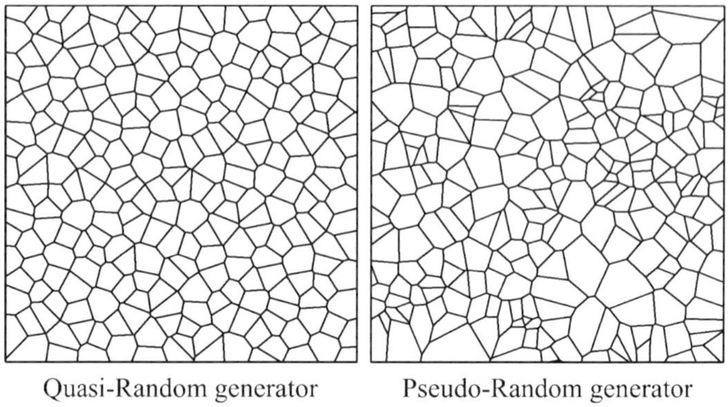

Fig. 1. Artificial microstructure generated by a quasi-random and a pseudo-random generator.

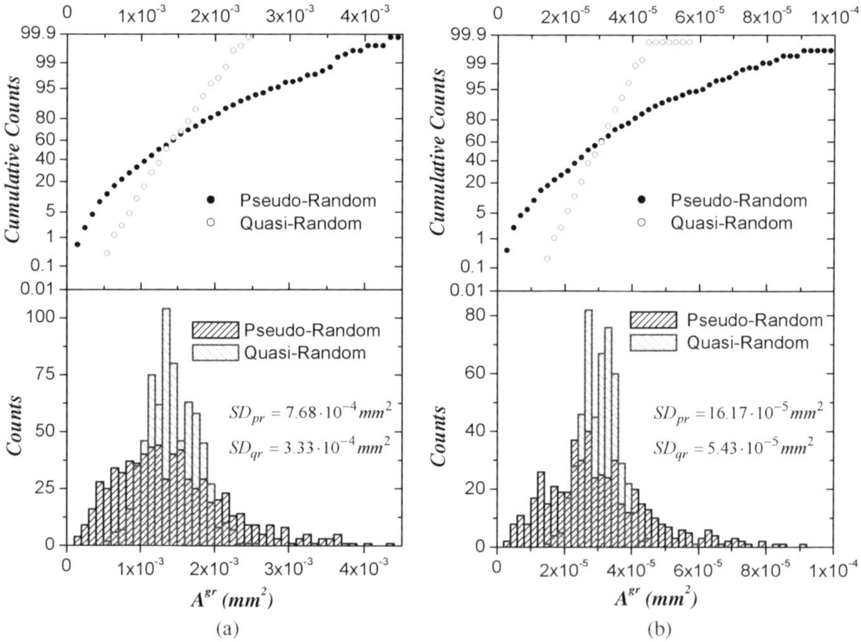

Fig. 2. Grain area distributions: (a) 700 grains, ASTM $G \simeq 6.45$[58] and (b) 500 grains, ASTM $G = 10.0$.[58]

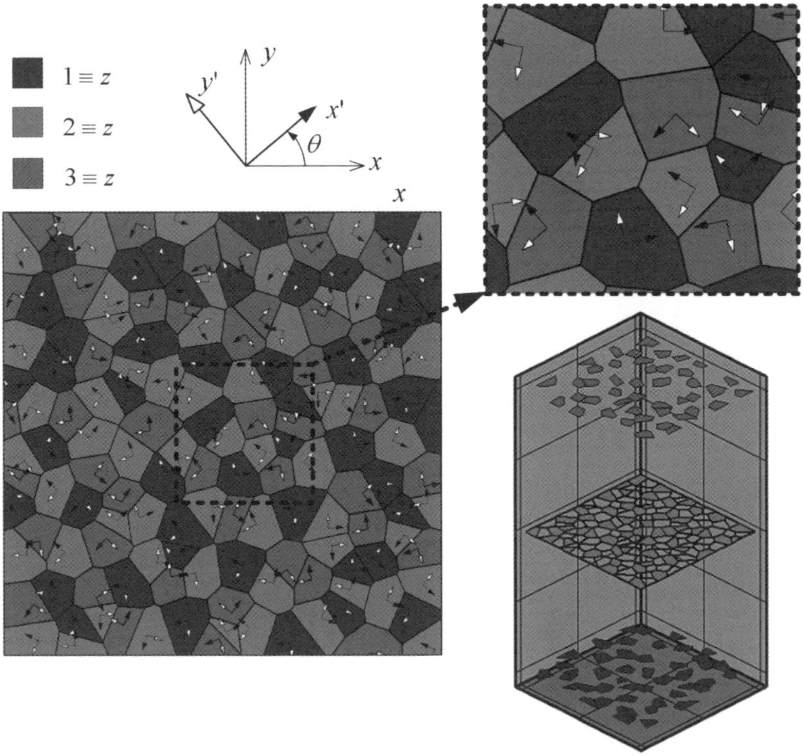

Fig. 3. Artificial microstructure generated by a quasi-random generator with randomly distributed material orientation for each grain.

of 500 grains with average grain area of $\bar{A}^{\mathrm{gr}} = 30.52 \cdot 10^{-6}\,\mathrm{mm}^2$ (ASTM $G = 12$[58]).

Figure 3 illustrates a randomly generated artificial microstructure, using the aforementioned method. Each grain is considered as a single crystal with orthotropic elastic behaviour and specific material orientation. Since the present study considers two-dimensional problems, to maintain the random character of the generated microstructure and the stochastic effects of each grain on the overall behaviour of the system, three different cases are considered for each grain.[46] Considering as xyz the geometry coordinate system and 123 the material coordinate system, three cases emerge in view of which of the three material axes coincide with the z-axis (out of plane) of the geometry; thus, *Case 1*: $1 \equiv z$, *Case 2*: $2 \equiv z$ and *Case 3*: $3 \equiv z$ (working plane is assumed the xy). Therefore, every generated grain is characterised by one of the aforementioned cases in a complete random manner. In Fig. 3, the different shades over each grain indicate the specific case. The three-dimensional view is provided to signify the random character of the grains selection for each case. However, in cases of cubic materials as the fcc and bcc metals, the characterisation of each grain as mentioned before is unnecessary as the material parameters are the same for every case. Moreover, to encounter the different orientation of each grain, a specific material orientation is given randomly to each grain (non-directional solidification is assumed). As Fig. 3 indicates, every grain orientation is randomly characterised by a counterclockwise angle θ off the x geometrical axis, where $0^\circ \leq \theta < 360^\circ$, that rotates the material coordinate system of each grain to a new position $x'y'$.

4. Microstructure Modelling

4.1. *Grain material modelling*

In the present study, an elastic-orthotropic model is used to describe the mechanical behaviour of the randomly created and orientated grains in a polycrystalline material. Hence, the constitutive relations combining the stresses and strains inside a specific grain are

$$\sigma_{ij} = c_{ijkl}\varepsilon_{kl}, \quad \varepsilon_{ij} = s_{ijkl}\sigma_{kl}, \tag{7}$$

where c_{ijkl} denotes the components of the stiffness tensor, which is inverse to the compliance tensor s_{ijkl}:

$$c_{ijkl}s_{klpq} = I_{ijpq}, \tag{8}$$

where $I_{ijpq} = (\delta_{ip}\delta_{jq} + \delta_{iq}\delta_{jp})/2$ is the fourth-order identity tensor, and δ_{ij} is the Kronecker delta function.

Using the concise Voigt notation to represent the elements of the elasticity tensor in the fixed basis, the compliance tensor is denoted by $\mathbf{S} = [S_{ij}]$, $i, j = 1, 2, \ldots, 6$, where $S_{12} = s_{1122}$, $S_{16} = s_{1112}$, $S_{44} = s_{2323}$, etc. For the case of an orthotropic material, having three mutually perpendicular symmetry planes, the compliance tensor unknown components are reduced to 9, since $S_{14} = S_{15} = S_{16} = 0$, $S_{24} = S_{25} = S_{26} = 0$, $S_{34} = S_{35} = S_{36} = 0$, $S_{45} = S_{46} = S_{56} = 0$.

In the case of two-dimensional problems, the compliance tensor for plane stress takes the following form:

$$\mathbf{S}' = \begin{bmatrix} S'_{11} & S'_{12} & S'_{16} \\ \ddots & S'_{22} & S'_{26} \\ sym & \ddots & S'_{66} \end{bmatrix}, \tag{9}$$

where the components of the above tensor S'_{ij}, $i,j = 1,2,6$ for two-dimensional problems, are taken from the compliance tensor S_{ij}, $i,j = 1,2,\ldots,6$ for three-dimensional problems, depending on which plane is normal to the plane that is modelled. For each of the three different cases explained in the previous section, Table 1 presents the corresponding S'_{ij} tensor compliance components. In the case where the material orientation axes coincide with the geometrical coordinate system, $S'_{16} = S'_{26} = 0$.

In the case of anisotropic elasticity, the fundamental solutions required in the boundary element method can be obtained for the two-dimensional case using the complex stress approach.[12] The fundamental solutions for the plane stress condition, for a source point:

$$z'_k = x'_1 + \mu_k x'_2 \tag{10}$$

in a complex plane with $k = 1,2$ and the field point defined by

$$z_k = x_1 + \mu_k x_2, \tag{11}$$

where $\mathbf{x}'$ and $\mathbf{x}$ are the cartesian coordinates of the source and the field points, respectively, are given as

$$U_{ij}(z'_k, z_k) = 2\Re[p_{j1}A_{i1}\ln(z_1 - z'_1) + p_{j2}A_{i2}\ln(z_2 - z'_2)], \tag{12}$$

$$T_{ij}(z'_k, z_k) = 2\Im\left[\frac{1}{(z_1 - z'_1)}q_{j1}(\mu_1 n_1 - n_2)A_{i1} + \frac{1}{(z_2 - z'_2)}q_{j2}(\mu_2 n_1 - n_2)A_{i2}\right], \tag{13}$$

Table 1. Corresponding compliance tensor components for the two-dimensional plane stress case.

S'_{ij}	$1 \equiv z$	$2 \equiv z$	$3 \equiv z$
S'_{11}	S_{22}	S_{11}	S_{11}
S'_{22}	S_{33}	S_{33}	S_{22}
S'_{12}	S_{23}	S_{13}	S_{12}
S'_{66}	S_{44}	S_{55}	S_{66}

where $\Re$ and $\Im$ denote the real and imaginary parts of the complex number in the square brackets, $\mathbf{n}$ is the outward normal unit vector, and

$$p_{ik} = \begin{bmatrix} S'_{11}\mu_k^2 + S'_{12} - S'_{16}\mu_k \\ S'_{12}\mu_k + S'_{22}/\mu_k - S'_{26} \end{bmatrix}, \tag{14}$$

$$q_{jk} = \begin{bmatrix} \mu_1 & \mu_2 \\ -1 & -1 \end{bmatrix}. \tag{15}$$

The complex coefficient A_{jk} are obtained after the solution of the following complex linear system:

$$\begin{bmatrix} 1 & -1 & 1 & -1 \\ \mu_1 & -\bar{\mu}_1 & \mu_2 & -\bar{\mu}_2 \\ p_{11} & -\bar{p}_{11} & p_{12} & -\bar{p}_{12} \\ p_{21} & -\bar{p}_{21} & p_{22} & -\bar{p}_{22} \end{bmatrix} \begin{Bmatrix} A_{j1} \\ \bar{A}_{j1} \\ A_{j2} \\ \bar{A}_{j2} \end{Bmatrix} = \begin{Bmatrix} \delta_{j2}/2\pi i \\ \delta_{j1}/2\pi i \\ 0 \\ 0 \end{Bmatrix}, \tag{16}$$

where δ_{ij} is the Kronecker delta function and μ_k are the complex or pure imaginary roots, which occur in conjugate pairs, μ_k and $\bar{\mu}_k$, of the characteristic equation:

$$S'_{11}\mu^4 - 2S'_{16}\mu^3 + (2S'_{12} + S'_{66})\mu^2 - 2S'_{26}\mu + S'_{22} = 0. \tag{17}$$

For the case of plain strain condition, the effective plain strain compliance tensor components must be used to calculate the complex μ_k and p_{ik}.[12] Table 2 presents these components for all three different cases described in the previous section for grain-level modelling.

Table 2. Effective compliance tensor components for the two-dimensional plain strain case.

$1 \equiv z$	$2 \equiv z$	$3 \equiv z$
$S'_{ij} = S_{kl} - \dfrac{S_{k1}S_{l1}}{S_{11}}$	$S'_{ij} = S_{kl} - \dfrac{S_{k2}S_{l2}}{S_{22}}$	$S'_{ij} = S_{kl} - \dfrac{S_{k3}S_{l3}}{S_{33}}$
$\begin{Bmatrix} i,j \\ k,l \end{Bmatrix} = \begin{Bmatrix} 1,2,6 \\ 2,3,4 \end{Bmatrix}$	$\begin{Bmatrix} i,j \\ k,l \end{Bmatrix} = \begin{Bmatrix} 1,2,6 \\ 1,3,5 \end{Bmatrix}$	$\begin{Bmatrix} i,j \\ k,l \end{Bmatrix} = \begin{Bmatrix} 1,2,6 \\ 1,2,6 \end{Bmatrix}$

4.2. A boundary cohesive element formulation

As already mentioned in the previous section, in a polycrystalline microstructure each grain is assumed to have a single crystal orthotropic behaviour of random orientation. Hence, in terms of the boundary element method, this system can be formulated as a multigrain-body system.[12] In this way, each grain can have any randomly specified elasticity parameters and material orientation.

Considering the microstructure illustrated in Fig. 3, two kinds of grains can be distinguished. The grains that are intersected by the domain boundary $\mathfrak{S}$ and from now on they will be referred as *domain boundary grains*, and the internal grains that are not intersected by $\mathfrak{S}$. The difference between them is that the internal grains have completely unknown boundary conditions while the others have both unknown and prescribed boundary conditions. For both cases the unknown boundary conditions will result from the solution of the problem and the use of tractions equilibrium and displacements compatibility conditions embedded along all the grain boundary interfaces. For each grain interface, a boundary of two neighbour grains, say A and B, tractions equilibrium and displacements compatibility is directly imposed, that is,

$$\tilde{\mathbf{t}}^I = \tilde{\mathbf{t}}_c^A = \tilde{\mathbf{t}}_c^B, \tag{18}$$

$$\delta\tilde{\mathbf{u}}^I = \tilde{\mathbf{u}}_c^A + \tilde{\mathbf{u}}_c^B, \tag{19}$$

where $\tilde{\mathbf{t}}^I$ and $\delta\tilde{\mathbf{u}}^I$ denote the interface tractions and relative displacement jump and the upper bar $(\tilde{\cdot})$ denotes values in the local coordinate system.

Each grain is bounded by a boundary S^H, where $H = 1, \ldots, N_g$ with N_g being the number of grains. The boundary of each grain is divided into the contact boundary S_c^H, indicating the contact with a neighbour grain boundary, and the free boundary S_{nc}^H, indicating the grain boundaries that coincide with the domain boundary $\mathfrak{S}$. Hence for every grain,

$$S^H = S_{nc}^H \cup S_c^H. \tag{20}$$

For the internal grains $S_{nc}^H = \varnothing$ and thus $S^H = S_c^H$. Therefore, S_{nc}^H exists only on boundary grains resulting to $\bigcup_{H=1}^{N_g}, S_{nc}^H = \mathfrak{S}$.

All the prescribed boundary conditions are transformed to the local coordinate system by

$$\tilde{\mathbf{t}}^p = \mathbf{R}\mathbf{t}^p, \quad \tilde{\mathbf{u}}^p = \mathbf{R}\mathbf{u}^p, \tag{21}$$

where $\mathbf{t}^p$, $\mathbf{u}^p$ denote the prescribed tractions and displacements, respectively, and $\mathbf{R}$ is a transformation matrix corresponding to the outward normal unit vector over the boundary S^H, given as

$$\mathbf{R} = \begin{bmatrix} -n_y & n_x \\ n_x & n_y \end{bmatrix}. \tag{22}$$

Each grain boundary S^H is rigidly rotated by an angle θ, to create a new *functional* grain boundary, that is, for $\forall \mathbf{x}_{S^H} \in S^H : \mathbf{x}_{S^H} \subset \mathbb{RR}^2$ a new *functional* grain boundary is created by

$$\tilde{\mathbf{x}}_{\tilde{S}^H} = \mathbf{R}_\theta \mathbf{x}_{S^H}, \tag{23}$$

where $\tilde{\mathbf{x}}_{\tilde{S}^H} \in \tilde{S}^H : \tilde{\mathbf{x}}_{\tilde{S}^H} \subset \mathbb{RR}^2$ denotes the new coordinates of every point on the *functional* grain boundary $\tilde{S}^H$ of each grain $H = 1, \ldots, N_g$, and $\mathbf{R}_\theta$ is the transformation matrix for the rigid rotation given as

$$\mathbf{R}_\theta = \begin{bmatrix} \cos\theta & \sin\theta \\ -\sin\theta & \cos\theta \end{bmatrix}. \tag{24}$$

The boundary integral equations are applied to the *functional* grain boundary, always corresponding to the local coordinate system. Hence, the fundamental solutions are transformed to the local coordinate system by

$$\tilde{\mathbf{T}} = \mathbf{T}\tilde{\mathbf{R}}, \quad \tilde{\mathbf{U}} = \mathbf{U}\tilde{\mathbf{R}}, \tag{25}$$

where $\mathbf{T}, \mathbf{U}$ denote the tractions and displacement fundamental solutions, respectively, and $\tilde{\mathbf{R}}$ is a transformation matrix corresponding to the outward normal unit vector over the *functional* boundary $\tilde{S}^H$, given as

$$\tilde{\mathbf{R}} = \begin{bmatrix} -\tilde{n}_y & \tilde{n}_x \\ \tilde{n}_x & \tilde{n}_y \end{bmatrix}, \tag{26}$$

where $\tilde{n}_x, \tilde{n}_y$ denote the components of the outward unit normal vector over the *functional* grain boundary.

The displacements integral equation[12] for each grain can now be written as

$$\begin{aligned} C_{ij}^H(\mathbf{x}')\tilde{u}_j^H(\mathbf{x}') &+ \fint_{\tilde{S}_{nc}^H} \tilde{T}_{ij}^H(\mathbf{x}',\mathbf{x})\tilde{u}_j^H(\mathbf{x})\,d\tilde{S}_{nc}^H + \fint_{\tilde{S}_c^H} \tilde{T}_{ij}^H(\mathbf{x}',\mathbf{x})\tilde{u}_j^H(\mathbf{x})\,d\tilde{S}_c^H \\ &= \int_{\tilde{S}_{nc}^H} \tilde{U}_{ij}^H(\mathbf{x}',\mathbf{x})\tilde{t}_j^H(\mathbf{x})\,d\tilde{S}_{nc}^H + \int_{\tilde{S}_c^H} \tilde{U}_{ij}^H(\mathbf{x}',\mathbf{x})\tilde{t}_j^H(\mathbf{x})\,d\tilde{S}_c^H, \end{aligned} \tag{27}$$

where $\tilde{u}_j^H$, $\tilde{t}_j^H$ are components of displacements and tractions, respectively, for each grain $H = 1, \ldots, N_g$, and C_{ij}^H is the so-called free-term. All components in Eq. (27) refer to the local coordinate system. In the case of internal grains, the first integral on the left- and the right-hand side of Eq. (27) vanishes since for these grains $\tilde{S}_{nc}^H = \varnothing$.

To solve Eq. (27), the *functional* boundaries $\tilde{S}_c^H$ and $\tilde{S}_{nc}^H$ of each grain $H = 1, \ldots, N_g$ are discretised into N_c^H and N_{nc}^H elements, respectively. Each element is composed of m_c^H and m_{nc}^H number of nodes for the grain boundary interfaces and the free grain boundaries, respectively. After the discretisation and using the point collocation method for solution, the final system of equations can be written in matrix form as

$$[\tilde{\mathbf{H}}_{nc}^H \quad \tilde{\mathbf{H}}_c^H] \begin{Bmatrix} \tilde{\mathbf{u}}_{nc}^H \\ \tilde{\mathbf{u}}_c^H \end{Bmatrix} = [\tilde{\mathbf{G}}_{nc}^H \quad \tilde{\mathbf{G}}_c^H] \begin{Bmatrix} \tilde{\mathbf{t}}_{nc}^H \\ \tilde{\mathbf{t}}_c^H \end{Bmatrix} \tag{28}$$

for $H = 1, \ldots, N_g$.

Combining and rearranging Eq. (28), for all grains $H = 1, \ldots, N_g$, and applying the interface boundary conditions (18) and (19), the final system of equations is obtained:

$$\begin{bmatrix} [& \mathbf{A} &] \\ & [\mathbf{0}] & [\mathbf{BC}] \end{bmatrix} \begin{Bmatrix} \tilde{\mathbf{x}} \\ \delta\tilde{\mathbf{u}}^I \\ \tilde{\mathbf{t}}^I \end{Bmatrix} = \begin{Bmatrix} \mathbf{R}\tilde{\mathbf{y}} \\ \mathbf{F} \end{Bmatrix}, \tag{29}$$

where the submatrices $\mathbf{A}$ and $\mathbf{R}$ are sparsed containing known integrals of the product of the shape functions, the Jacobians and the fundamental fields. Submatrix $\mathbf{A}$ also contains the interface boundary conditions (18) and (19). The vectors $\tilde{\mathbf{x}}$ and $\tilde{\mathbf{y}}$ denote the unknown boundary conditions and the prescribed boundary values along the domain boundary $\mathfrak{S}$, respectively. The submatrix $\mathbf{BC}$ contains all the interface conditions for the grain facets, corresponding to $\delta\tilde{\mathbf{u}}^I$ and $\tilde{\mathbf{t}}^I$, while the submatrix $\mathbf{F}$ contains the right-hand sides of these interface conditions. The size of the final system is substantially reduced by directly imposing the tractions equilibrium (18) and displacements compatibility conditions (19), on the $\tilde{\mathbf{H}}_c^H$ and $\tilde{\mathbf{G}}_c^H$ submatrices of each grain, Eq. (28), inside the submatrix $\mathbf{A}$, instead of imposing them conventionally on the interface tractions $\tilde{\mathbf{t}}^I$ and the displacement discontinuities $\delta\tilde{\mathbf{u}}^I$. The above methodology reduces substantially the size of the final system.

4.3. *Grain discretisation*

In the present study the artificial microstructure is discretised using constant subparametric elements, that is, linear elements for the geometry and constant elements for the field unknowns. Only the grain boundaries are discretised due to the proposed boundary element formulation. Hence, the overall size of the system is substantially reduced compared with the FE formulations presented in the past.[8,46–49,53–56] Two are the main reasons for using constant elements in the present study. The first motivation for using constant elements is that all field unknowns, these are interface tractions and displacement discontinuities, are located at the center of these elements and not at the edges; thus, problems at triple points (points where three grains meet) are automatically avoided. The other reason for using constant elements is that analytical integration can be carried out over each element. In this way, numerical integration is avoided and furthermore singularities that exist when the source point $\mathbf{x}'$ coincides with the field point $\mathbf{x}$ are treated in the best way. As a result the proposed formulation becomes faster and more accurate.

As it was demonstrated before, the generated microstructure is composed of grains with various sizes and shapes. Consequently, the grain boundary interface lengths may vary significantly from place to place. If a constant number of elements were assumed per grain boundary side, it would result to a great variation of the grain boundary elements size. In order to avoid this, a smoothing technique is used in the present study. Initially the average length, denoted by $\bar{L}^f$, of all grain boundary facets is evaluated for the microstructure in consideration. Then all the grain boundary facets lengths, L^f, are compared one by one with the average length $\bar{L}^f$. The number of elements that each facet will be discretised is given by

$$N^f = \bar{n}^f \frac{L^f}{\bar{L}^f}, \tag{30}$$

where $N^f \in I_n : I_n = \{1, \ldots, n\}$ is the integer number closer to the real number resulting on the right-hand side and $\bar{n}^f \in I_n : I_n = \{1, \ldots, n\}$ is an input parameter, denoting the number of elements that the average length grain interface will be discretised. In this way, the number of elements over each grain boundary interface is increased or reduced depending on the size of the facet, resulting to a smoother discretisation

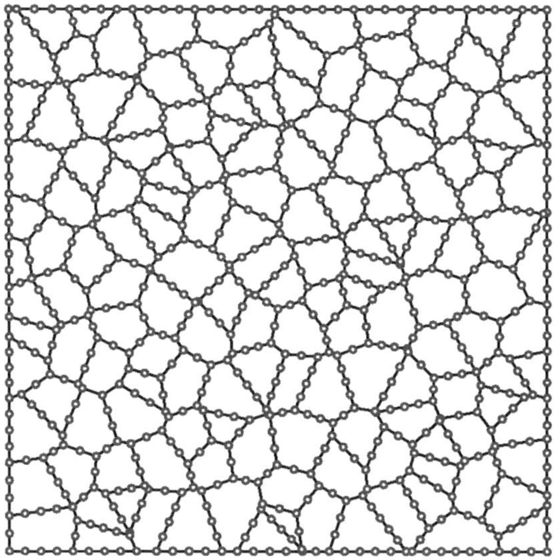

Fig. 4. Artificial microstructure discretisation into grain boundary elements (150 grains, $\bar{n}^f = 2$).

of the artificially generated microstructure. Figure 4 illustrates an artificial microstructure created by a quasi-random point generator, of 150 grains, using $\bar{n} = 2$.

In the cohesive modelling of cracks, two main factors influence the element size independency and reproducibility of the solution, as it was investigated by Tomar *et al.*[39] and Espinosa and Zavatierri.[47] Firstly, to obtain an accurate resolution of the fields near the crack-tips, the element size, $2L_e$, must be small enough to accurately resolve the stress distribution inside the cohesive zones. Hence, $2L_e \ll L_{\mathcal{CZ}}$, where $L_{\mathcal{CZ}}$ denotes the cohesive zone length. The second factor results from the macroscopic stiffness reduction due to the cohesive separation along the element boundaries in the case where the initial stiffness of the cohesive surfaces is finite. In the proposed formulation, the second factor does not affect the solution process, since in the formulation zero displacement discontinuities are enforced directly, Eq. (19) $\delta\bar{\mathbf{u}}^I = 0$, for all the undamaged interface node pairs. However, the first factor has an important role on the accuracy of the results. For a linearly softening cohesive law, as the one used in the present study, an approximation of the cohesive zone size $L_{\mathcal{CZ}}$ at the crack tip is given by Rice,[59] after Tomar *et al.*[39] and Espinosa and

Zavattieri,[47] as

$$L_{\mathcal{CZ}} = \frac{\pi}{2}\left(\frac{K_{IC}}{T_{\max}}\right)^2, \tag{31}$$

or in the case of cohesive relations derived from a potential ϕ as in the following equation[39]:

$$L_{\mathcal{CZ}} = \frac{9\pi E}{32(1-v^2)}\frac{\phi}{T_{\max}^2}, \tag{32}$$

where K_{IC} denotes the fracture toughness of the material in Mode I, for plane strain conditions, E is the modulus of elasticity, v is the Poisson ratio and $T_{\max}$ denotes the strength of the cohesive grain boundary pair under pure normal separation.[39] As already mentioned, it is necessary for the grain boundary interface element size, $2L_e$, to satisfy the inequality: $2L_e \ll L_{\mathcal{CZ}}$. Therefore, to ensure solution convergence and mesh independence for all the examples simulated in the present study, using $\bar{n} = 2$ in Eq. (30) for the considered average grain sizes, the resulting grain boundary elements size was always $(L_{\mathcal{CZ}}/2L_e) > 10$.

5. Grain Boundary Interface

In polycrystalline materials, grain boundary interfaces require special care for the evolution of intergranular microfracture to be simulated. In the present study three different states of a grain boundary interface are distinguished:

(i) *Potential crack zone*, denoted by $\mathcal{PC}$, where all grain boundaries are considered as undamaged interfaces. In this zone, tractions equilibrium, Eq. (18) and displacements compatibility, Eq. (19), where $\delta\bar{\mathbf{u}}^I = 0$, are directly implemented in the formulation. In this way any penetration or separation of the grain interfaces is not allowed.

(ii) *Free crack zone*, denoted by $\mathcal{FC}$, where a complete intergranular microcrack has separated two grains along their interface and two new surfaces have been formed. In this zone the newly formed surfaces act independently and frictional contact conditions are introduced in the formulation, as explained later.

(iii) *Cohesive zone*, denoted by $\mathcal{CZ}$, where the grain boundary is partially damaged and a microcrack starts forming. This is the process zone where cohesive interface laws are introduced to model the interaction of the local tractions and displacement on the partially damaged grain boundary interface.

Cohesive zone modelling is being increasingly used in recent years to simulate fracture process in a variety of materials. Since Barenblatt[60] and Dugdale[61] proposed the concept of cohesive modelling, many other models have been proposed over the last decades. For a review over the chronological evolution of the proposed cohesive models, the readers are referred to Refs. 62 and 63. Cohesive modelling is ideal for modelling interfaces where materials with different properties are met, since it avoids the singular crack fields very close to the crack tip. In the present study, the linear cohesive law initially proposed by Ortiz and Pandolfi[18] for fully mixed mode fracture is adopted and coupled with the grain boundary element formulation, to simulate mixed mode intergranular microcracking evolution in polycrystalline brittle materials.

In our formulation, the displacements compatibility conditions (19), where $\delta\bar{\mathbf{u}}^I = 0$, are directly implemented resulting in the cancellation of any penetration or separation of the grain boundary interfaces. Hence, no relative displacement discontinuities exist until some damage is initiated along an interface. In this way, difficulties with the initial slope of the bilinear cohesive law extensively used in the FEM are avoided.[39,46] However, to initiate damage in the BEM formulation, considering mixed mode failure criteria, all the information must be gathered by the interface tractions. Therefore, an effective traction is introduced, over all grain boundary interface node pairs $i = 1, \ldots, M_c : i \in \mathcal{PC}$, given as

$$t^{I,\text{eff}} = \left[\langle t_n^I \rangle^2 + \left(\frac{\beta}{\alpha} t_t^I \right)^2 \right]^{1/2}, \tag{33}$$

where t_n^I, t_t^I are the normal and tangential components of the interface traction $\tilde{\mathbf{t}}^I$; β and α assign different weights to the sliding and opening mode and $\langle \cdot \rangle$ denotes the McCauley bracket defined as $\langle x \rangle = \max\{0, x\}$ $x \in \mathbb{R}$. Damage is initiated once the effective traction, $t^{I,\text{eff}}$, exceeds a maximum traction, denoted as $T_{\max}$; hence, $t^{I,\text{eff}} \geq T_{\max}$.

Once damage has initiated on a specific grain boundary node pair, say i_o, it is assumed that this pair enters the cohesive zone; that is, $i_o \in \mathcal{CZ}$. Following Ortiz and Pandolfi,[18] an effective opening displacement is introduced, which accounts for both opening (Mode I) and sliding (Mode II) separation, given as

$$d = \left[\left(\frac{\delta u_n^I}{\delta u_n^{I,\mathrm{cr}}} \right)^2 + \beta^2 \left(\frac{\delta u_t^I}{\delta u_t^{I,\mathrm{cr}}} \right)^2 \right]^{1/2}, \tag{34}$$

where δu_n^I, δu_t^I are the normal and tangential relative displacements of the interface and $\delta u_n^{I,\mathrm{cr}}, \delta u_t^{I,\mathrm{cr}}$ are critical values at which interface failure takes place in the case of pure Mode I and pure Mode II, respectively. The parameter β assigns different weights to the sliding and normal opening displacements. Its physical meaning is that β is a measure of the strain energy release rate ratio considering Mode I and II; thus, $\beta^2 = (G_{II}/G_I)$,[46] and $G_I = \frac{1}{2}\delta u_n^{I,\mathrm{cr}} T_{\max}$. The effective opening displacement d in Eq. (34), or else *damage parameter* for the present study, takes values $d \in [0, 1]$, where $d = 0$ means completely undamaged and $d = 1$ means completely failed. In this case, a complete microcrack has been formed and the specific grain boundary node pair i_o now becomes $i_o \in \mathcal{FC}$.

Following Espinosa and Zavattieri,[46] a potential of the following form is assumed:

$$\phi(\delta u_n^I, \delta u_t^I) = \delta u_n^{I,\mathrm{cr}} \int_0^d t^{I,\mathrm{eff}}(d')\, dd' = T_{\max} \frac{\delta u_n^{I,\mathrm{cr}}}{2} d(2-d), \tag{35}$$

where

$$t^{I,\mathrm{eff}}(d) = T_{\max}(1-d). \tag{36}$$

Figure 5 illustrates the variation of the effective interface traction with respect to the effective opening displacement d, Eq. (36), considering both the opening and the sliding of the interface.

The normal and tangential components of the traction acting on the interface in the fracture process zone are given by

$$t_n^I = \frac{\partial \phi}{\partial \delta u_n^I} = T_{\max} \frac{\delta u_n^I}{\delta u_n^{I,\mathrm{cr}} d}(1-d), \tag{37}$$

$$t_t^I = \frac{\partial \phi}{\partial \delta u_t^I} = T_{\max} \frac{\alpha \delta u_t^I}{\delta u_t^{I,\mathrm{cr}} d}(1-d), \tag{38}$$

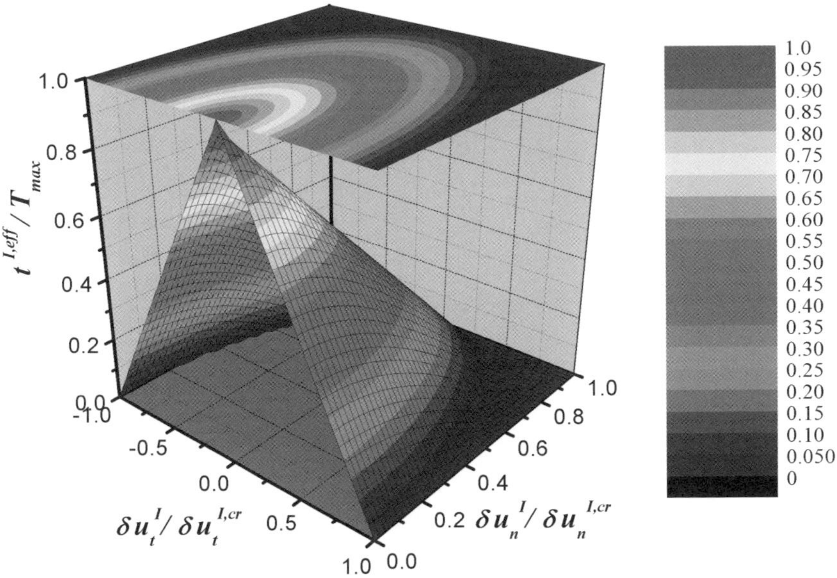

Fig. 5. Variation of the effective interface traction with respect to the opening and sliding of the grain boundary interface.

where $\alpha = \beta^2(\delta u_n^{I,\mathrm{cr}}/\delta u_t^{I,\mathrm{cr}})$. The variation of the (a) normal and (b) tangential components of the interface cohesive traction is illustrated in Fig. 6 with respect to the relative opening and sliding of the interface.

Owing to the irreversibility of the interface cohesive law, unloading–reloading in the range $0 \leq d < d^*$ is given by

$$t_n^I = T_{\max}\frac{\delta u_n^I}{\delta u_n^{I,\mathrm{cr}} d^*}(1 - d^*), \tag{39}$$

$$t_t^I = T_{\max}\frac{\alpha\delta u_t^I}{\delta u_t^{I,\mathrm{cr}} d^*}(1 - d^*). \tag{40}$$

The term d^* denotes the last effective opening displacement where unloading took place.

In the case of pure sliding mode that is under compressive tractions acting on the interface due to the impenetrability condition $\delta u_n^I = 0$, only Eqs. (38) and (40) are implemented concerning the tangential interface tractions.

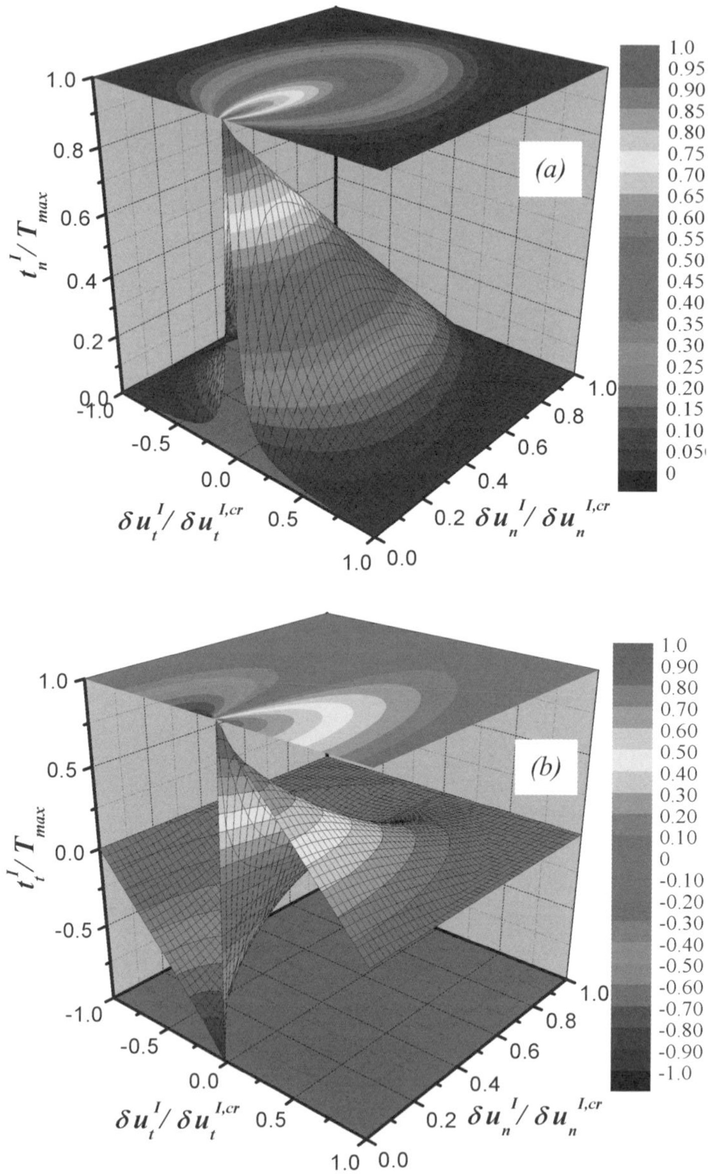

Fig. 6. (a) Variation of the normal and (b) tangential component of the interface cohesive traction.

Table 3. Contact constraints for different contact modes.

Separation	Stick	Slip
$t_n^I = 0$	$\delta u_n^I = 0$	$\delta u_n^I = 0$
$t_t^I = 0$	$\delta u_t^I = 0$	$t_t^I \pm \mu t_n^I = 0$

In the case where a cohesive element, introduced inside a damaged grain boundary interface, has completely failed, that is, $d = 1$, a completely free microcrack is introduced by separating the specific grain boundaries node pair. Once a microcrack has formed, the two free surfaces of the microcrack can come into contact, slide or separate. Upon interface failure,[64] the equivalent nodal tangential tractions are computed using Coulomb's frictional law.[67] Therefore, a full frictional contact analysis is introduced in the proposed formulation to model such effects. Table 3 presents the boundary constraints introduced in submatrix **BC** of the final system of equations (62), to model effects due to contact, sliding or separation of the crack free surfaces. Tables 4 and 5 present the criteria that are employed to check for any contact mode or status violation (for convention with the local coordinate system along the grain interfaces, $\delta u_n^I > 0$: penetration).

It is worth noting that all the aforementioned interface laws can be implemented directly in the submatrix **BC** of the final system of equations (62). This is a great advantage of the proposed boundary element formulation, since the introduction of the cohesive elements and later of the free microcracks do not affect the size of the final system. This is due to the fact that all the interface laws can be directly implemented as local boundary conditions along the grain boundaries of the microstructure, by coupling the local tractions and relative displacement discontinuities through the interface laws. The system becomes non-linear only when interface elements exist along grain boundaries that are in the loading case (unloading/reloading), since the interpretation of equations (37) and (38) is required. For all other cases, the system is fully linear.

6. Microcracking Evolution Algorithm

The algorithm starts by creating the grain boundary element mesh, as it was demonstrated. Zero displacement discontinuities are enforced in the formulation to simulate the undamaged yet microstructure. The initial solution of the problem is performed for a unit load by using a special sparse solver. Since the coefficient matrix of the final system of equations (62) is highly sparsed, using a sparse solver speeds up the solution process several times comparing with ordinary solvers. Then the inversed coefficient matrix is stored and the main algorithm for simulating microcracking evolution starts.

Initially, since the whole system is still linear, the load λ, at which damage will be initiated for the first time in a grain boundary interface node pair, is evaluated. The specific node pair is assumed to enter the $\mathcal{CZ}$ zone and for an additional small increment $\Delta\lambda$ of the load λ, $\lambda + \Delta\lambda$, the problem is resolved by just multiplying the right-hand vector with the inversed coefficient matrix. Checks for each node pair $i = 1, \ldots, M_c : i \in \mathcal{PC} \cup \mathcal{CZ} \cup \mathcal{FC}$ are performed considering if the node pair is undamaged, $i \in \mathcal{PC}$, partially damaged, $i \in \mathcal{CZ}$ or completely damaged, $i \in \mathcal{FC}$. If it is undamaged, the effective traction $t^{I,\mathrm{eff}}$ is compared with the $T_{\max}$, and if $t^{I,\mathrm{eff}} \geq T_{\max}$ then damage is initiated. In the case where the specific pair was already in the $\mathcal{CZ}$ zone, two subcases are distinguished considering if it is in the loading or in the irreversible unloading–reloading state. For both cases, checks are performed for the monotony of the effective opening displacement. Finally, in the case where the specific pair are completely damaged, that is, $i \in \mathcal{FC}$, a frictional contact analysis routine is employed to ensure no contact mode or status violations will occur. Once the required checks are performed for all grain boundary interface node pairs, the system is re-solved. To ensure convergence in every step, checks for any violations over the state of the interface node pairs are performed again. In the case of no violations, the program outputs the intermediate step data and continues to the next step.

To update the inverse coefficient matrix, the Sherman–Morisson update formula is used.[65,66] Since the changes of the coefficient matrix to include the different interface conditions are taking place in a small part of the coefficient matrix, inside submatrix **BC** equation (62), updating the inverse matrix speeds up the iterative solution several times.[36] Hence, only the interface node pairs that require updates are considered instead of re-solving the whole system. For every load step of the algorithm that is the outer loop, the solution is re-calculated several times by simply multiplying the inverse coefficient matrix by the right-hand side vector of the system. For the pairs that are in the cohesive zone and in loading state, the normal and tangential components of the discontinues displacements are coupled with the corresponding traction components, in submatrix **BC**, by implementing Eqs. (37) and (38) and assuming the previous step effective opening displacement d for each pair, that is, the inner loop. After every re-solution of the problem, a new damage factor d for each pair $i \in \mathcal{CZ}$ is evaluated and introduced in submatrix **BC**. The criteria for detecting any violations over all interface node pairs after the re-solution of the problem, include also criteria for checking the convergence of the resulting effective opening displacement d after comparing it with the previous step evaluation.

In the case where an interface node pair has completely failed, a free microcrack is introduced by separating the node pair, as already mentioned. For all these completely damaged pairs, a frictional contact analysis is performed to model possible contact, sliding or separation of the crack surfaces. Initially, checks are performed for any contact mode violation. Table 4 presents the criteria used to ensure that a specific crack pair is still in separation or contact mode. The aforementioned criteria are checked over all free crack node pairs, $i = 1, \ldots, M_c : i \in \mathcal{FC}$. If the pair is in contact, checks are performed for any contact status violation, Table 5. Coulomb's frictional law[67] is used to model the frictional forces developed due to compressive sliding, where μ denotes the friction coefficient. Any contact violation results in an update of the corresponding components of the submatrix **BC** in Eq. (62), according to the boundary constraints listed in Table 3.

Table 4. Contact mode violation check criteria.

Assumption/decision	Stick	Slip
Stick	$t_t^I < \mu t_n^I$	$t_t^I \geq \mu t_n^I$
Slip	$t_t^I \cdot \delta u_t^I > 0$	$t_t^I \cdot \delta u_t^I \leq 0$

Table 5. Contact status violation check criteria.

Assumption/decision	Separation	Contact
Separation	$\delta u_n^I < 0$	$\delta u_n^I \geq 0$
Contact	$t_n^I > 0$	$t_n^I \leq 0$

6.1. *Non-local approach*

To ensure mesh independency and reproducibility of the numerical results, a non-local approach must be introduced in order to avoid the pathological localisation of microdamage in the macro-scale. Generally, a non-local approach consists of replacing a specific variable by its non-local weighted volume averaged counterpart.[23,27,28] The choice of the variable to be averaged is arbitrary to some extent. However, the new non-local model must exactly agree with the standard modelling approach, as long as the material behaviour remains elastic.

In the proposed multi-scale boundary element formulation, the local degradation of the material stiffness due to the microdamage evolution is modelled by introducing in the macro-scale the decremental stress, $\dot{\sigma}^D$, which results from the initiation and propagation of microcracks inside each RVE, in the micro-scale. However, this stress component cannot be replaced directly by its non-local counterpart. To overcome this, the following technique is introduced. For every domain point, $i = 1, \ldots, M$, that has been assigned an RVE for monitoring the microscopic behaviour, the non-local macro-strain $\hat{\dot{\varepsilon}}^M(\mathbf{X}')$ is evaluated after every macroscopic solution, by considering the macro-strains in the neighbourhood of this point, as follows:

$$\begin{aligned} \hat{\dot{\varepsilon}}^M(\mathbf{X}') &= \int_V a(\mathbf{X}', \mathbf{X})\dot{\varepsilon}^M(\mathbf{X})\, dV(\mathbf{X}), \\ a(\mathbf{X}', \mathbf{X}) &= a_o(\mathbf{X}', \mathbf{X}) \left\{ \int_V a_o(\mathbf{X}', \xi)\, dV(\xi) \right\}^{-1}, \end{aligned} \tag{41}$$

and $a_o(\mathbf{X}', \mathbf{X})$ in the present work is taken to be the Gauss distribution function, given for the two-dimensional case as

$$a_o(\mathbf{X}', \mathbf{X}) = \exp\left(-\frac{2|\mathbf{X} - \mathbf{X}'|^2}{l^2} \right), \tag{42}$$

where l denotes the material characteristic length, which measures the heterogeneity scale of the material.[27]

This non-local macro-strain is used to evaluate the periodic boundary conditions to be assigned to the corresponding point $\mathbf{X}'$, RVE, as it will be described in a later section. After solution of the specific micromechanics problem with the prescribed boundary values defined by $\hat{\dot{\varepsilon}}^M$, the volume average total stress $\hat{\bar{\dot{\sigma}}}^t$ is evaluated using averaging theorems presented in the following section. From this stress, the decremental component that is used as initial stress in the boundary element formulation is evaluated as $\hat{\dot{\sigma}}^D = \hat{\dot{\sigma}}^e - \hat{\dot{\sigma}}^t$, where $\hat{\dot{\sigma}}^e$ is the elastic stress in case of no-microdamage; the upper hat $(\hat{\cdot})$ denotes that these stresses resulted by the non-local macro-strain that corresponded to the specific point $\mathbf{X}'$ in the macro-continuum the RVE was assigned.

However, the aforementioned decremental component of stress, $\hat{\dot{\sigma}}^D$, can not be directly implemented in the boundary integral equation (1), since it corresponds to the non-local strain field and not to the local one. At this point a macro-damage coefficient is introduced, denoted by D_{ij}, given by the subdivision of the decremental stress by the non-local elastic stress, resulting in

$$D_{ij}(\mathbf{X}') = 1 - \hat{\dot{\sigma}}^t_{ij}(\mathbf{X}')[\hat{\dot{\sigma}}^e_{ij}(\mathbf{X}')]^{-1} \tag{43}$$

where no summations are implied for the repeated indices i, j and $D_{ij} = D_{ji}$ due to the symmetry of the strain and stress tensors. In the case where $D_{ij} = 0$, no damage has taken place, where in cases of $D_{ij} = 1$ the macro-continuum is completely damaged and a macro-crack (fracture) must be introduced.

In the context of the proposed boundary element method for the macro-continuum, to implement the aforementioned damage, a local decremental stress is evaluated by

$$\dot{\sigma}^D_{ij}(\mathbf{X}') = D_{ij} C^M_{ijkl} \dot{\varepsilon}^M_{kl}(\mathbf{X}') \tag{44}$$

where C^M_{ijkl} denotes the fourth-order elasticity stiffness tensor of the macro-continuum and, again, no summation is implied for the repeated indices ij.

At this stage it should be noted that some attention must be paid to cases of loading an RVE by a strain tensor of the form $\{\varepsilon_{11}, \varepsilon_{22}, \varepsilon_{12}\} = \{0, a, 0\}$, where $a \in \mathbb{R}$. In this case, damage is expected to appear along the 11-*direction* (for an isotropic material without defects). Therefore, the aforementioned damage coefficient D_{ij} should describe the developed damage due to loading on the 22-*direction*; that is, $0 < D_{22} \leq 1$. However, due to the Poisson effect, the developed average stress component on the 11-*direction* will also be lower than the undamaged (linear elastic) component on the same direction. Consequently, Eq. (43) would give a damage coefficient D_{11}, which is artificial, since no damage has occurred on this direction and is due to the Poisson effect. For the case of a perfect homogeneous isotropic material, this artificial damage is always equal to the actual one. For the general case of a polycrystalline material composed of randomly orientated anisotropic grains, as in the present work, this artificial damage appears to be lower, or in the range of the actual one.

7. Definitions: Averaging Theorems

As pointed out before, an RVE represents the microstructure of an infinitesimal material neighbourhood for a point in a macro-continuum mass. Hence, the stress and strain fields corresponding to the macro-scale will be referred to as macro-stress/strain and will be denoted by a superscript M, as σ^M and ε^M, respectively. On the other hand, the stress, strain fields corresponding to the RVEs (that is the micro-scale), will be referred as micro-stress/strain and denoted by a superscript m, as σ^m and ε^m, respectively. In multi-scale mechanics, averaging theorems and quantities is required in order to transfer information through the different scales.[34,35] Therefore, every averaged quantity referring to the RVEs will be denoted by an upper bar, that is, $\bar{\sigma}^m, \bar{\varepsilon}^m$ for the volume average microstress and micro-strain, respectively. As pointed out before, a rate problem is regarded here, where the field unknowns are denoted by an upper dot, that is, $\dot{\sigma}, \dot{\varepsilon}$ for the stresses and strains, respectively. Moreover, as infinitesimal deformations are considered in the present work, it should be noted that the average micro-stress/strain rates equal the rate of change of the average micro-stress/strain,[34] that is, $\bar{\dot{\sigma}}^m = \dot{\bar{\sigma}}^m, \bar{\dot{\varepsilon}}^m = \dot{\bar{\varepsilon}}^m$.

As a benchmark problem in the present work, a polycrystalline brittle material is considered, which is susceptible to intergranular fracture. Assume now the RVE illustrated in Fig. 7. This RVE represents the microstructure of a polycrystalline brittle material and is composed of randomly distributed and orientated single crystal anisotropic elastic grains. It was produced by the Poisson–Voronoi tessellation method, which is extensively used in the literature for modelling polycrystalline materials in a random manner.[14,37] Each grain is assumed to have a randomly assigned material orientation, defined by an angle θ subtended from the x geometrical axis, where $0° \leq \theta < 360°$ (non-directional solidification is assumed). Since the present study considers two-dimensional problems,

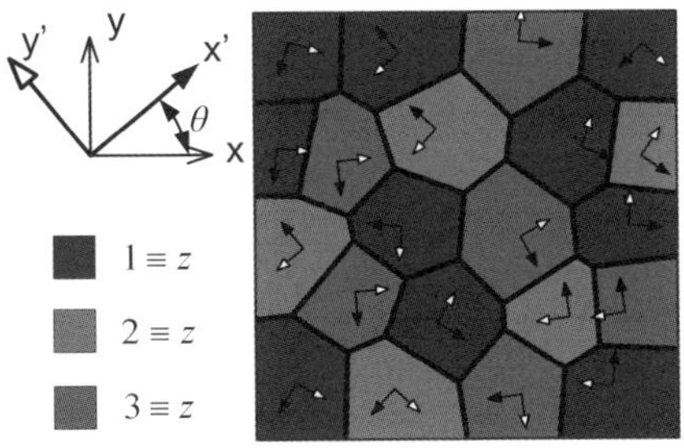

Fig. 7. Artificial microstructure with randomly distributed material orientation for each grain.

to maintain the random character of the generated microstructure and the stochastic effects of each grain on the overall behaviour of the system, three different cases are considered for each grain in view of which material axis is normal to the plane,[46] i.e. *Case 1*: $1 \equiv z$, *Case 2*: $2 \equiv z$ and *Case 3*: $3 \equiv z$ (working plane is assumed the xy).

Since every grain is assumed to have a general anisotropic mechanical behaviour, the RVE would behave in a linear elastic manner as long as the interfaces are still intact. Each grain $H : H = 1, \ldots, N_g$, where N_g denotes the total number of grains in the RVE, has a volume denoted by V^H and a surface denoted by S^H. Therefore, the volume of the RVE, V^m, is given by

$$V^m = \bigcup_{H=1}^{N_g} V^H. \tag{45}$$

The boundary of each grain is divided into the contact boundary S_c^H, indicating the contact with a neighbour grain boundary, and the free boundary S_{nc}^H, indicating the grain boundaries that coincide with the boundary of the RVE, S^m. Hence for every grain,

$$S^H = S_{nc}^H \cup S_c^H. \tag{46}$$

For the internal grains $S_{nc}^H = \varnothing$ and thus $S^H = S_c^H$. Therefore, S_{nc}^H exists only on the RVE boundary grains resulting to

$$S^m = \bigcup_{H=1}^{N_g} S_{nc}^H, \tag{47}$$

where S^m denotes the boundary of an RVE.

Let us assume the overall collection of all grain boundary interfaces within an RVE to be denoted by S_{pc}^m and given as

$$S_{pc}^m = \frac{1}{2} \bigcup_{H=1}^{N_g} S_c^H . \tag{48}$$

Along this path, potential intergranular microcracks may be initiated and propagated. In general, the overall properties of this RVE are strongly affected by the morphology and the material orientation of its grains and the condition of all its grain boundary interfaces S_{pc}^m. These grain boundary interfaces may be undamaged, partially damaged and completely damaged. The latter deflects intergranular cracks that can propagate along the grain boundaries. This type of debonding consumes mechanical energy and leads to greater toughness. Therefore, its effect on the overall behaviour of the RVE must be considered when averaging theorems are used. The overall volume average micro-stress $\bar{\dot{\sigma}}_{ij}^{m,t}$ of an RVE composed by grains can be given as

$$\bar{\dot{\sigma}}_{ij}^{m,t} = \frac{1}{V^m} \int_{V^m} \dot{\sigma}_{ij}^m \, dV^m = \frac{1}{V^m} \sum_{H=1}^{N_g} \int_{V^H} \dot{\sigma}_{ij}^H \, dV^H , \tag{49}$$

and since the stress tensor is divergence-free,[34] using the divergence theorem and considering Eq. (46)

$$\bar{\dot{\sigma}}_{ij}^{m,t} = \frac{1}{V^m} \sum_{H=1}^{N_g} \left\{ \int_{S_{nc}^H} x_i^H \dot{t}_j^H \, dS_{nc}^H + \int_{S_c^H} x_i^H \dot{t}_j^H \, dS_c^H \right\} , \tag{50}$$

where $\dot{t}_j = \dot{\sigma}_{ij} n_i$ denotes the surface tractions.

Consider now that the debonding of the grain boundaries can be modelled as displacement discontinuities, $\delta\dot{\mathbf{u}}^I$, and tractions jumps, $\delta\dot{\mathbf{t}}^I$. However, to ensure equilibrium, tractions jumps must always vanish. In other words, in cases of partially damaged boundaries or closed cracks the local tractions must cancel each other and in cases of completely formed opened cracks their surfaces must be traction free. Hence, by using the definition of the RVE boundary, Eq. (47), the overall volume average stress, $\bar{\dot{\sigma}}_{ij}^{m,t}$, can be evaluated by

$$\bar{\dot{\sigma}}_{ij}^{m,t} = \frac{1}{V^m} \int_{S^m} x_i^m \dot{t}_j^m \, dS^m , \tag{51}$$

where $x_i^m, \dot{t}_j^m$ represent the position vectors of the points lying on the RVE boundary and their tractions, respectively.

In terms of strains, the volume average strain, $\bar{\dot{\varepsilon}}_{ij}^m$, can be evaluated in a similar manner as

$$\bar{\dot{\varepsilon}}_{ij}^m = \frac{1}{2V^m}\int_{V^m}(\dot{u}_{i,j}^m + \dot{u}_{j,i}^m)\,dV^m = \frac{1}{2V^m}\sum_{H=1}^{N_g}\int_{V^H}(\dot{u}_{i,j}^H + \dot{u}_{j,i}^H)\,dV^H, \tag{52}$$

and by using again the divergence theorem and Eq. (46), leads to

$$\bar{\dot{\varepsilon}}_{ij}^m = \frac{1}{2V^m}\sum_{H=1}^{N_g}\left\{\int_{S_{nc}^H}(\dot{u}_i^H n_j^H + \dot{u}_j^H n_i^H)\,dS_{nc}^H + \int_{S_c^H}(\dot{u}_i^H n_j^H + \dot{u}_j^H n_i^H)\,dS_c^H\right\}. \tag{53}$$

Considering now small deformations, for two adjacent grains A and B over an interface the displacement discontinuities are defined as $\delta\dot{\mathbf{u}}^I = \dot{\mathbf{u}}^A - \dot{\mathbf{u}}^B$, in global coordinates, and the outward normal unit vectors of each grain are $\mathbf{n}^A$ and $\mathbf{n}^B = -\mathbf{n}^A$, respectively. The volume average strain can be evaluated after using Eqs. (47) and (48) by

$$\bar{\dot{\varepsilon}}_{ij}^m = \frac{1}{2V^m}\left\{\int_{S^m}(\dot{u}_i^m n_j^m + \dot{u}_j^m n_i^m)\,dS^m + \int_{S_{pc}^m}(\delta\dot{u}_i^I n_j^A + \delta\dot{u}_j^I n_i^A)\,dS_{pc}^m\right\}. \tag{54}$$

Transforming the displacement discontinuities from global, $\delta\dot{\mathbf{u}}^I$, to local, $\delta\tilde{\dot{\mathbf{u}}}^I$, coordinates, the opening gap $\delta\tilde{\dot{u}}_n^I$ and the sliding gap $\delta\tilde{\dot{u}}_t^I$ along the damaged interfaces can be used directly for evaluating the volume average strain. The transformation of the displacement discontinuities is given by

$$\delta\dot{u}_i^I = R_{ik}\delta\tilde{\dot{u}}_k^I, \tag{55}$$

where R_{ik} denotes the transformation tensor. Finally, Eq. (54) can be written as

$$\bar{\dot{\varepsilon}}_{ij}^m = \frac{1}{2V^m}\left\{\int_{S^m}(\dot{u}_i^m n_j^m + \dot{u}_j^m n_i^m)\,dS^m + \int_{S_{pc}^m}(R_{ik}\delta\tilde{\dot{u}}_k^I n_j^A + R_{jk}\delta\tilde{\dot{u}}_k^I n_i^A)\,dS_{pc}^m\right\}. \tag{56}$$

In the case of perfect grain boundary interfaces, the displacement discontinuities vanish, i.e. $\delta\tilde{\mathbf{u}}^I = \mathbf{0}$, and therefore the last term of the above equation vanishes too. On the other hand, when the interfaces are imperfect, partially damaged and/or completely damaged (cracked), displacement discontinuities exist, i.e. $\delta\tilde{\mathbf{u}}^I \neq \mathbf{0}$, and therefore an additional strain appears due to the presence of microcracks and partially damaged interfaces. This additional strain is represented by the last term in Eq. (56) and provides a *correction* to the effective volume average strain due to the possible discontinuity of the displacements on a grain boundary interface that has been partially damaged or cracked.[11,34,38] It should be noted that for the sliding component of the displacement discontinuities, $\delta\tilde{u}_t^I$, both positive and negative values may be considered to model the two way sliding of the grain boundary interfaces. However, for the normal opening component, $\delta\tilde{u}_n^I$, only opening is considered, that is, negative values for convention with the definition of the outward normal unit vectors of the grains. This is because the impenetrability conditions are enforced in the contact detection algorithm to ensure the non-penetration of the cracked grain boundaries.[21] Moreover, the detailed contact history of every interface crack is being recorded throughout the incremental process, in order for the internal friction effect on the sliding and the sticking of the crack interfaces to be considered in evaluating the volume average strain.

Generally, in a multi-scale method, the macro-stress $\dot{\sigma}^M$ and macro-strain $\dot{\varepsilon}^M$ tensors corresponding to a point $\mathbf{X}^M$ in the macro-continuum can be evaluated directly by the volume average micro-stress $\dot{\bar{\sigma}}^m$ and micro-strain $\dot{\bar{\varepsilon}}^m$ over the RVE, which represents the microstructure of the infinitesimal material neighbourhood at point $\mathbf{X}^M$. On the contrary, the macro-stress/strain can provide the boundary conditions for the RVE.[34]

7.1. *RVE boundary conditions*

The accurate estimation of the overall response of an RVE is of great importance in a multi-scale modelling and is directly related to the applied type of boundary conditions. In order to be able to use the averaging theorems presented in previous section, for transferring information through the scales, four types of boundary conditions can be used; these are uniform tractions, uniform displacements, mixed boundary conditions and periodic boundary conditions.[5,34,40,41]

The first case of the aforementioned boundary conditions, uniform tractions, does not provide all the required information for a numerical analysis, since rigid body motion will be inevitable.

The uniform displacement boundary conditions can be applied directly on the RVE boundary, considering the macrostrain $\dot{\varepsilon}^M$ at the domain point $\mathbf{X}'$ in the macro-continuum[34]:

$$\dot{u}_i^{m,o} = \dot{\varepsilon}_{ij}^M x_j^m, \tag{57}$$

where $\mathbf{x}^m$ denotes the position vector of every point on the domain boundary S^m of an RVE, i.e. $\mathbf{x}^m \in S^m$. By applying uniform displacements boundary conditions on the RVE, an underestimation of the mechanical properties of the RVE is achieved.[5] However, in the present case where intergranular cracks may run up to the RVE boundaries, uniform displacement boundary conditions are overconstraining the response of the RVE in excess loading that would result in excess microdamage. This is due to the fact that the applied displacements are always a linear translation of the square boundaries of the RVE and therefore they overconstraint crack propagation close to the RVE's boundaries.

The mixed boundary conditions would not overconstraint the crack propagation, however are not applicable in the present case since they require the RVE to have at least orthotropic behaviour and the mixed uniform boundary data must exclude shear stresses or strains.[40]

To date, the periodic boundary conditions (PBC) are usually preferred since they provide the most reasonable estimates of mechanical properties of heterogeneous materials, even in cases where the microstructure is not periodic.[5,6] To apply the PBC, the RVE boundary S^m is separated into left, right, top and bottom parts, as Fig. 8 illustrates, and for the two-dimensional case, the following conditions are applied:

$$\dot{u}_i^R = \dot{u}_i^L + \dot{\varepsilon}_{ij}^M (x_j^2 - x_j^1) \quad \text{and} \quad \dot{u}_i^T = \dot{u}_i^B + \dot{\varepsilon}_{ij}^M (x_j^4 - x_j^1) \tag{58}$$

$$\dot{t}_i^R = -\dot{t}_i^L \quad \text{and} \quad \dot{t}_i^T = -\dot{t}_i^B \tag{59}$$

where $\mathbf{u}^s$ and $\mathbf{t}^s$ for $s = \{T, B, R, L\}$ represent the applied displacements and tractions, respectively, on the top, bottom, right and left side of the RVE boundary. The position vectors of the vertices 1,2 and 4, as Fig. 8 illustrates, are denoted by $\mathbf{x}^i, i = \{1, 2, 4\}$. In the present case where all field unknowns in the micro-scale are referred to the local coordinates, Eq. (27),

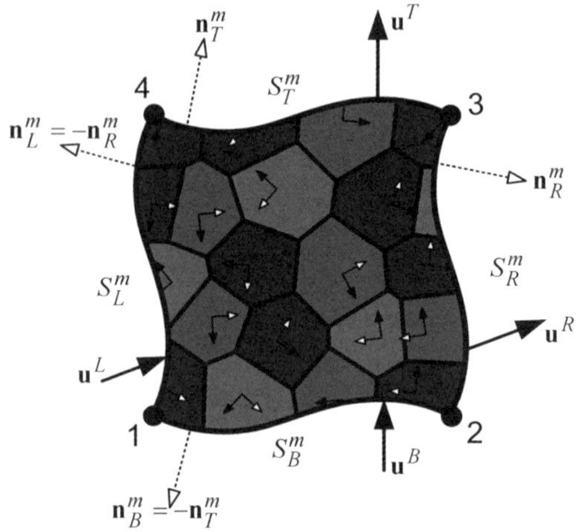

Fig. 8. Schematic representation of a typical RVE under periodic boundary conditions.

the PBCs take the following form:

$$\tilde{\dot{u}}_i^R + \tilde{\dot{u}}_i^L = \delta\dot{x}_i^{R-L} \quad \text{and} \quad \tilde{\dot{u}}_i^T + \tilde{\dot{u}}_i^B = \delta\dot{x}_i^{T-B} \tag{60}$$

$$\tilde{\dot{t}}_i^R = \tilde{\dot{t}}_i^L \quad \text{and} \quad \tilde{\dot{t}}_i^T = \tilde{\dot{t}}_i^B \tag{61}$$

where $\delta\dot{x}_i^{R-L} = (R_{ij}^R)^{-1}\dot{\varepsilon}_{jk}^M(x_k^2 - x_k^1)$, $\delta\dot{x}_i^{T-B} = (R_{ij}^T)^{-1}\dot{\varepsilon}_{jk}^M(x_k^4 - x_k^1)$ and $\mathbf{R}^R$ and $\mathbf{R}^T$ are the right and top side rotation matrices, respectively.

However, closer examination of Eqs. (60) and (61) shows that these boundary conditions cannot be directly implemented into the BEM, as they are constraint equations instead of prescribed boundary values as in the case of uniform displacements, Eq. (57). In other words, the prescribed boundary conditions are obtained from the final solution of the RVE. Hence, there are no initial prescribed conditions but boundary constraints that increase the size of the final system of Eqs. (62). In order to implement the aforementioned periodic boundary conditions in the presented boundary cohesive grain element formulation, without increasing the final system of equations, the PBCs, Eqs. (60) and (61), are directly implemented in the coefficient submatrix [**A**], Eq. (62), and the unknown boundary values are now the displacements and tractions of the right and top RVE boundary sides. To be more precise, considering Eq. (62), the part of submatrix [**A**] that corresponds to the RVE boundary unknown values would take the following form:

$$\begin{bmatrix} [& \mathbf{A} &] \\ [\mathbf{0}] & [\mathbf{BC}] & \end{bmatrix} \begin{Bmatrix} \tilde{\mathbf{x}} \\ \delta\tilde{\mathbf{u}}^I \\ \tilde{\mathbf{t}}^I \end{Bmatrix} = \begin{Bmatrix} \mathbf{R}\tilde{\mathbf{y}} \\ \mathbf{F} \end{Bmatrix} \tag{62}$$

where the submatrices $\mathbf{H}^s, \mathbf{G}^s$, for $s = \{T, B, R, L\}$, contain known integrals of the products of the Jacobian and the anisotropic tractions and displacement fundamental solutions, respectively, corresponding to the RVE boundary nodes.

The general condition for applying the aforementioned PBC is that the discretisation of the RVE boundary on opposite sides must coincide. Therefore, the grain boundary mesh generator must place the same number of elements at same locations on opposite sides, for the PBC to be directly implemented. Fortunately, in the framework of boundary element methods, such implementations of the mesh are relatively easy to achieve. Moreover, considering Fig. 8, rigid body motions can be eliminated by requiring $\dot{\mathbf{u}}^k = \mathbf{0}$ for either $k = \{1, 2, 4\}$.[42]

8. Micro–Macro Interface

8.1. *Coupling with macro-BEM*

Considering now the case where the RVE boundary conditions are defined by a macro-strain $\dot{\varepsilon}^M$. In the absence of any partially damaged, cracked grain boundary interface, the corresponding overall volume average stress $\dot{\bar{\sigma}}_{ij}^{m,t}$ associated with the prescribed macro-strain would be equal to

$$\dot{\sigma}_{ij}^{m,\mathrm{el}} = C_{ijkl}^{m} \dot{\varepsilon}_{kl}^{M}, \tag{63}$$

where the term $\dot{\sigma}_{ij}^{m,\mathrm{el}}$ denotes the corresponding average elastic stress, related to the prescribed macro-strain and C_{ijkl}^{m} is the fourth-order elasticity tensor corresponding to the RVE. If the RVE is sufficiently large so that even though is composed of randomly distributed and orientated single crystal anisotropic grains, its overall mechanical behaviour is isotropic due to the homogenisation[5,21] and equal to the macro-continuum (if the macro-continuum is assumed to be isotropic). In this case, Eq. (63) can be used directly by replacing the RVE elasticity tensor with the macro-continuum elasticity tensor C_{ijkl}^{M}. Nevertheless, the elastic average stress can always be computed by the averaging theorem, Eq. (51), for each RVE by considering no damage at the grain boundary interfaces.

Owing to the presence of partially damaged and cracked grain boundary interfaces, the volume average micro-stress is not in general equal to Eq. (63). Nevertheless, the total volume average micro-stress is defined by

$$\bar{\dot{\sigma}}_{ij}^{m,t} = \dot{\sigma}_{ij}^{m,\text{el}} - \bar{\dot{\sigma}}_{ij}^{m,D}, \tag{64}$$

where $\bar{\dot{\sigma}}^{m,D}$ denotes the decrement in the overall stress, due to the presence of cracked and damaged grain boundary interfaces.

Taking into account Eq. (51) for the evaluation of the overall volume average stress over an RVE, the additional stress term in the above equation can be evaluated as

$$\bar{\dot{\sigma}}_{ij}^{m,D} = \dot{\sigma}_{ij}^{m,\text{el}} - \frac{1}{V^m} \int_{S^m} x_i^m \dot{t}_j^m \, dS^m. \tag{65}$$

This component of stress is considered as initial stress for the macro-continuum boundary element formulation presented previously. When no microdamage has taken place, the last term in Eq. (65) is equal to $\dot{\sigma}^{m,\text{el}}$ and therefore the initial stress component vanishes. Hence, the macro-continuum is still in the elastic regime without any damage. On the other hand, when the RVE is completely broken and cannot carry any more load, the last term in Eq. (65) vanishes and the decremental component of stress equals the fully elastic. In the macro-continuum BE formulation, this initial stress completely cancels the elastic and therefore the macro-material stiffness has completely degraded at that point.

8.2. *Coupling with macro-FEM*

In the case where the macro-continuum is being modelled with a domain numerical method, like the FEM, an RVE can be assigned at every integration point or centroid of an element. Degradation of the RVE stiffness due to possible initiation and propagation of microcracks can be modelled directly by assuming a new stiffness tensor, C_{ijkl}^D, that correlates the total volume average micro-stress with the prescribed macro-strain, i.e.

$$\bar{\dot{\sigma}}_{ij}^{m,t} = C_{ijkl}^D \dot{\varepsilon}_{kl}^M. \tag{66}$$

To this extent and considering Eq. (64), the overall average stress over an RVE can be evaluated in terms of strains as

$$\bar{\dot{\sigma}}_{ij}^{m,t} = C_{ijkl}^m \dot{\varepsilon}_{kl}^M + C_{ijkl}^m \bar{\dot{\varepsilon}}_{kl}^{m,D}, \tag{67}$$

where $\dot{\bar{\sigma}}_{ij}^{m,D} = -C_{ijkl}^{m}\dot{\bar{\varepsilon}}_{kl}^{m,D}$, and $\dot{\bar{\varepsilon}}_{kl}^{m,D}$ denotes the additional strain component due to the presence of microcracks.[34]

Considering now Eq. (56), this additional volume average strain component can be evaluated by

$$\dot{\bar{\varepsilon}}_{ij}^{m,D} = \frac{1}{2V^m}\left\{\int_{S_{pc}^m}(R_{ik}\delta\dot{\tilde{u}}_k^I n_j^A + R_{jk}\delta\dot{\tilde{u}}_k^I n_i^A)\, dS_{pc}^m\right\}. \tag{68}$$

Following Kouznetsova[41] and considering the periodic boundary conditions for an RVE, the final system of the proposed micromechanics BEM Eq. (62), can be rearranged in terms of the displacement discontinuities as

$$\begin{bmatrix}\mathbf{K}^1 & \mathbf{K}^2\\ \mathbf{K}^3 & \mathbf{K}^4\end{bmatrix}\begin{Bmatrix}\dot{\tilde{\mathbf{x}}}\\ \delta\dot{\tilde{\mathbf{u}}}^I\end{Bmatrix} = \begin{Bmatrix}\mathbf{P}\dot{\varepsilon}^M\\ \mathbf{0}\end{Bmatrix}, \tag{69}$$

where $\mathbf{P} = -\mathbf{H}^B(\mathbf{R}^T)^{-1}\delta\mathbf{x}^{4-1} - \mathbf{H}^L(\mathbf{R}^R)^{-1}\delta\mathbf{x}^{2-1}$ and $\mathbf{K}^1 = \mathbb{R}^{m\times m}$, $\mathbf{K}^2 = \mathbb{R}^{m\times n}$, $\mathbf{K}^3 = \mathbb{R}^{n\times m}$, $\mathbf{K}^4 = \mathbb{R}^{n\times n}$ denote submatrices.

At the end of a microstructural increment, where a converged state has been achieved, a third-order tensor L_{ijk} can be evaluated that relates directly the displacement discontinuities with the prescribed macro-strains, i.e. $\delta\dot{\tilde{u}}_i^I(\mathbf{X}) = L_{ijk}(\mathbf{X})\dot{\varepsilon}_{jk}^M$, where $L_{ijk} = [K_{il}^3(K_{lp}^1)^{-1}K_{pn}^2 - K_{in}^4]^{-1}K_{nm}^3(K_{ms}^1)^{-1}P_{sjk}$ and $L_{ijk} = L_{ikj}$.

Using now the relation between the displacement discontinuities and the prescribed macro-strain, Eq. (68) takes the form

$$\dot{\bar{\varepsilon}}_{ij}^{m,D} = J_{ijkl}\dot{\varepsilon}_{kl}^M, \tag{70}$$

where J_{ijkl} is a fourth-order tensor with symmetries $J_{ijkl} = J_{jikl} = J_{ijlk}$ given by

$$J_{ijkl} = \frac{1}{2V^m}\left\{\int_{S_{pc}^m}(R_{im}L_{mkl}n_j^A + R_{jm}L_{mkl}n_i^A)\, dS_{pc}^m\right\}. \tag{71}$$

Finally, the damaged stiffness tensor is obtained by substituting Eqs. (71) and (70) into Eq. (67) and considering Eq. (66). The resulting expression must be valid for any constant symmetric macro-strain,[34] given by

$$C_{ijkl}^{D} = C_{ijkl} - C_{ijmn} J_{mnkl}. \tag{72}$$

From the above expression, the damaged stiffness matrices, in the context of the FEM, are evaluated, depending if the specific RVE is assigned to an integration point or the centroid of a macro-FE.

9. Multiprocessing Algorithm

The proposed multi-scale boundary element method is a parallel processing formulation that requires special attention during the implementation, in order to be efficient and robust. Each micromechanics simulation, that is, each RVE, is assumed to be an individual subprogram that runs separately and in parallel with all the other micromechanics programs and the macromechanics main program. Since the proposed formulation is an incremental solution method, for every micromechanics simulation the inverse coefficient matrix of the final system of equations, Eq. (62), must be stored. As micro-damage progresses and therefore the interface boundary conditions are changing, the coefficient matrix of each micromechanics simulation would dynamically change. Therefore, throughout the simulation only updates of the inversed matrix should be made in order to reduce the computational effort of repeated inversion of the coefficient matrix. For more details on the implementation of the micromechanics, the readers are referred to Ref. 21. The macromechanics main program controls all the micromechanics programs. The macro-program starts all the micro-programs and gives them the green flag for reading its output. Once all the micro-programs have finished, the macro-program reads their outputs and processes them. When the micro-programs are running, the macro-program is placed on pause and *vice versa*.

Once all the RVE subprograms have started, built the BE mesh and inverted their main coefficient matrix, the critical macro-load λ where micro-damage will be initiated in the first RVE, for the first time, is evaluated. This is done directly since the whole system remains fully linear elastic and saves computational effort of incrementing the macro-load in the linear elastic regime. The incremental scheme starts by increasing step-by-step the applied macro-load. The macro-continuum is being solved and the macro-strains are evaluated for every domain point that has assigned an RVE to represent the corresponding microstructure.

Parallel processing of every RVE micromechanics starts by applying the new periodic boundary conditions. When all of them have finished, the main program reads their outputs, i.e. the decremental component of stress, and evaluates the right-hand vector to encounter the possible microdamage. After resolving the macro-continuum, the convergence is checked by evaluating the macro internal energy at each internal loop k, by $U^{M,k} = \int_{V^M} \sigma_{ij}^M \varepsilon_{ij}^M \, dV^M$, and enforcing the following tolerance: $100 \cdot |\frac{U^{M,k} - U^{M,k-1}}{U^{M,k}}| \leq 0.1\%$. If the prescribed tolerance has not been reached, the macro-strains are re-evaluated considering the previous macro-microdamage state and the micromechanics subprograms resolve the RVEs for the new boundary conditions. When convergence is achieved, the intermediate results are printed and another macro-load increment is applied.

In continuum damage models, a macro-crack is represented by a region of completely damaged material. However, this completely damaged region should be excluded from the macro-continuum formulation, since the governing equations are meaningless as the material has no stiffness there. Moreover, in non-local formulations as the one used here, the large strains due to the complete loss of the material stiffness would lead to wrong estimates of the non-local averaged strains. Additionally, by excluding this region from the macro-continuum formulation, the assigned completely damaged RVEs are also excluded, resulting in savings in computational time and storage. By excluding this completely damaged region, a new internal or external boundary is specified and boundary conditions are applied. In order to do so, the macro-continuum is remeshed and the local solution is remapped onto the new mesh.[43] However, this is rather complicated and interpolation errors will be inevitably introduced. Moreover, in the case of multi-scale modelling the macro-positions where the RVEs are assigned cannot change during the solution process. In this chapter, after following Peerlings *et al.*[31] who proposed the following remeshing method in the context of the FEM, the completely damaged macro-cells, that is, the assigned RVEs macro-points and their neighbourhood, are removed from the macro-continuum and the additional newly formed macro-boundary is being discretised using quadratic boundary elements. To ensure smooth transition and crack propagation and, on the other hand, to avoid numerical singularities, a critical damage factor is specified, i.e. $D^* = 0.999$. The criterion for removing a completely damaged cell was chosen to be $\max\{D_{11}, D_{22}, D_{12}\} \geq D^*$.

10. Multi-Scale Damage Simulations

Multi-scale damage simulations are performed using the proposed method for a polycrystalline Al_2O_3 ceramic material. At the micro-scale, multiple intergranular crack initiation and propagation under mixed-mode failure conditions is considered. Moreover, the random grain distribution, morphology and orientation is also taken into account. In cases of fully cracked grain boundary interfaces, a fully frictional contact analysis is performed to allow for sliding, sticking and separation of the crack's surfaces. The mesh independency of the proposed formulation is addressed. Additionally, comparisons with the FEM are made in order to investigate the different modelling philosophies. Several examples are illustrated to conclude the study.

Figure 9 illustrates a schematic representation of the problem solved here. A polycrystalline Al_2O_3 is subjected to three-point bending, at the macro-scale, by applying displacement control. The expected non-linear macro-region is assigned a number of domain points and on each point an RVE is handed over. Two cases are investigated: (a) initially the same

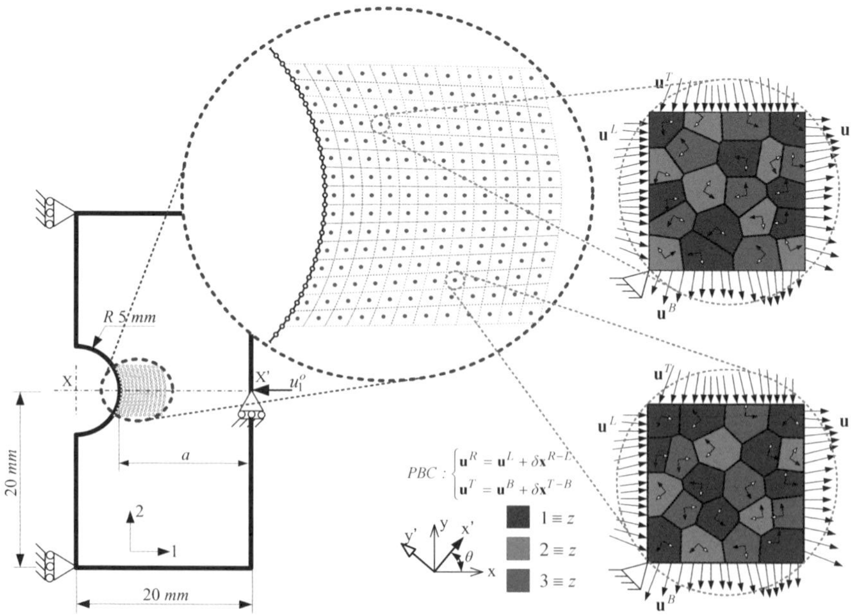

Fig. 9. Schematic representation of the multi-scale problem.

RVE is considered for every macro-domain point and (b) a randomly picked different RVE is assigned to each point to investigate heterogeneous microstructures with possible defects, randomly distributed in the macro-domain. The RVEs are randomly generated by Voronoi tesselations as described.[21] The single crystal elastic constants of Al_2O_3 considered here are $C_{11} = 496.8\,\text{GPa}$, $C_{33} = 498.1\,\text{GPa}$, $C_{44} = 147.4\,\text{GPa}$, $C_{12} = 163.6\,\text{GPa}$, $C_{13} = 110.9\,\text{GPa}$, $C_{14} = -23.5\,\text{GPa}$.[44] The fracture toughness of the material $K_{IC} = 4\,\text{MPam}^{1/2}$, $T_{\max} = 500\,\text{MPa}$, $\alpha = \beta = 1$ and plain strain conditions were assumed. The RVEs were composed by 21 grains, randomly distributed with random material orientation, of average grain size ASTM $G = 10$ ($\bar{A}^{\text{gr}} = 126\,\mu\text{m}^2$, $\bar{d}^{\text{gr}} = 11.2\,\mu\text{m}$[58]). The interface internal friction coefficient was assumed to be $\mu = 0.2$. The macro-continuum elastic properties were $E = 415.0\,\text{GPa}$, for the elastic moduli and $\nu = 0.24$ for the Poisson ratio. The non-local material's characteristic length was set to $l = 1.5\,\text{mm}$.

The macro-continuum was modelled using 65 quadratic boundary elements and 228 domain points and therefore 228 RVEs. The macro-continuum was also modelled using the FE commercial software ABAQUS.[45] To compare directly the results from both macro-formulations, the expected non-linear region was modelled in exactly the same manner in both numerical methods. The FEM model was created using quadratic quadrilateral elements in order to match exactly the BEM model in the non-linear region, and the rest was discretised using quadratic triangular elements. In order to investigate the influence of modelling the damage, which the micro feeds the macro, using the initial stress approach in the context of the BEM, two different formulations were considered in the case of macro-FEM. The first one is to consider the damage as an initial decremental stress that softens the material locally, as exactly the same as in the case of the proposed boundary element formulation. The second formulation is to directly implement the new damaged material stiffness, as cracks initiate and propagate in the micro-scale. In both cases it was assumed that the damage is uniformly distributed inside an FE in order to make a direct comparison with the BEM and to avoid partially damaged elements.[31] Figure 10 illustrates the different meshes used in the case of macro-BEM and macro-FEM.

The results from the macro-BEM/FEM comparison are illustrated in Fig. 11, where the dimensionless macro-stress component σ_{22} in front of the hole, along the cross section X–X', Fig. 9, is presented. The first frame shot, (i), illustrates the stress state when no-damage has appeared yet, i.e.

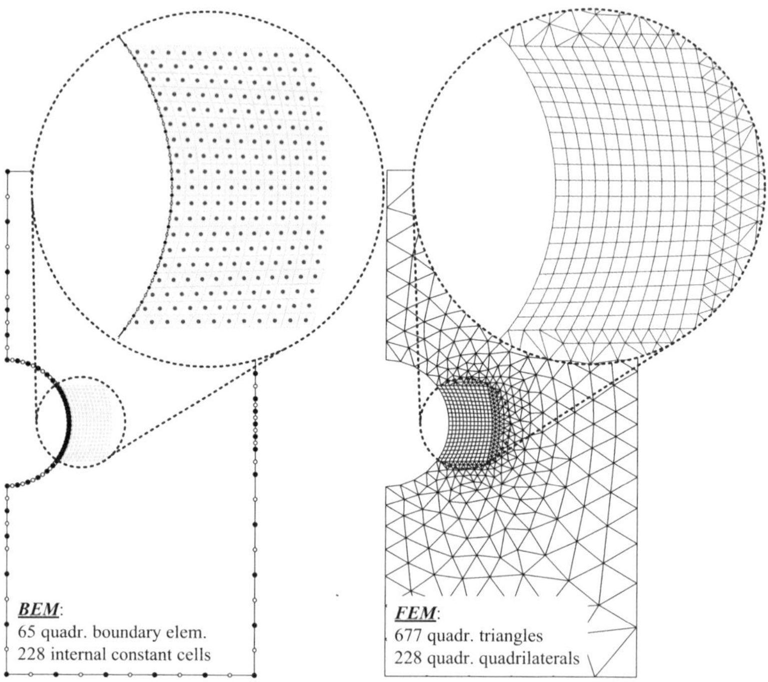

Fig. 10. Macro-BEM mesh and macro-FEM mesh, used in the present study.

still in the fully elastic regime. The next two frame shots illustrate some damage, due to partially damaged and cracked grain boundary interfaces in the micro-scale, which reduce the stiffness of the macro-continuum and therefore less stress can be sustained over this area. The elastic BEM stress curve is also presented as a dash-dotted line for comparison. The last frame shot is the increment just before a macro-crack will be initiated. As the initial stress FEM approach is denoted by dashed line while the damaged stiffness FEM approach is denoted by dash-dotted line. It can be seen that both macro-FEM results are very close and moreover the proposed macro-BEM formulation is in good agreement with both macro-FEMs.

Figure 12 illustrates the two different domain discretisations that were used in the present study to investigate the mesh independency of the proposed formulation. The same exact region in front of the hole was assigned 120 points for cell mesh A (61 quadratic boundary elements) and 228 points for cell mesh B (65 quadratic boundary elements). The same

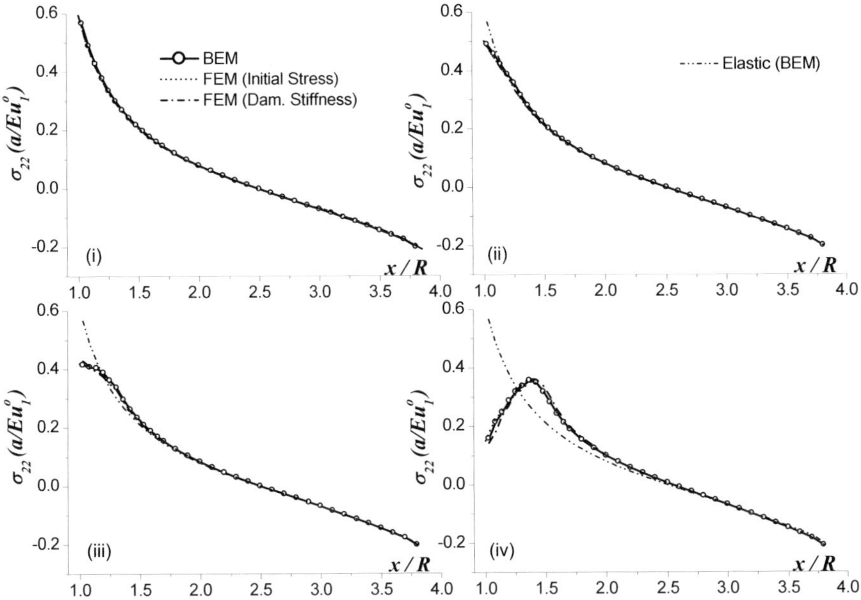

Fig. 11. Comparison between a macro-BEM and a macro-FEM formulation in the context of the proposed multi-scale damage modelling.

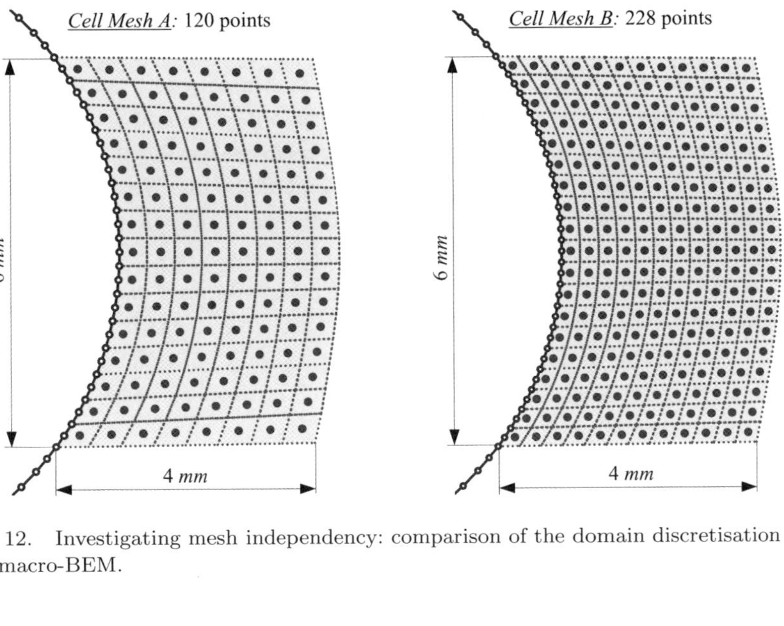

Fig. 12. Investigating mesh independency: comparison of the domain discretisation for the macro-BEM.

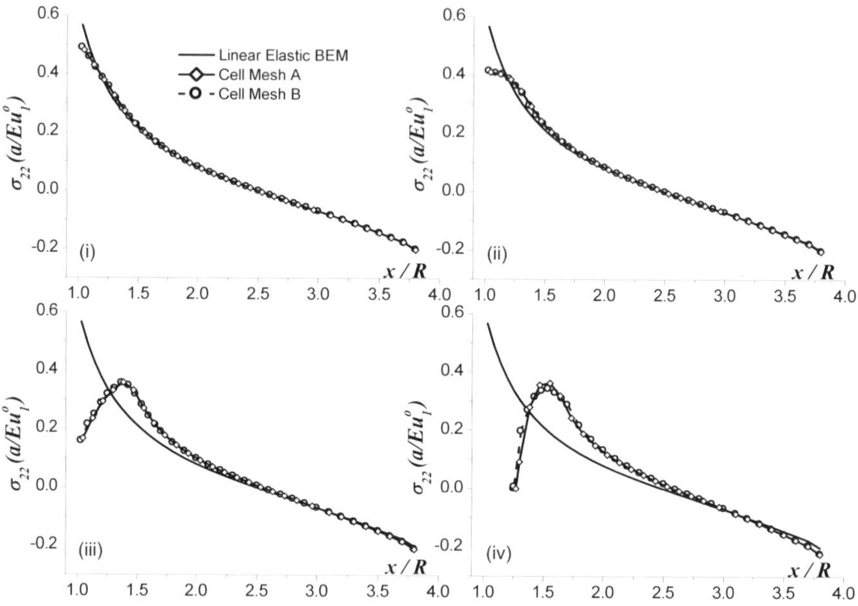

Fig. 13. Dimensionless stress component along X–X' cross section: comparison between different domain discretisations for the macro-BEM.

characteristic length in the integral non-local model was kept for both cases and the same RVE was assigned at each macro-domain point. Figure 13 illustrates the resulting dimensionless stress component in front of the hole. It can be seen that the proposed formulation, with the non-local approach for the macro-continuum, does not suffer from severe localisation of the damage that eventually leads to mesh-dependent results. In Fig. 13, frame shot (iii) corresponds to the last increment just before a macro-crack is initiated, while in the last frame a macro-crack has already been initiated. The corresponding frame shots of Fig. 13 macro-damage patterns, due to microcracking evolution, for both mesh cases, are illustrated in Fig. 14. Even though the damage patterns are represented in a discrete manner (uniform damage distribution over each cell), both mesh cases give similar macro-damage pattern.

Figure 15 illustrates the macro-damage evolution for the case of cell mesh A density, Fig. 12, but with additional domain discretisation. The new discretisation is composed of 180 macro-domain points with the same corresponding RVEs. Even though between the previous cell mesh A example and the current example there is an increase of +50% more RVEs,

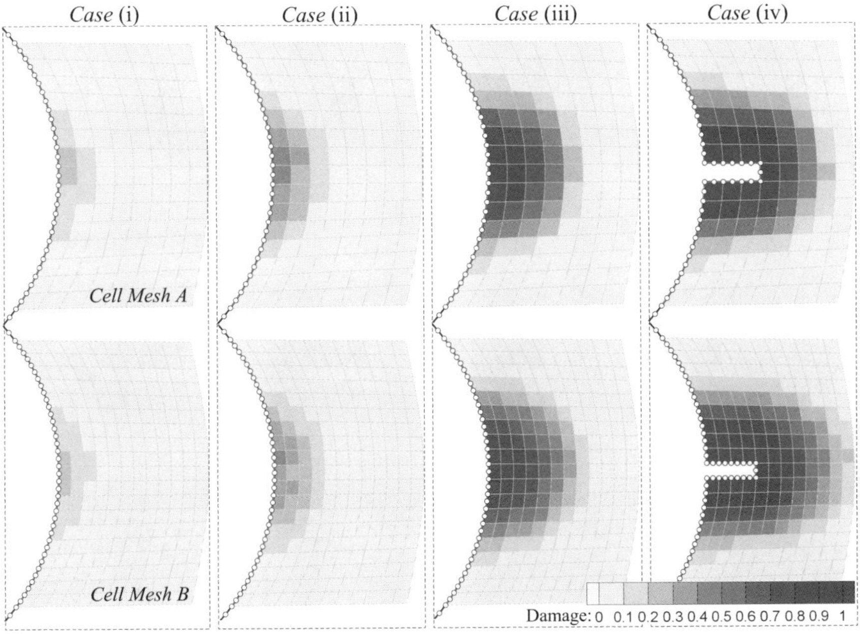

Fig. 14. Macro-damage patterns for different domain discretisations.

until case (iv) in Fig. 14 and case (vi) in Fig. 15, which corresponds at the same macro-load increment, the computational effort was only 9% higher. This is due to the proposed multi-scale boundary element formulation, where as long as the RVEs remain undamaged, only a matrix–vector multiplication is performed to finalise the increment. Figure 16 illustrates the evolution of the dimensionless internal macro-stress along the X–X' cross section at the fracture load. The curves correspond to the damage patterns illustrated in Fig. 15.

Consider now the case that most of the engineering materials are in general heterogeneous at a certain scale. From the definition of the RVE,[34] it represents the microstructure at the infinitesimal material neighbourhood around a macro-point and moreover it should statistically represent the microstructure of the macro-continuum. Therefore, it could be argued that a material may have different microstructure in different areas of the macro-continuum, with certain defects or not. In this case, the selected RVE must represent in the same sense the microstructure of the material at the specific region. For this reason and to demonstrate the capability of the proposed

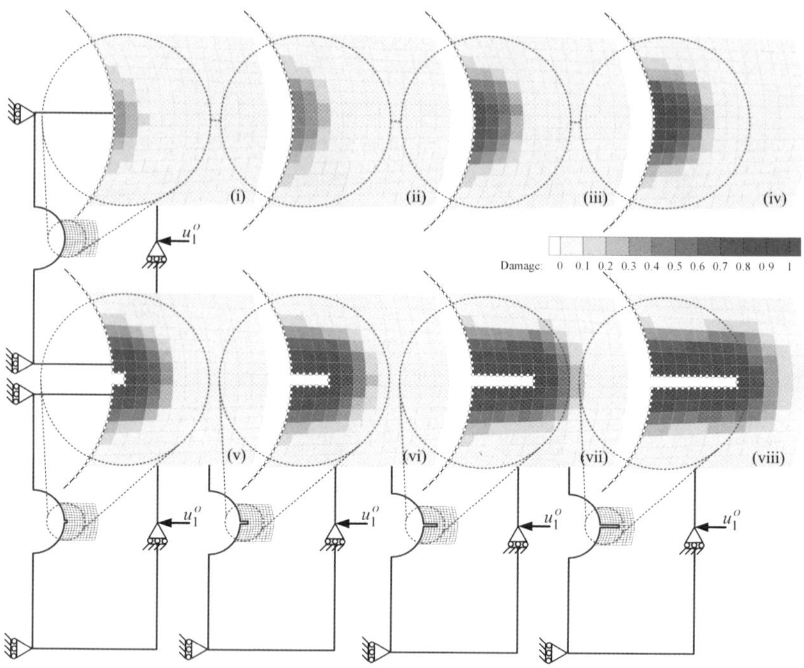

Fig. 15. Macro-damage evolution.

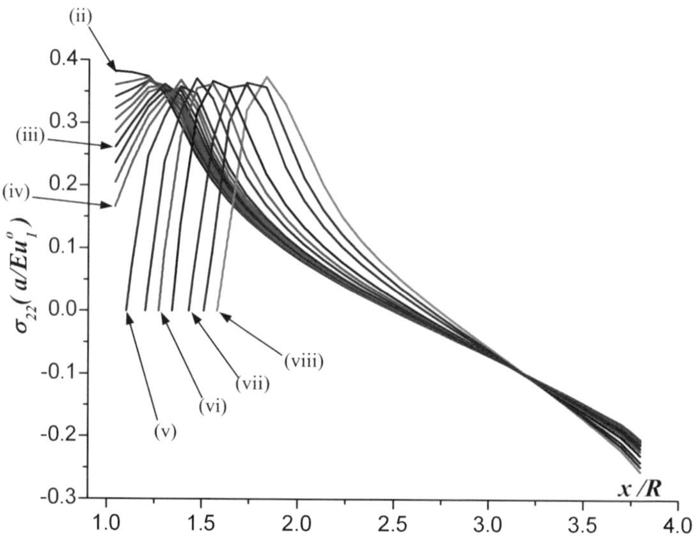

Fig. 16. Evolution of the dimensionless internal stress σ_{22} component along the X–X' cross section.

method to deal with such heterogeneous problems, the next examples consist of randomly distributed different RVEs for the macro-domain points. A set of eight RVE-grain morphologies and distributions are produced and assigned randomly to the macro-domain points. Even though the different RVE-grain morphologies are eight, each RVE has a unique grain material orientation, randomly distributed. In this way, a mixture of microstructure morphologies is randomly distributed at specific macro-points in the continuum, with the same average grain size, to study influence of microstructural variation.

Two sets of different RVEs were created and simulated with the proposed method. Figure 17 illustrates the damage evolution of the first set. It can be seen that the damage at the macro-scale, which is due to the intergranular fracture evolution in the micro-scale, is not fully symmetric. Moreover at early stages, i.e. frames (ii)–(iii), the highest damage is not exactly at the boundary of the hole but slightly inside of the

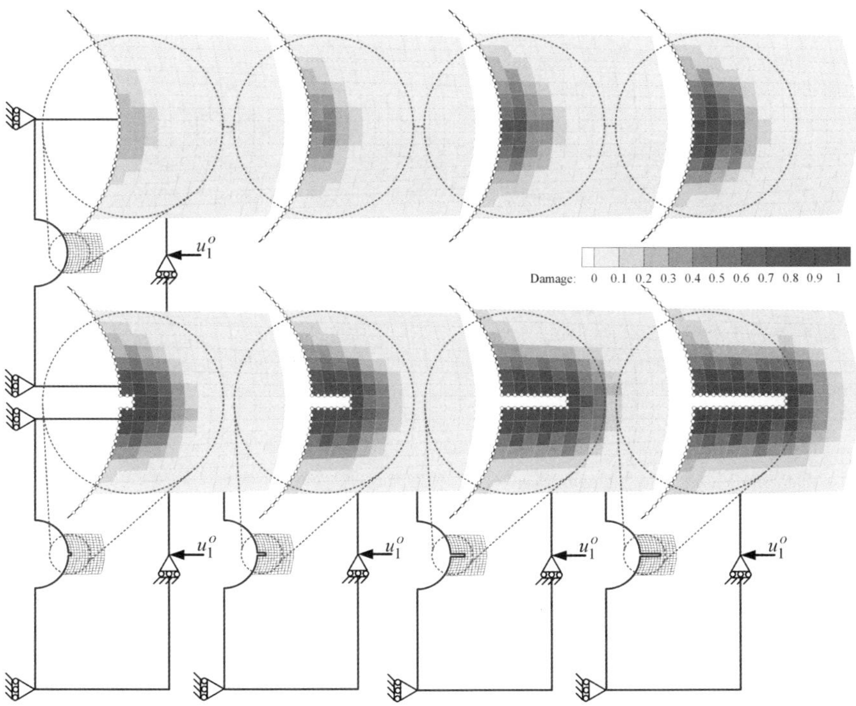

Fig. 17. Damage evolution at the macro-continuum for randomly distributed different RVEs: Set 1.

boundary. Both phenomena are due to the fact that some RVEs are more susceptible to fracture than others. Therefore, some areas of the macro-continuum are being damaged faster than what it was expected with classic continuum theory. The capability to model efficiently such phenomena is important in terms of modelling materials with variable properties through their thickness, such as coated and generally surface-treated material. The micro-damage evolution inside the corresponding RVEs is illustrated in Figs. 18 and 19. Figure 18 illustrates the microstructural state just at the initiation of the macro-crack, while Fig. 19 at a specific moment

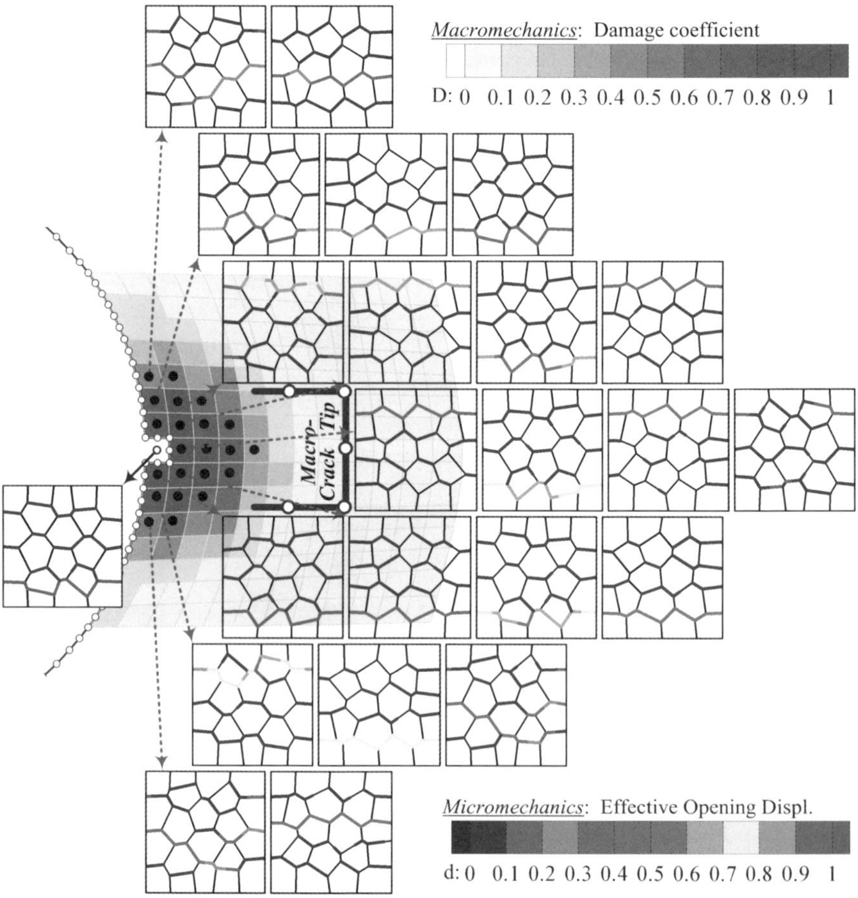

Fig. 18. Intergranular fracture evolution at the micro-scale for frame shot (v) of Fig. 17.

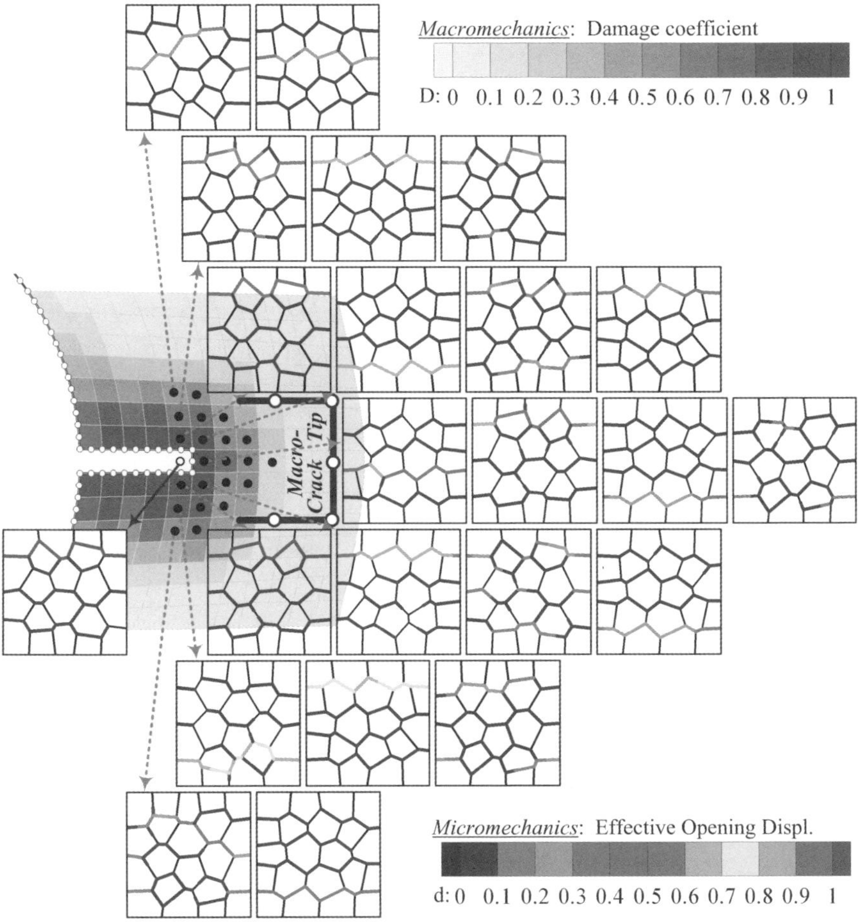

Fig. 19. Intergranular fracture evolution at the micro-scale for frame shot (vii) of Fig. 17.

after the macro-crack has propagated. In these figures, the progression of microcracking in front of the macro-crack tip is illustrated. This is in agreement with experimental findings where in front and around the crack tip, microcracks are formed, propagate and coalescence in order to form a macro-crack.[15] The damage evolution of the second set is illustrated in Fig. 20 and the corresponding state at the micro-scale at the initiation and after some propagation of the macro-crack is illustrated in Figs. 21 and 22, respectively. Comparing the damage evolution of the two sets,

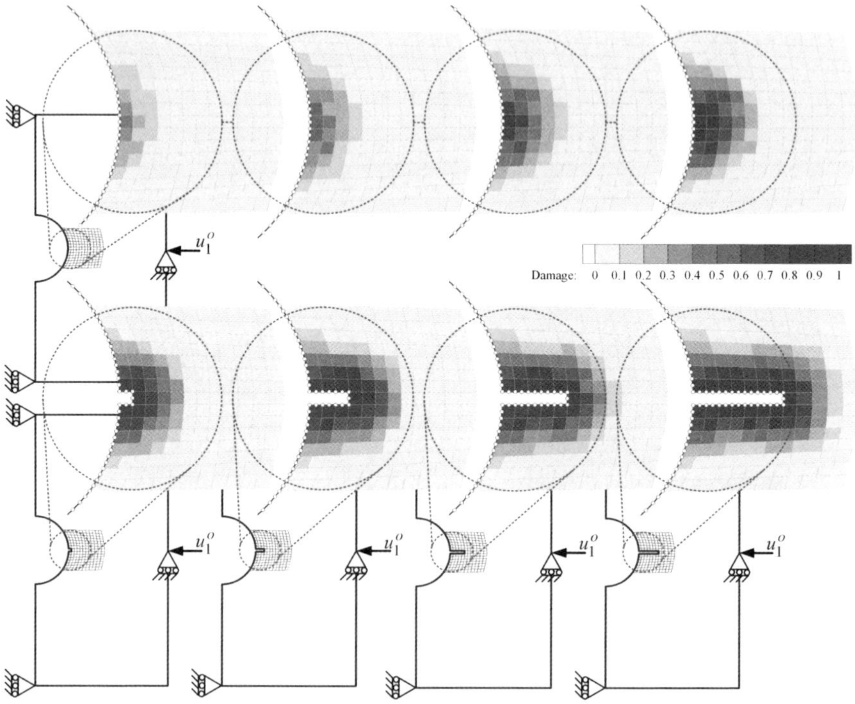

Fig. 20. Damage evolution at the macro-continuum for randomly distributed different RVEs: Set 2.

Figs. 17 and 20, a slight difference of the macro-damage response can be seen. This is due to the microstructural difference that is illustrated in Figs. 18 and 19 comparing with Figs. 21 and 22. It must be noted that even though the corresponding microstructures of the macro-continuum were different, the fracture macro-loads differed by only 1.2% between the two random examples.

11. Conclusions

A multi-scale boundary element formulation and its effective numerical implementation for modelling damage are proposed for the first time. Information about the constitutive behaviour of a polycrystalline material at the macro-continuum are obtained by the micro-scale using averaging

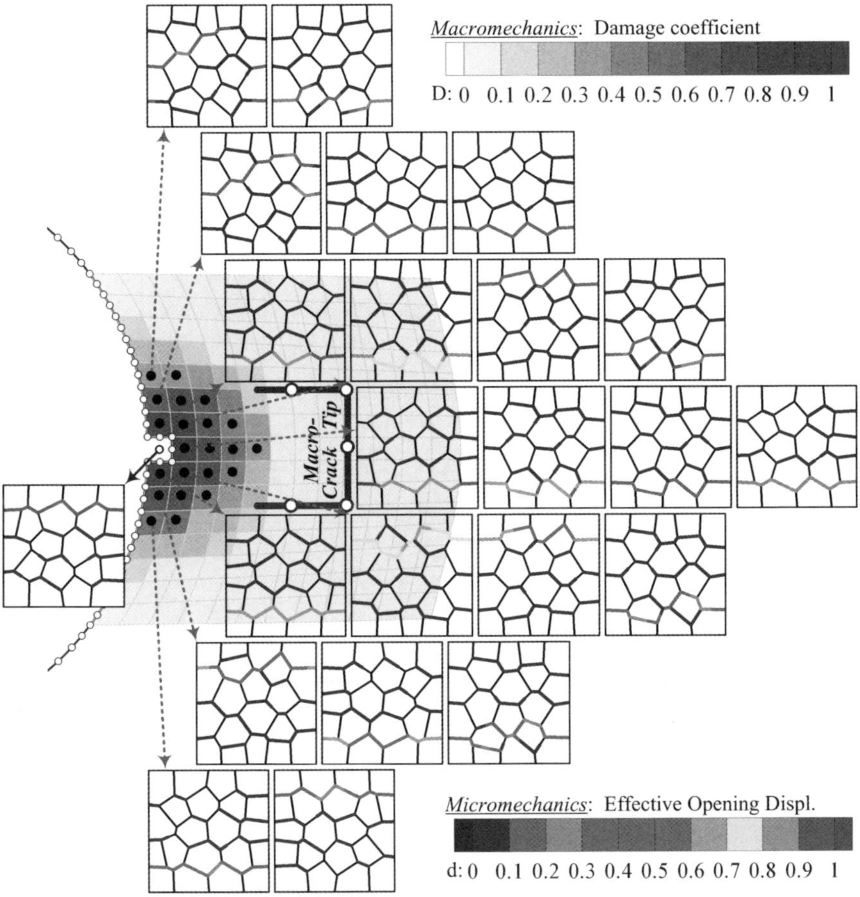

Fig. 21. Intergranular fracture evolution at the micro-scale for frame shot (v) of Fig. 20.

theorems in a multiprocessing manner. Both macro-continuum and micro-scale are modelled using the BEM. An approach for coupling the micro-BEM with the macro-FEM is also proposed. An integral non-local approach is employed for avoiding the pathological localisation of microdamage at the macro-scale. At the micro-scale, after considering a random distribution, morphology and orientation of the grains, multiple intergranular crack initiation and propagation under mixed-mode failure conditions was modelled. A fully frictional contact analysis was used to allow for crack surfaces to come into contact, slide, stick or separate.

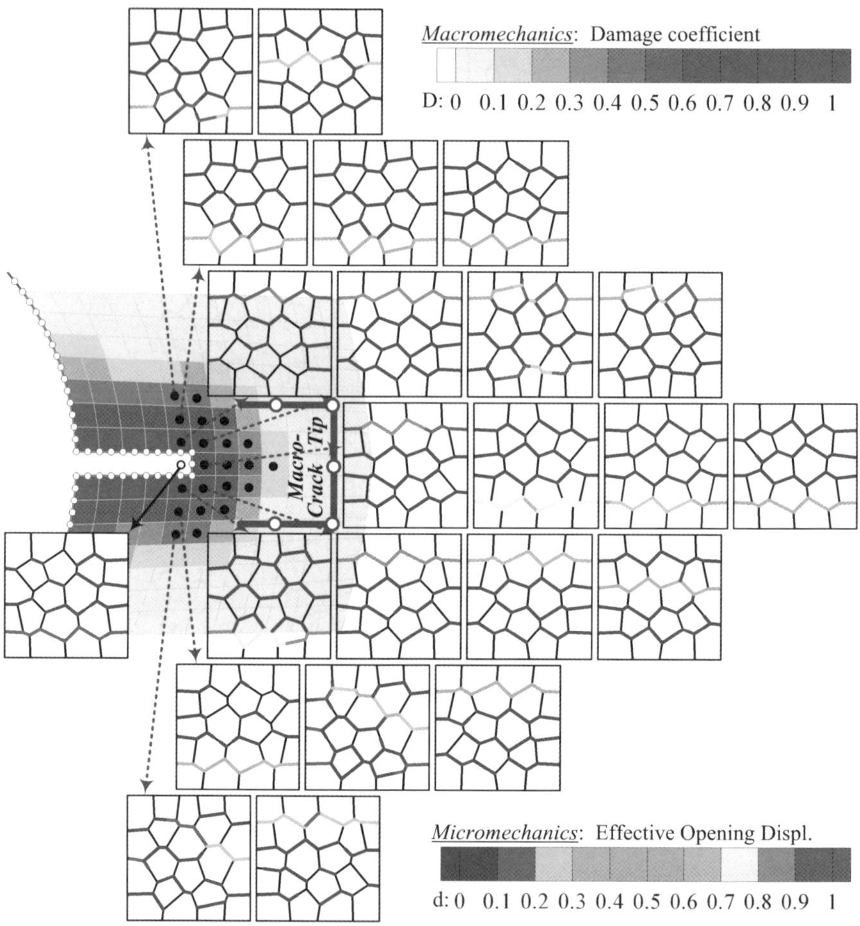

Fig. 22. Intergranular fracture evolution at the micro-scale for frame shot (vii) of Fig. 20.

Different numerical examples for a polycrystalline Al_2O_3 were investigated in order to demonstrate the accuracy of the proposed method. Mesh independency of the results was achieved due to the non-local approach used at the macro-scale. Comparing the proposed method with two macro-FEM models, one using an initial stress approach and another with a damaged stiffness tensor approach, good agreement was also obtained. Cases of inhomogeneous materials were also investigated by randomly assigning RVEs with variations in the microstructure.

The analysis demonstrates that the proposed method can be considered as a promising tool for future modelling of heterogeneous materials or materials with microstructural variation through their thickness.

References

1. L. M. Kachanov, On the time to failure under creep conditions, *Izv. Akad. Nauk. SSSR Otd. Tekhn. Nauk.* **8**, 26–31 (1958).
2. J. Lemaitre and J.-L. Chaboche, *Mechanics of Solid Materials* (Cambridge University Press, Cambridge, 1990).
3. J. Lemaitre, *A Course on Damage Mechanics*, 2nd edn. (Springer, Berlin, 1996).
4. S. Ghosh, K. Lee and S. Moorthy, Two scale analysis of heterogeneous elastic–plastic materials with asymptotic homogenisation and Voronoi cell finite element model, *Comput. Meth. Appl. Mech. Eng.* **132**, 63–116 (1996).
5. K. Terada, M. Hori, T. Kyoya and N. Kikuchi, Simulation of the multi-scale convergence in computational homogenization approaches, *Int. J. Solids Struct.* **37**, 2285–2311 (2000).
6. V. Kouznetsova, M. G. D. Geers and W. A. M. Brekelmans, Multi-scale constitutive modelling of heterogeneous materials with a gradient-enhanced computational homogenization scheme, *Int. J. Numer. Meth. Eng.* **54**, 1235–1260 (2002).
7. V. Kouznetsova, M. G. D. Geers and W. A. M. Brekelmans, Multi-scale second-order computational homogenization of multi-phase materials: A nested finite element solution strategy, *Comput. Meth. Appl. Mech. Eng.* **193**, 5525–5550 (2004).
8. J. D. Clayton and D. L. McDowell, Finite polycrystalline elastoplasticity and damage: Multiscale kinematics, *Int. J. Solids Struct.* **40**, 5669–5688 (2003).
9. P. Ladevèze, Multiscale modelling and computational strategies for composites, *Int. J. Numer. Meth. Eng.* **60**, 233–253 (2004).
10. S. Ghosh, L. Kyunghoon and P. Raghavan, A multi-level computational model for multi-scale damage analysis in composite and porous materials, *Int. J. Solids Struct.* **38**, 2335–2385 (2001).
11. P. Raghavan and S. Ghosh, A continuum damage mechanics model for unidirectional composites undergoing interfacial debonding, *Mech. Mater.* **37**, 955–979 (2005).
12. M. H. Aliabadi, *The Boundary Element Method*, Applications in Solids and Structures, Vol. 2, Wiley, London, 2002.
13. M. H. Aliabadi, A new generation of boundary element methods in fracture mechanics, *Int. J. Fract.* **86**, 91–125 (1997).
14. A. G. Crocker, P. E. J. Flewitt and G. E. Smith, Computational modelling of fracture in polycrystalline materials, *Int. Mater. Rev.* **50**, 99–124 (2005).
15. R. W. Rice, Mechanical properties of ceramics and composites: Grain and particle effects (Marcel Dekker, New York, 2000).

16. C. J. McMahon Jr., Hydrogen-induced intergranular fracture of steels, *Eng. Fract. Mech.* **68**, 773–788 (2001).
17. E. P. George, C. T. Liu, H. Lin and D. P. Pope, Environmental embrittlement and other causes of brittle grain boundary fracture in Ni_3Al, *Mater. Sci. Eng. A* **92/93**, 277–288 (1995).
18. M. Ortiz and A. Pandolfi, Finite-deformation irreversible cohesive elements for three-dimensional crack-propagation analysis, *Int. J. Numer. Meth. Eng.* **44**, 1267–1282 (1999).
19. V. Tvergaard, Effect of fibre debonding in a whisker-reinforced metal, *Mater. Sci. Eng. A Struct. Mater. Prop. Microstruct. Process. A* **125**, 203–213 (1990).
20. X.-P. Xu and A. Needleman, Numerical simulations of dynamic crack growth along an interface, *Int. J. Fract.* **74**, 289–324 (1995–1996).
21. G. K. Sfantos and M. H. Aliabadi, A boundary cohesive grain element formulation for modelling intergranular microfracture in polycrystalline brittle materials, *Int. J. Numer. Meth. Eng.* **69**, 1590–1626 (2007).
22. G. K. Sfantos and M. H. Aliabadi, Multi-scale boundary element modelling of material degradation and fracture, *Comput. Meth. Appl. Mech.* **196**, 1310–1329 (2007).
23. Z. P. Bažant, T. Belytschko and T. P. Chang, Continuum theory for strain-softening, *J. Eng. Mech.* **110**, 1666–1692 (1984).
24. R. de Borst, L. J. Sluys, H.-B. Mühlhaus and J. Pamin, Fundamental issues in finite element analysis of localization of deformation, *Eng. Comput.* **10**, 99–121 (1993).
25. J. H. P. de Vree, W. A. M. Brekelmans and M. A. J. van Gils, Comparison of nonlocal approaches in continuum damage mechanics, *Comput. Struct.* **55**, 581–588 (1995).
26. A. Needleman and V. Tvergaard, Mesh effects in the analysis of dynamic ductile crack growth, *Eng. Fract. Mech.* **47**, 75–91 (1994).
27. M. Jirásek, Nonlocal models for damage and fracture: Comparison of approaches, *Int. J. Solids Struct.* **35**, 4133–4145 (1998).
28. Z. P. Bažant and M. Jirásek, Nonlocal integral formulations of plasticity and damage: Survey of progress, *J. Eng. Mech.* **128**, 1119–1149 (2002).
29. H.-B. Mühlhaus and E. C. Aifantis, A variational principle for gradient plasticity, *Int. J. Solids Struct.* **28**, 845–857 (1991).
30. R. de Borst, J. Pamin, R. H. J. Peerlings and L. J. Sluys, On gradient-enhanced damage and plasticity models for failure in quasi-brittle and frictional materials, *Comput. Mech.* **17**, 130–141 (1995).
31. R. H. J. Peerlings, W. A. M. Brekelmans, R. de Borst and M. G. D. Geers, Gradient-enhanced damage modelling of high-cycle fatigue, *Int. J. Numer. Meth. Eng.* **49**, 1547–1569 (2000).
32. J. Sládek, V. Sládek and Z. P. Bažant, Non-local boundary integral formulation for softening damage, *Int. J. Numer. Meth. Eng.* **57**, 103–116 (2003).
33. F.-B. Lin, G. Yan, Z. P. Bažant and F. Ding, Nonlocal strain-softening model of quasi-brittle materials using the boundary element method, *Eng. Anal. Bound. Elem.* **26**, 417–424 (2002).

34. S. Nemat-Nasser and M. Hori, *Micromechanics: Overall Properties of Heterogeneous Materials* (Elsevier Science, Amsterdam, 1999).
35. S. Nemat-Nasser, Averaging theorems in finite deformation plasticity, *Mech. Mater.* **31**, 493–523 (1999).
36. W. H. Press, S. A. Teukolsky, W. T. Vetterling and B. P. Flannery, *Numerical Recipes in Fortran, The Art of Scientific Computing*, 2nd edn. (Cambridge University Press, Cambridge, 1994).
37. A. Okabe, B. Boots, K. Sugihara and S. N. Chiu, *Spatial Tessellations: Concepts and Applications of Voronoi Diagrams*, 2nd edn. (Wiley, Chichester, 2000).
38. C. J. Lissenden and C. T. Herakovich, Numerical modeling of damage development and viscoplasticity in metal matrix composites, *Comput. Meth. Appl. Mech. Eng.* **126**, 289–303 (1995).
39. V. Tomar, Z. Jun and Z. Min, Bounds for element size in a variable stiffness cohesive finite element model, *Int. J. Numer. Meth. Eng.* **61**, 1894–1920 (2004).
40. S. Hazanov, Hill condition and overall properties of composites, *Arch. Appl. Mech.* **68**, 385–394 (1998).
41. V. Kouznetsova, Computational homogenization for the multi-scale analysis of multi-phase materials, PhD dissertation, University of Technology, Eindhoven, The Netherlands (2002).
42. O. van der Sluis, P. J. G. Schreurs, W. A. M. Brekelmans and H. E. H. Meijer, Overall behaviour of heterogeneous elastoviscoplastic materials: Effect of microstructural modelling, *Mech. Mater.* **32**, 449–462 (2000).
43. B. Patzák and M. Jirásek, Adaptive resolution of localized damage in quasi-brittle materials, *J. Eng. Mech.* **130**, 720–732 (2004).
44. J. B. Wachtman Jr., W. E. Tefft, D. G. Lam Jr. and R. P. Stinchfield, Elastic constants of synthetic single crystal corundum at room temperature, *J. Res. Nat. Bur. Stand.* **64**, 213–228 (1960).
45. *ABAQUS 6.5 Documentation* (ABAQUS Inc., USA, 2004).
46. H. D. Espinosa and P. D. Zavattieri, A grain level model for the study of failure initiation and evolution in polycrystalline brittle materials. Part I: Theory and numerical implementation, *Mech. Mater.* **35**, 333–364 (2003).
47. H. D. Espinosa and P. D. Zavattieri, A grain level model for the study of failure initiation and evolution in polycrystalline brittle materials. Part II: Numerical examples, *Mech. Mater.* **35**, 365–394 (2003).
48. Y. J. Wei and L. Anand, Grain-boundary sliding and separation in polycrystalline metals: Application to nanocrystalline fcc metals, *J. Mech. Phys. Solids* **52**, 2587–2616 (2004).
49. J. Zhai, V. Tomar and M. Zhou, Micromechanical simulation of dynamic fracture using the cohesive finite element, *J. Eng. Mater. Tech. Trans. ASME* **126**, 179–191 (2004).
50. B.-N. Kim, S. Wakayama and M. Kawahara, Characterization of 2-dimensional crack propagation behavior by simulation and analysis, *Int. J. Fract.* **75**, 247–259 (1996).

51. M. Grah, K. Alzebdeh, P. Y. Sheng, M. D. Vaudin, K. J. Bowman and M. Ostoja-Starzewski, Brittle intergranular failure in 2D microstructures: Experiments and computer simulations, *Acta Mater.* **44**, 4003–4018 (1996).
52. N. Sukumar, D. J. Srolovitz, T. J. Baker and J. H. Prevost, Brittle fracture in polycrystalline microstructures with the extended finite element method, *Int. J. Numer. Meth. Eng.* **56**, 2015–2037 (2003).
53. S. Weyer, A. Fröhlich, H. Riesch-Oppermann, L. Cizelj and M. Kovac, Automatic finite element meshing of planar voronoi tessellations, *Eng. Fract. Mech.* **69**, 945–958 (2002).
54. Y. Liu, Y. Kageyama and S. Murakami, Creep fracture modeling by use of continuum damage variable based on Voronoi simulation of grain boundary cavity, *Int. J. Mech. Sci.* **40**, 147–158 (1998).
55. K. S. Zhang, D. Zhang, R. Feng and M. S. Wu, Microdamage in polycrystalline ceramics under dynamic compression and tension, *J. Appl. Phys.* **98**, 023505 (2005).
56. K. S. Zhang, M. S. Wu and R. Feng, Simulation of microplasticity-induced deformation in uniaxially strained ceramics by 3-D Voronoi polycrystal modeling, *Int. J. Plast.* **21**, 801–834 (2005).
57. R. L. Mullen, R. Ballarini, Y. Yin and A. H. Heuer, Monte Carlo simulation of effective elastic constants of polycrystalline thin films, *Acta Mater.* **45**, 2247–2455 (1997).
58. *ASTM E112-96 (Reapproved 2004): Standard Test Methods for Determining Average Grain Size* (ASTM International, 2004).
59. J. R. Rice, Mathematical analysis in the mechanics of fracture, in *Fracture*, Vol. 2, ed. H. Liebowitz (Academic Press, New York, 1968), pp. 191–311.
60. G. I. Barenblatt, On equilibrium cracks formed in brittle fracture. General concepts and hypotheses. Axisymmetric cracks, *J. Appl. Math. Mech.* **23**, 622–636 (1959).
61. D. S. Dugdale, Yielding of steel sheets containing slits, *J. Mech. Phys. Solids* **8**, 100–104 (1960).
62. N. Chandra, H. Li, C. Shet and H. Ghonem, Some issues in the application of cohesive zone models for metal-ceramic interfaces, *Int. J. Solids Struct.* **39**, 2827–2855 (2002).
63. C. Shet and N. Chandra, Analysis of energy balance when using cohesive zone models to simulate fracture processes, *J. Eng. Mat. Technol. Trans. ASME* **124**, 440–450 (2002).
64. H. D. Espinosa, P. D. Zavattieri and G. L. Emore, Adaptive FEM computation of geometric and material nonlinearities with application to brittle failure, *Mech. Mater.* **29**, 275–305 (1998).
65. J. Sherman and W. J. Morrison, Adjustment of an inverse matrix corresponding to a change in one element of a given matrix, *Ann. Math. Statist.* **21**, 124–127 (1950).
66. J. Sherman and W. J. Morrison, Adjustment of an inverse matrix corresponding to changes in the elements of a given column or a given row of the original matrix, *Ann. Math. Statist.* **20**, 621 (1949).
67. K. L. Johnson, *Contact Mechanics* (Cambridge University Press, Cambridge, 2001).

68. J. J. Bellante, H. Kahn, R. Ballarini, C. A. Zorman, M. Mehregany and A. H. Heuer, Fracture toughness of polycrystalline silicon carbide thin films, *Appl. Phys. Lett.* **86**, 071920 (2005).
69. J. F. Shackelford and W. Alexander, *CRC Materials Science and Engineering Handbook*, 3rd edn. (CRC Press, London, 2000).
70. T. Ohji, Y. Yamauchi, W. Kanematsu and S. Ito, Tensile rupture strength and fracture defects of sintered silicon carbide, *J. Amer. Ceramic Soc.* **72**, 688–690 (1989).
71. J. Kubler, Weibull characterization of four hipped/posthipped engineering ceramics between room temperature and 1500°C, EMPA Swiss Federal Laboratories for Materials Testing and Research, EMPA-Nr. 129747 (1992), pp. 1–88.
72. S. F. Pugh, The fracture of brittle materials, *Br. J. Appl. Phys.* **18**, 129–162 (1967).
73. P. A. Klerck, E. J. Sellers and D. R. J. Owen, Discrete fracture in quasi-brittle materials under compressive and tensile stress states, *Comput. Meth. Appl. Mech. Eng.* **193**, 3035–3056 (2004).
74. M. Yumlu and M. U. Ozbay, Study of the behaviour of brittle rocks under plane strain and triaxial loading conditions, *Int. J. Rock Mech. Mining Sci.* **32**, 725–733 (1995).
75. J. D. Poloniecki and T. R. Wilshaw, Determination of surface crack size densities in glass, *Nature Phys. Sci.* **229**, 226–227 (1971).
76. A. S. Jayatilaka and K. Trustrum, Statistical approach to brittle fracture, *J. Mater. Sci.* **12**, 1426–1430 (1977).

Appendix

The fundamental solutions used in the boundary integral equations are given as

$$U_{ij}(\mathbf{x}',\mathbf{x}) = c_1\left\{c_2 \ln\left(\frac{1}{r}\right)\delta_{ij} + r_{,i}r_{,j}\right\},$$

$$T_{ij}(\mathbf{x}',\mathbf{x}) = c_3\{r_{,m}n_m(c_4\delta_{ij} + 2r_{,i}r_{,j}) - c_4(r_{,i}n_j - r_{,j}n_i)\},$$

$$E_{ijk}(\mathbf{x}',\mathbf{X}) = -\frac{c_1}{r}\{c_4(r_{,j}\delta_{ik} + r_{,k}\delta_{ij}) - r_{,i}\delta_{jk} + 2r_{,i}r_{,j}r_{,k}\},$$

$$D^{\varepsilon}_{ijk}(\mathbf{X}',\mathbf{x}) = \frac{c_1}{r}\{c_4(r_{,i}\delta_{jk} + r_{,j}\delta_{ik}) - r_{,k}\delta_{ij} + 2r_{,i}r_{,j}r_{,k}\},$$

$$\begin{aligned} S^{\varepsilon}_{ijk}(\mathbf{X}',\mathbf{x}) = -\frac{c_3}{r^2}\{&2r_{,m}n_m\{r_{,k}\delta_{ij} + \nu(r_{,i}\delta_{jk} + r_{,j}\delta_{ik}) - 4r_{,i}r_{,j}r_{,k}\} \\ &+ n_i(c_4\delta_{jk} + 2\nu r_{,j}r_{,k}) + n_j(c_4\delta_{ik} + 2v r_{,i}r_{,k}) \\ &- n_k c_4(\delta_{ij} - 2r_{,i}r_{,j})\}, \end{aligned}$$

$$W^{\varepsilon}_{ijkl}(\mathbf{X}',\mathbf{X}) = \frac{c_1}{r^2}\{2\nu(\delta_{li}r_{,j}r_{,k} + \delta_{ik}r_{,j}r_{,l} + \delta_{lj}r_{,k}r_{,i} + \delta_{jk}r_{,l}r_{,i}) + 2\delta_{kl}r_{,i}r_{,j}$$
$$+ c_4(\delta_{jk}\delta_{li} + \delta_{lj}\delta_{ik}) - \delta_{ij}(\delta_{kl} - 2r_{,k}r_{,l}) - 8r_{,i}r_{,j}r_{,k}r_{,l}\},$$

$$\dot{g}^{\varepsilon}_{ij}(\mathbf{X}') = \frac{\pi c_1}{2}\{\dot{\sigma}^{D}_{mm}(\mathbf{X}')\delta_{ij} - 2c_2\dot{\sigma}^{D}_{ij}(\mathbf{X}')\},$$

$$D^{\sigma}_{ijk}(\mathbf{X}',\mathbf{x}) = -\frac{c_3}{r}\{c_4(r_{,i}\delta_{jk} + r_{,j}\delta_{ik} - r_{,k}\delta_{ij}) + r_{,i}r_{,j}r_{,k}\},$$

$$S^{\sigma}_{ijk}(\mathbf{X}',\mathbf{x}) = \frac{c_3^2}{c_1 r^2}\{2r_{,m}n_m[c_4\delta_{ij}r_{,k} + \nu(r_{,j}\delta_{ik} + r_{,i}\delta_{jk}) - 4r_{,i}r_{,j}r_{,k}]$$
$$+ 2\nu(n_i r_{,i}r_{,k} + n_j r_{,i}r_{,k}) + c_4(2n_k r_{,i}r_{,j} + n_j\delta_{ik} + n_i\delta_{jk})$$
$$- (1-4\nu)n_k\delta_{ij}\},$$

$$W^{\sigma}_{ijkl}(\mathbf{X}',\mathbf{X}) = -\frac{c_3}{r^2}\{c_4(\delta_{li}\delta_{jk} + \delta_{ik}\delta_{lj} - \delta_{ij}\delta_{kl} + 2\delta_{ij}r_{,k}r_{,l}) + 2\delta_{kl}r_{,i}r_{,j}$$
$$+ 2\nu(\delta_{li}r_{,j}r_{,k} + \delta_{ik}r_{,j}r_{,l} + \delta_{lj}r_{,k}r_{,i} + \delta_{jk}r_{,l}r_{,i})$$
$$- 8r_{,i}r_{,j}r_{,k}r_{,l}\},$$

$$\dot{g}^{\sigma}_{ij}(\mathbf{X}') = \frac{\pi c_3}{2}\{2\dot{\sigma}^{D}_{ij}(\mathbf{X}') - c_2\dot{\sigma}^{D}_{mm}(\mathbf{X}')\delta_{ij}\},$$

where $c_1 = 1/(8\pi\mu(1-\nu))$, $c_2 = 3-4\nu$, $c_3 = -1/(4\pi(1-\nu))$ and $c_4 = 1-2\nu$. Moreover, $r = \sqrt{(r_i r_i)}$, $r_i = x_i - x'_i$, $r_{,i} = r_i/r$, $r_{,m}n_m = \partial r/\partial n$ and

$$\delta_{ij} = \begin{cases} 1 & \text{if } i = j \\ 0 & \text{if } i \neq j \end{cases}$$

denotes the Kronecker delta function. The Poisson ratio is denoted by ν and the shear modulus by μ.

NON-UNIFORM TRANSFORMATION FIELD ANALYSIS: A REDUCED MODEL FOR MULTISCALE NON-LINEAR PROBLEMS IN SOLID MECHANICS

Jean-Claude Michel* and Pierre Suquet

L.M.A./C.N.R.S. 31 Chemin Joseph Aiguier
13402, Marseille, Cedex 20, France
**michel@lma.cnrs-mrs.fr*

This chapter is devoted to the *Non-uniform Transformation Field Analysis* which is a reduction technique introduced in the realm of Multiscale Problems in Non-linear Solid Mechanics to achieve scale transition for materials exhibiting a non-linear behaviour. It is indeed well recognised that the non-linearity introduces a strong coupling between the problems at different scales which, in full rigour, remain coupled.

To avoid the computational cost of the scale coupling, reduced models have been developed. To improve on the predictions of Transformation Field Analysis where the plastic strain field is assumed to be uniform in each domain, the authors (Michel and Suquet[18]) have proposed another reduced model, called the *Non-uniform Transformation Field Analysis*, where the plastic strain fields follow shape functions which are not piecewise uniform.

The model is presented for individual phases exhibiting an elasto-viscoplastic behaviour. A brief account on the reduction technique is given first. Then the time-integration of the model at the level of a macroscopic material point is performed by means of a numerical scheme.

This reduced model is applied to structural problems. The implementation of the model in a Finite Element code is discussed. It is shown that the model predicts accurately the effective behaviour of non-linear composite materials with just a few internal variables. Another worth-noting feature of the method is that the local stress and strain fields can be determined simply by postprocessing the output of the structural (macroscopic) computation performed with the model. The flexibility and accuracy of the method are illustrated by assessing the lifetime of a plate subjected to cyclic four-point bending. Using the distribution in the structure of the energy dissipated locally in the matrix by viscoplasticity as fatigue indicator, the lifetime prediction for the structure is seen to be in good agreement with large-scale computations taking into account all heterogeneities.

1. Introduction

A common engineering practice in the analysis of composite structures is to use *effective* or *homogenised* material properties instead of taking into account all details of the individual phase properties and geometrical arrangement (fibre and matrix in the case of a fibre-reinforced composite). These effective properties are sometimes difficult to measure and this difficulty has motivated the development of mathematical *homogenisation* techniques which provide a rational way of deriving *effective* material properties directly from those of the individual constituents and from their arrangement or *microstructure*. A further interest of such predictive schemes is that the material or geometrical parameters can be varied easily which opens the way for tailoring new materials for a given application. Although homogenisation has been developed for both periodic[25] or random composites,[20] the present study is focused on periodic composites.

Periodic homogenisation of *linear* properties of composites is now well-established and the reader is referred to Bensoussan *et al.*[2] or Sanchez-Palencia[25] for the general theory, and to Suquet[29] or Guedes and Kikuchi[10] (among others) for computational aspects. The central theoretical result for linear properties is that, provided that the scales are well separated, the linear effective properties of a composite are completely determined by solving a finite number of unit-cell problems. These unit-cell problems are solved once for all and their resolution yields the effective properties of the composite. Then the analysis of a structure comprised of such a composite material can be performed using these effective linear properties. In summary, for linear problems, the analysis consists of two completely independent steps, a homogenisation step at the unit-cell level only, and a standard structural analysis performed at the structure level only.

In comparison, the situation for non-linear composites is more complicated. For composites governed by a single non-quadratic but strictly convex potential (elastic potential or dissipation potential), homogenisation results can be established to define an effective behaviour, deriving from an effective potential (provided that the scales are well separated). However, except in very specific cases, this effective potential cannot be found by solving a small, or even a finite, number of unit-cell problems. To *each* macroscopic stress or strain state corresponds a unit-cell problem which has to be solved independently of the unit-cell problem for a different macroscopic state. Therefore, although there exists a homogenised behaviour for the composite, the rigorous analysis of a composite structure consists of two *coupled* computational problems: (1) a structural problem where the (unknown) effective constitutive relations express the relations between the microscopic stress and strain fields solution of the second problem; (2) a unit-cell problem whose loading conditions are imposed by the (unknown) macroscopic stress or strain (or their rates).

Exactly the same type of complication occurs when the composite is made of individual constituents governed by two potentials, free-energy and dissipation potential, accounting for reversible and irreversible processes, respectively. The most common examples of such materials are elasto-viscoplastic or elasto-plastic materials. It has long been recognised by Rice,[21] Mandel,[14] or Suquet[28] that the exact description of the effective constitutive relations of such composites requires the determination of all microscopic plastic strains *at the unit-cell level.* For structural computations, the consequence of this theoretical result is that the number of integration points required in the analysis is equal to the product of the number of integration points at all scales, which is prohibitively large. With the increase in computational power, numerical strategies for solving these coupled problems have been proposed (see Feyel and Chaboche[7] or Terada and Kikuchi[32] for instance) but are so far limited by the formidable size of the corresponding problems.

In order to derive constitutive models of the effective behaviour of composites which are both useable and reasonably accurate, one has to resort to approximations. The *Transformation Field Analysis* (TFA) originally proposed by Dvorak and Benveniste[5] is an elegant way of reducing the number of macroscopic internal variables by assuming the *microscopic* fields of internal variables to be piecewise uniform. It has been extended by Fish *et al.*[8] to periodic composites using asymptotic expansions. Assuming the eigenstrains to be uniform within each individual constituent, Fish *et al.*[8] derived an approximate scheme which they called, for a two-phase material, the "two-point homogenisation scheme". The original scheme and this extended scheme have been incorporated successfully in structural computations.[6,9,11] However, it has been noticed[4,16,30] that the application of the TFA to two-phase systems may require, in certain circumstances, a subdivision of each individual phase into several (and sometimes numerous) subdomains to obtain a satisfactory description of the effective behaviour of the composite. The need for a finer subdivision of the phases stems from the intrinsic non-uniformity of the plastic strain field which can be highly heterogeneous even within a single material phase. As a consequence, the number of internal variables needed to achieve a reasonable accuracy in the effective constitutive relations, although finite, is prohibitively high.

In order to reproduce accurately the actual effective behaviour of the composite, it is important to capture correctly the heterogeneity of the plastic strain field. This observation has motivated the introduction in Refs. 16 and 18 of *non-uniform transformation fields*. More specifically the (visco)plastic strain within each phase is decomposed on a finite set of *plastic modes* which can present large deviations from uniformity. An approximate effective model for the composite can be derived from this decomposition where the internal variables are the components of the (visco)plastic strain field on the (visco)plastic modes. This theory is called the *Non-uniform Transformation Field Analysis* (NTFA). For two-phase composites (non-linear matrix and elastic fibres), comparison of the classical TFA, and of the NTFA with numerical simulations of the response of a unit-cell under monotone or cyclic loadings, has shown the accuracy of the NTFA.[18] The present study is devoted to the presentation of the NTFA and to its implementation into a macroscopic structural Finite Element (FE) analysis. It will be shown that the NTFA not only provides accurate predictions for the effective behaviour of composite materials, which is its initial goal, but also provides an accurate approximation of the local fields which are the quantities of interest in predicting the lifetime of structures.

2. Structural Problems with Multiple Scales

2.1. *Homogenisation and two-scale expansions*

Structures made of composite materials naturally involve two very different length-scales. The largest scale (the *macroscopic* scale) is related to the structure itself and is characterised by length L (Fig. 1). The second and smallest scale (the *microscopic* scale) is related to the size of the heterogeneities in the composite material (typically the fibre scale in fibre-reinforced structures). The typical length at this scale is denoted by d. In the fibre-reinforced laminates, d is of the order of the fibre diameter, whereas L is typically related to the thickness, or length, of the layered structure. When the scales are "well-separated", i.e. when the ratio $\eta = d/L$ is small ($\eta \ll 1$), one can expect all details about the microstructure to be "smeared out". In other words, the response of the structure at the macroscale can be computed by replacing the very contrasted physical properties of

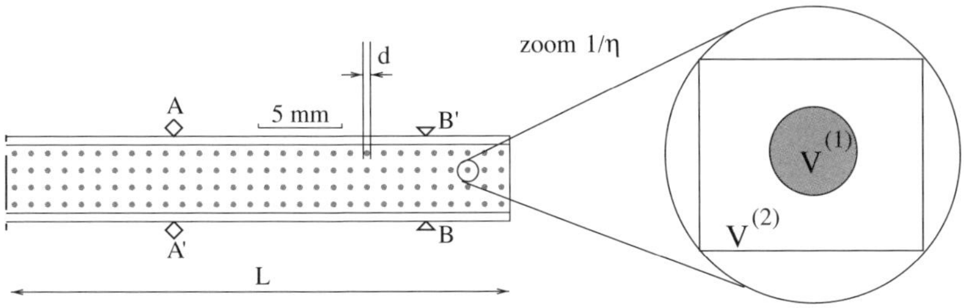

Fig. 1. Composite structure (left) and unit-cell (right).

the individual constituents by *effective or homogenised properties* (at the macroscale).

The aim of the mathematical theories of *homogenisation* is to determine exactly or to bound these effective properties from the information available, often partially, on the individual constituents themselves and on their arrangement (microstructure). However, if the effective properties are sufficient for the analysis performed in the linear range (stiffness of a composite structure, few first eigenfrequencies...) where the structure responds macroscopically as a whole, in many problems of engineering interest it is essential to take into consideration not only averaged fields, or effective properties, but also full *local fields*. Damage or fracture for instance are dramatically dependent on the local details of the strain or stress fields. The procedure by which the local fluctuations of fields are reconstructed from their macroscopic average is sometimes called *localisation*, and one important objective of the present approach is to propose a simple localisation rule for strain and stress fields.

The microstructure of periodic composites is completely known as soon as the geometry of a single unit-cell V is prescribed. For such composites, homogenisation results can be obtained heuristically by means of two-scale expansions making use of the fact that the parameter $\eta = d/L$ is small and that the geometry (and therefore the fields) are periodic at the microscopic scale (Sanchez-Palencia[24] and Bensoussan *et al.*[2]). Rigorous mathematical techniques have been developed to establish convergence theorems which usually confirm that homogenisation results obtained by asymptotic expansions usually hold true (see for instance Tartar[31]).

A brief reminder (by no means exhaustive) about two-scale expansions is given now. A function f defined on the macroscopic structure has variations at the two different spatial scales and can be denoted as $f(\boldsymbol{X}, \boldsymbol{x})$ to highlight this dependence on both variables, where $\boldsymbol{X}$ denotes the macroscopic spatial variable (structural scale), whereas $\boldsymbol{x}$ denotes the microscopic variable (at the unit-cell level). A dependence of a function on the microscopic variable $\boldsymbol{x}$ corresponds to fast oscillations of this function at the macroscopic scale, whereas a dependence on the macroscopic variable $\boldsymbol{X}$ corresponds to slower variations at the structural level.

The scale ratio η is finite and different from 0 in the actual structure (even though it is convenient mathematically to consider that it tends to 0). Therefore, all mechanical fields (stress, strain, displacement, etc.) in the actual structure depend on this ratio. For instance, the displacement field and the stress field in the actual structure will be denoted by $\boldsymbol{u}^\eta$ and $\boldsymbol{\sigma}^\eta$. The homogenised relations are obtained by taking the limit of $\boldsymbol{u}^\eta$ and $\boldsymbol{\sigma}^\eta$ as η goes to 0 and by studying the set of equations satisfied by these limits. These limits can be determined by means of two-scale expansions. For any function f^η defined on the composite structure with finite scale ratio η, its two-scale expansion is defined as

$$f^\eta(\boldsymbol{X}) = \sum_{j=0}^{+\infty} \eta^j f^j\left(\boldsymbol{X}, \frac{\boldsymbol{X}}{\eta}\right), \tag{1}$$

where, by virtue of the periodicity of the microstructure, all functions $f^k(\boldsymbol{X}, \boldsymbol{x})$ are periodic with respect to the microscopic variable $\boldsymbol{x}$. Therefore, for a macroscopic point $\boldsymbol{X}$, the argument $\boldsymbol{X}/\eta$ of the functions f^j, denotes the location of $\boldsymbol{X}$ in the unit-cell at the microscopic scale.

Let us recall that, setting $g^\eta(\boldsymbol{X}) = g\left(\frac{\boldsymbol{X}}{\eta}\right)$ where g is periodic over the unit-cell, the limit of g^η as η goes to 0 is the average of g over the unit-cell. The convergence is weak (only in average) and not pointwise. Consequently, the limit of f^η as η goes to 0 is the average with respect to $\boldsymbol{x}$ of the zeroth order term in the expansion (1):

$$\lim_{\eta \to 0} f^\eta(\boldsymbol{X}) = \bar{f}^0(\boldsymbol{X}) = \frac{1}{|V|} \int_V f^0(\boldsymbol{X}, \boldsymbol{x})\, d\boldsymbol{x}.$$

The homogenised (or effective) relations for the composite are therefore the relations between the limits as η goes to 0 of the fields $\boldsymbol{\sigma}^\eta$ and $\boldsymbol{\varepsilon}^\eta$, or equivalently between the averages of the zeroth order terms in the expansion of the stress field and strain field (or strain-rate field), and additional internal variables $\boldsymbol{\alpha}$, depending on the constitutive relations of the individual constituents which remain to be specified (see Sec. 2.2).

To understand how these zeroth order terms behave, one has to expand the unknown displacement, strain and stress fields $\boldsymbol{u}^\eta, \boldsymbol{\varepsilon}^\eta$, and $\boldsymbol{\sigma}^\eta$ in powers of η, after due account of the equations satisfied by these fields. In addition to the constitutive equations (to be specified), these equations are the compatibility equations and the equilibrium equations:

$$\varepsilon_{ij}^\eta = \frac{1}{2}\left(\frac{du_i^\eta}{dX_j} + \frac{du_j^\eta}{dX_i}\right), \quad \frac{d\sigma_{ij}^\eta}{dX_j} + F_i = 0, \tag{2}$$

where $\boldsymbol{F}$ denote the body forces applied to the structure. The derivation of a two-scale function $f(\boldsymbol{X}, \boldsymbol{x})$ which is periodic with respect to $\boldsymbol{x}$ with $\boldsymbol{x} = \boldsymbol{X}/\eta$ is performed according to the chain-rule:

$$\frac{d}{d\boldsymbol{X}} = \frac{\partial}{\partial \boldsymbol{X}} + \frac{1}{\eta}\frac{\partial}{\partial \boldsymbol{x}}.$$

Applying this derivation rule to the double-scale expansion of $\boldsymbol{u}^\eta$, $\boldsymbol{\varepsilon}^\eta$ and $\boldsymbol{\sigma}^\eta$:

$$\left.\begin{aligned}
\boldsymbol{u}^\eta(\boldsymbol{X}) &= \boldsymbol{u}\left(\boldsymbol{X}, \frac{\boldsymbol{X}}{\eta}\right) = \sum_{k=0}^{\infty} \eta^k \boldsymbol{u}^k(\boldsymbol{X}, \boldsymbol{x}),\\
\boldsymbol{\varepsilon}^\eta(\boldsymbol{X}) &= \boldsymbol{\varepsilon}\left(\boldsymbol{X}, \frac{\boldsymbol{X}}{\eta}\right) = \sum_{k=0}^{\infty} \eta^k \boldsymbol{\varepsilon}^k(\boldsymbol{X}, \boldsymbol{x}),\\
\boldsymbol{\sigma}^\eta(\boldsymbol{X}) &= \boldsymbol{\sigma}\left(\boldsymbol{X}, \frac{\boldsymbol{X}}{\eta}\right) = \sum_{k=0}^{\infty} \eta^k \boldsymbol{\sigma}^k(\boldsymbol{X}, \boldsymbol{x}),
\end{aligned}\right\} \tag{3}$$

one obtains the expansion of the compatibility and equilibrium equations in powers of η:

$$\left.\begin{aligned}
&\text{Order } -1: && \boldsymbol{\varepsilon}_{\boldsymbol{x}}(\boldsymbol{u}^0) = 0, \quad \operatorname{div}_{\boldsymbol{x}} \boldsymbol{\sigma}^0 = 0,\\
&\text{Order } 0: && \boldsymbol{\varepsilon}^0 = \boldsymbol{\varepsilon}_{\boldsymbol{X}}(\boldsymbol{u}^0) + \boldsymbol{\varepsilon}_{\boldsymbol{x}}(\boldsymbol{u}^1),\\
& && \operatorname{div}_{\boldsymbol{X}} \boldsymbol{\sigma}^0 + \operatorname{div}_{\boldsymbol{x}} \boldsymbol{\sigma}^1 + \boldsymbol{F} = 0,\\
& && \boldsymbol{\sigma}^0, \boldsymbol{\varepsilon}^0, \text{ and } \boldsymbol{\alpha}^0 \text{ satisfy the constitutive relations.}
\end{aligned}\right\} \tag{4}$$

Similar equations corresponding to higher-order terms in the expansions can be obtained in the same way. The operators $\boldsymbol{\varepsilon}_{\boldsymbol{x}}$ and $\operatorname{div}_{\boldsymbol{x}}$ in (4) stand for the deformation and divergence operators with respect to the microscopic variable $\boldsymbol{x}$ (with similar conventions for these operators with respect to the macroscopic variable $\boldsymbol{X}$). The constitutive equations of the phases may involve internal variables, in which case the zeroth order terms of these internal variables also enter the relations between $\boldsymbol{\sigma}^0$ and ε^0.

It follows from the first equation of the first line in (4) that $\boldsymbol{u}^0(\boldsymbol{X},\boldsymbol{x}) = \boldsymbol{u}^0(\boldsymbol{X})$. $\boldsymbol{u}^0$ has no dependence on the microscopic variable (no fast oscillations in the displacement field). In addition, taking the average over the unit-cell of the first equation at order 0 (second line in (4)), and taking into account the fact that the average of the gradient of a periodic function vanishes identically, one obtains that

$$\varepsilon_{\boldsymbol{X}}(\boldsymbol{u}^0)(\boldsymbol{X}) = \bar{\varepsilon}^0(\boldsymbol{X}), \tag{5}$$

where an overlined letter denotes an averaged quantity

$$\bar{\varepsilon}^0(\boldsymbol{X}) = \left\langle \varepsilon^0(\boldsymbol{X},.)\right\rangle \quad \text{with } \langle f\rangle = \frac{1}{|V|}\int_V f(\boldsymbol{x})\, d\boldsymbol{x}.$$

In other words, the macroscopic strain $\varepsilon_{\boldsymbol{X}}(\boldsymbol{u}^0)$ is the average over the unit-cell of the zeroth order term in the expansion of the strain field ε^η.

Unlike the displacement field, the zeroth order terms $\boldsymbol{\sigma}^0$ and ε^0 of the stress and strain fields have microscopic fluctuations (i.e. they depend on both the macroscopic and the microscopic variables). It follows from the second equation in the first line of (4) that $\boldsymbol{\sigma}^0$ is self-equilibrated at the microscopic scale, whichever body forces $\boldsymbol{F}$ are applied to the structure at the macroscopic scale. Taking the average over the unit-cell of the third line in (4), and noting that the average of the divergence of a periodic field vanishes identically, one finds that the average stress $\bar{\boldsymbol{\sigma}}^0 = \left\langle \boldsymbol{\sigma}^0\right\rangle$ satisfies the macroscopic equilibrium equations:

$$\operatorname{div}_{\boldsymbol{X}} \bar{\boldsymbol{\sigma}}^0 + \langle \boldsymbol{F}\rangle = 0. \tag{6}$$

The two equations (5) and (6) are valid irrespective of the constitutive behaviour of the phases. The homogenised, or effective, constitutive relations relate the average stress $\bar{\boldsymbol{\sigma}}^0$ and the average strain $\bar{\varepsilon}^0$. The determination of these relations requires, in principle, a complete knowledge of the fields $\boldsymbol{\sigma}^0$ and ε^0 with all their microscopic fluctuations. The dependence of these fields on the macroscopic variable $\boldsymbol{X}$ is known by solving the equilibrium problem for the structure subjected to the imposed macroscopic loading and where the effective constitutive relations are used for the composite material. Their dependence on the microscopic variable is known by solving the so-called *local problem* (or unit-cell problem), where the macroscopic variable $\boldsymbol{X}$ is only a parameter and will be omitted for clarity:

$$\left.\begin{array}{c}\varepsilon^0(\boldsymbol{x}) = \bar{\varepsilon}^0 + \varepsilon_{\boldsymbol{x}}(\boldsymbol{u}^1(\boldsymbol{x})) \text{ in } V \quad \text{where } \boldsymbol{u}^1 \text{ is periodic,}\\ \mathrm{div}_{\boldsymbol{x}}\, \boldsymbol{\sigma}^0 = 0 \text{ in } V, \quad \boldsymbol{\sigma}^0 \cdot \boldsymbol{n} \quad \text{antiperiodic on } \partial V,\\ \boldsymbol{\sigma}^0, \varepsilon^0, \text{and } \boldsymbol{\alpha}^0 \text{ are related by the constitutive equations of the phases.}\end{array}\right\} \tag{7}$$

The antiperiodicity condition for the traction $\boldsymbol{\sigma}^0 \cdot \boldsymbol{n}$ on ∂V is derived from the periodicity of $\boldsymbol{\sigma}^0$ and the antiperiodicity of $\boldsymbol{n}$ on opposite sides of the unit-cell V. The first line in (7) can be replaced by $\langle \varepsilon^0 \rangle = \bar{\varepsilon}^0$ and periodicity conditions. The constitutive relations of the phases have to be specified in order to further exploit these relations. For simplicity, the zeroth order terms ε^0 and $\boldsymbol{\sigma}^0$ will simply be denoted by ε and $\boldsymbol{\sigma}$ in the rest of the paper and the dependence on the variable $\boldsymbol{X}$ will be omitted in the rest of this section.

2.2. *Individual constituents*

As already noted, the microstructure of periodic composites is completely specified by the knowledge of a unit-cell V, which plays, for periodic media, a role parallel to that of a *representative volume element* (rve) in homogenisation theories for random media. The unit-cell V is occupied by N homogeneous phases $V^{(r)}$ with characteristic function $\chi^{(r)}(\boldsymbol{x})$ and volume fraction $c^{(r)}$:

$$\chi^{(r)}(\boldsymbol{x}) = \begin{cases} 1 & \text{if } \boldsymbol{x} \in V^{(r)}, \\ 0 & \text{otherwise,} \end{cases} \qquad c^{(r)} = \langle \chi^{(r)} \rangle.$$

The average of a field f over the unit-cell V and over each individual phase $V^{(r)}$ is denoted by overlined letters $\bar{f}$ and $\bar{f}^{(r)}$:

$$\bar{f} = \langle f \rangle = \sum_{r=1}^{N} c^{(r)} \bar{f}^{(r)}, \quad \bar{f}^{(r)} = \langle f \rangle_r = \frac{1}{|V^{(r)}|} \int_{V^{(r)}} f(\boldsymbol{x})\, d\boldsymbol{x}.$$

The composite structures of interest for this study may be subjected to thermomechanical loadings. Therefore, the validity of the constitutive relations of the individual constituents must cover a wide range of temperature and strain-rates. For simplicity, attention will be restricted here to isotropic materials.

We shall adopt in the sequel a viscoplastic model with non-linear kinematic hardening proposed by Chaboche,[3] generalising the Armstrong–Fredericks constitutive relations:

$$\left.\begin{aligned}
&\boldsymbol{\sigma} = \boldsymbol{L} : (\varepsilon - \varepsilon_{\mathsf{vp}}),\\
&\dot{\varepsilon}_{\mathsf{vp}} = \frac{3}{2}\dot{p}\frac{\boldsymbol{s} - \boldsymbol{\mathfrak{X}}}{(\sigma - \mathfrak{X})_{\mathsf{eq}}}, \quad \dot{p} = \dot{\varepsilon}_0 \left[\frac{((\sigma - \mathfrak{X})_{\mathsf{eq}} - R)^+}{\sigma_0}\right]^n,\\
&\dot{\boldsymbol{\mathfrak{X}}} = \frac{2}{3}H\dot{\varepsilon}_{\mathsf{vp}} - \xi\boldsymbol{\mathfrak{X}}\dot{p}, \quad R = R(p),
\end{aligned}\right\} \tag{8}$$

where $(.)^+$ denotes the McCauley bracket (positive part):

$$A^+ = A \text{ if } A \geq 0, \quad A^+ = 0 \text{ if } A \leq 0.$$

When the phases are isotropic, their elastic properties are characterised by a bulk modulus k and a shear modulus G. Kinematic hardening effects are characterised by the back-stress $\boldsymbol{\mathfrak{X}}$, whereas isotropic hardening manifests itself through the dependence of the yield stress $R(p)$ on the cumulated viscoplastic strain p defined as $\dot{p} = (2/3\dot{\varepsilon}_{\mathsf{vp}} : \dot{\varepsilon}_{\mathsf{vp}})^{1/2}$. To simplify notations it is useful to introduce the viscoplastic potential:

$$\psi(\boldsymbol{\mathcal{A}}, R) = \frac{\sigma_0\dot{\varepsilon}_0}{n+1}\left[\frac{(\mathcal{A}_{\mathsf{eq}} - R)^+}{\sigma_0}\right]^{n+1}, \tag{9}$$

by means of which the second line of the constitutive relations (8) can be written as

$$\dot{\varepsilon}_{\mathsf{vp}} = \frac{\partial\psi}{\partial\boldsymbol{\mathcal{A}}}(\boldsymbol{\sigma} - \boldsymbol{\mathfrak{X}}, R), \quad \dot{p} = -\frac{\partial\psi}{\partial R}(\boldsymbol{\sigma} - \boldsymbol{\mathfrak{X}}, R). \tag{10}$$

The model (8) (and subsequent refinements which will not be considered here) is commonly used in the analysis of the lifetime of metallic or polymeric structures under repeated thermomechanical loadings (see Samrout *et al.*[23] and Amiable *et al.*[1] among others). The material parameters of the model, namely the elastic moduli $\boldsymbol{L}$, the rate-sensitivity exponent n, the flow-stress σ_0, the isotropic hardening function $R(p)$, the kinematic hardening modulus H, and the spring-back coefficient ξ, are strongly temperature-dependent. For simplicity, thermal loadings and thermal strains will not be considered in the present analysis, but the strong temperature-dependence of the material parameters will be accounted for. For instance, the rate-sensitivity exponent n can vary from 5 to 20 for Aluminum alloys when the temperature varies from 20°C to 500°C. The method proposed here will make use of certain objects, called plastic modes, identified at a given temperature but used over the whole range of temperature with the appropriate material parameters. In other words, these plastic modes do not need to be identified at each temperature.

2.3. *Unit-cell problem: Effective response of heterogeneous materials*

As seen in Sec. 2.1, the first-order terms of the stress and strain field solve a unit-cell problem (also called the *local problem*) consisting of the equilibrium and compatibility equations (7) and the constitutive relations (10). All material properties are assumed to be uniform in each individual phase:

$$\boldsymbol{L}(\boldsymbol{x}) = \sum_{r=1}^{N} \boldsymbol{L}^{(r)} \chi^{(r)}(\boldsymbol{x}), \quad \psi(\boldsymbol{x}, \boldsymbol{\mathcal{A}}, R) = \sum_{r=1}^{N} \psi^{(r)}(\boldsymbol{\mathcal{A}}, R) \chi^{(r)}(\boldsymbol{x}).$$

The overall stress $\bar{\boldsymbol{\sigma}}$ and the overall strain $\bar{\varepsilon}$ are the averages of their local counterparts $\boldsymbol{\sigma}$ and ε (for simplicity the dependence on the macroscopic variable $\boldsymbol{X}$ of all fields will be omitted):

$$\bar{\boldsymbol{\sigma}} = \langle \boldsymbol{\sigma} \rangle, \quad \bar{\varepsilon} = \langle \varepsilon \rangle. \tag{11}$$

The homogenised effective relations are the relations between the macroscopic stress $\bar{\boldsymbol{\sigma}}$ (and its time-derivatives) and the overall strain $\bar{\varepsilon}$ (and its time-derivatives).

To find these relations, a history of macroscopic strain $\bar{\varepsilon}(t)$ is prescribed on a time interval $[0, T]$ generating a time-dependent local stress field $\boldsymbol{\sigma}(\boldsymbol{x}, t)$. Its average $\bar{\boldsymbol{\sigma}}(t)$ is the macroscopic stress whose history is therefore related to the history of $\bar{\varepsilon}(t)$.

The local problem to be solved to determine $\boldsymbol{\sigma}(t)$ reads:

$$\left.\begin{aligned} \boldsymbol{\sigma}(\boldsymbol{x}, t) &= \boldsymbol{L}(\boldsymbol{x}) : (\varepsilon(\boldsymbol{x}, t) - \varepsilon_{\mathsf{vp}}(\boldsymbol{x}, t)), \\ \dot{\varepsilon}_{\mathsf{vp}}(\boldsymbol{x}, t) &= \frac{\partial \psi}{\partial \boldsymbol{\mathcal{A}}}(\boldsymbol{\sigma}(\boldsymbol{x}, t) - \boldsymbol{\mathfrak{X}}(\boldsymbol{x}, t), R(\boldsymbol{x}, t)), \\ \operatorname{div}_{\boldsymbol{x}} \boldsymbol{\sigma}(\boldsymbol{x}, t) &= 0, \quad \langle \varepsilon(t) \rangle = \bar{\varepsilon}(t), \quad \text{boundary conditions.} \end{aligned}\right\} \tag{12}$$

In view of the local periodicity of the structure, periodic boundary conditions are assumed on the boundary of the unit-cell.

The average of the local stress field $\boldsymbol{\sigma}(\boldsymbol{x}, t)$ is the macroscopic stress response of the composite to a prescribed history of macroscopic strain $\bar{\varepsilon}(t)$. Unfortunately, except in very specific situations (e.g. laminates), these effective relations for nonlinear materials cannot be given in closed form. They are accessible only numerically, along a prescribed path. An important consequence of this observation for the computational analysis of a composite structure, is that the macroscopic and microscopic levels are intimately coupled. At the structural level, the macroscopic strain $\bar{\varepsilon}(\boldsymbol{X}, t)$ is a function of position, and a problem similar to (12) has to be solved at every macroscopic point $\boldsymbol{X}$ or, in a computational analysis, at every macroscopic integration point. As pointed out by Fish and Shek,[8] history data has to be updated at a number of integration points equal to the product of the numbers of integration points at all scales at each time increment.

To avoid the computational difficulty associated with the coupling of scales, approximations are introduced to render the resolution of the local problem (12) possible in closed form or amenable to simple algebra.

2.4. *An auxiliary elasticity problem*

Before introducing approximate resolution schemes for the local problem (12), it is important to emphasise that the stress and strain fields are solutions to *a linear elasticity problem* on the unit-cell when the fields of internal variables are known. Indeed, assuming that the viscoplastic part of the strain is prescribed, the stress and strain fields in the rve solve the following linear elastic problem, with appropriate boundary conditions (for simplicity, the time dependence of the fields is omitted):

$$\boldsymbol{\sigma}(\boldsymbol{x}) = \boldsymbol{L}(\boldsymbol{x}) : (\varepsilon(\boldsymbol{x}) - \varepsilon_{\mathsf{vp}}(\boldsymbol{x})), \quad \operatorname{div} \boldsymbol{\sigma}(\boldsymbol{x}) = 0, \quad \langle \varepsilon \rangle = \bar{\varepsilon}. \tag{13}$$

Assume that $\varepsilon_{\mathsf{vp}}(\boldsymbol{x})$ is known. It plays the role of a thermal strain in thermoelasticity when the temperature is prescribed, or that of a *transformation strain* in phase transformation problems.

The solution of (13) can be expressed in terms of $\varepsilon_{\mathsf{vp}}$ and $\bar{\varepsilon}$ by a straightforward application of the superposition principle. Consider first the case where $\varepsilon_{\mathsf{vp}}$ is identically $\mathbf{0}$. Problem (13) is then a standard elasticity problem and its solution can be expressed by means of the elastic strain-localisation tensor $\boldsymbol{A}(\boldsymbol{x})$ as

$$\varepsilon(\boldsymbol{x}) = \boldsymbol{A}(\boldsymbol{x}) : \bar{\varepsilon}. \tag{14}$$

Consider next the case where $\bar{\varepsilon} = 0$ and $\varepsilon_{\mathsf{vp}}(\boldsymbol{x})$ is arbitrary. Problem (13) can then be written as an elasticity problem with eigenstress (sometimes called polarisation stress) $\boldsymbol{\tau}(\boldsymbol{x}) = -\boldsymbol{L}(\boldsymbol{x}) : \varepsilon_{\mathsf{vp}}(\boldsymbol{x})$;

$$\boldsymbol{\sigma}(\boldsymbol{x}) = \boldsymbol{L}(\boldsymbol{x}) : \varepsilon(\boldsymbol{x}) + \boldsymbol{\tau}(\boldsymbol{x}), \quad \operatorname{div} \boldsymbol{\sigma}(\boldsymbol{x}) = 0, \quad \langle \varepsilon \rangle = \mathbf{0}. \tag{15}$$

Introducing the non-local elastic Green operator $\boldsymbol{\Gamma}(\boldsymbol{x}, \boldsymbol{x}')$ of the non-homogeneous elastic medium, the solution of (15) can be expressed as

$$\varepsilon(\boldsymbol{x}) = -\boldsymbol{\Gamma} * \boldsymbol{\tau}(\boldsymbol{x}), \quad \text{where} \quad \boldsymbol{\Gamma} * \boldsymbol{\tau}(\boldsymbol{x}) \stackrel{\text{def}}{=} \frac{1}{|V|} \int_V \boldsymbol{\Gamma}(\boldsymbol{x}, \boldsymbol{x}') : \boldsymbol{\tau}(\boldsymbol{x}')\, d\boldsymbol{x}'. \tag{16}$$

The superposition principle applied to (14) and (16) gives that the solution of (13) reads as

$$\varepsilon(\boldsymbol{x}) = \boldsymbol{A}(\boldsymbol{x}) : \bar{\varepsilon} + \frac{1}{|V|} \int_V \boldsymbol{D}(\boldsymbol{x}, \boldsymbol{x}') : \varepsilon_{\mathsf{vp}}(\boldsymbol{x}')\, d\boldsymbol{x}' = \boldsymbol{A}(\boldsymbol{x}) : \bar{\varepsilon} + \boldsymbol{D} * \varepsilon_{\mathsf{vp}}(\boldsymbol{x}), \tag{17}$$

where the non-local operator $\boldsymbol{D}(\boldsymbol{x}, \boldsymbol{x}') = \boldsymbol{\Gamma}(\boldsymbol{x}, \boldsymbol{x}') : \boldsymbol{L}(\boldsymbol{x}')$ gives the strain at point $\boldsymbol{x}$ created by a transformation strain at point $\boldsymbol{x}'$.

3. Non-Uniform Transformation Field Analysis (NTFA)

3.1. *Motivation: Approximate resolution of the local problem*

The *Transformation Field Analysis* (TFA), originally developed by Dvorak[6] (see also references herein), is based on the assumption that the viscoplastic strains are *uniform* within each individual domain $V^{(r)}$:

$$\varepsilon_{\mathsf{vp}}(\boldsymbol{x}, t) = \sum_{r=1}^{N} \bar{\varepsilon}_{\mathsf{vp}}^{(r)}(t) \chi^{(r)}(\boldsymbol{x}). \tag{18}$$

The determination of the field $\varepsilon_{\mathsf{vp}}(\boldsymbol{x})$ is therefore reduced to the determination of the tensorial variables $\bar{\varepsilon}_{\mathsf{vp}}^{(r)}$, $r = 1, \ldots, N$. Using this decomposition, the macroscopic stress reads as

$$\bar{\boldsymbol{\sigma}} = \sum_{r=1}^{N} c^{(r)} \bar{\boldsymbol{\sigma}}^{(r)}, \quad \bar{\boldsymbol{\sigma}}^{(r)} = \langle \boldsymbol{\sigma} \rangle_r = \boldsymbol{L}^{(r)} : (\bar{\varepsilon}^{(r)} - \bar{\varepsilon}_{\mathsf{vp}}^{(r)}), \tag{19}$$

where

$$\bar{\varepsilon}^{(r)} = \langle \varepsilon \rangle_r = \boldsymbol{A}^{(r)} : \bar{\varepsilon} + \sum_{s=1}^{N} \boldsymbol{D}^{(rs)} : \bar{\varepsilon}_{\mathsf{vp}}^{(s)}, \quad r = 1, \ldots, N, \tag{20}$$

and

$$\boldsymbol{A}^{(r)} = \langle \boldsymbol{A} \rangle_r, \tag{21}$$

$$\boldsymbol{D}^{(rs)} = \frac{1}{c^{(r)}} \frac{1}{|V|} \frac{1}{|V|} \int_V \int_V \chi^{(r)}(\boldsymbol{x}) \boldsymbol{\Gamma}(\boldsymbol{x}, \boldsymbol{x}') : \boldsymbol{L}(\boldsymbol{x}') \chi^{(s)}(\boldsymbol{x}')\, d\boldsymbol{x}'\, d\boldsymbol{x}. \tag{22}$$

The evolution of $\bar{\boldsymbol{\varepsilon}}_{\mathsf{vp}}^{(r)}$ is governed by the constitutive relations of the individual phases applied to the average stresses and thermodynamic forces on the phases. Assuming that these constitutive relations take the form (8) (or (10)), with material properties labelled by the phase r, the evolution equations for the generalised variables $\bar{\boldsymbol{\varepsilon}}_{\mathsf{vp}}^{(r)}$ read as:

$$\left.\begin{aligned} \dot{\bar{\boldsymbol{\varepsilon}}}_{\mathsf{vp}}^{(r)} &= \frac{\partial \psi^{(r)}}{\partial \boldsymbol{\mathcal{A}}}(\bar{\boldsymbol{\sigma}}^{(r)} - \bar{\boldsymbol{\mathfrak{X}}}^{(r)}, \bar{R}^{(r)}), \quad \dot{\bar{p}}^{(r)} = -\frac{\partial \psi^{(r)}}{\partial R}(\bar{\boldsymbol{\sigma}}^{(r)} - \bar{\boldsymbol{\mathfrak{X}}}^{(r)}, \bar{R}^{(r)}), \\ \dot{\bar{\boldsymbol{\mathfrak{X}}}}^{(r)} &= \frac{2}{3} H^{(r)} \dot{\bar{\boldsymbol{\varepsilon}}}_{\mathsf{vp}}^{(r)} - \xi^{(r)} \bar{\boldsymbol{\mathfrak{X}}}^{(r)} \dot{\bar{p}}^{(r)}, \quad \bar{R}^{(r)} = R^{(r)}(\bar{p}^{(r)}). \end{aligned}\right\} \tag{23}$$

When a prescribed path $\bar{\boldsymbol{\varepsilon}}(t)$, $t \in [0, T]$ is prescribed in the space of macroscopic strains, the corresponding history of the average strains $\bar{\boldsymbol{\varepsilon}}^{(r)}(t)$ and viscoplastic strains $\bar{\boldsymbol{\varepsilon}}_{\mathsf{vp}}^{(r)}(t)$ in each phase can be obtained by integrating in time the systems of differential Eqs. $(19)_2$, (20), and (23).

A nice feature of the TFA is that its implementation is relatively easy. However, applying the TFA to two-phase systems using plastic strains which are uniform in each phase yields predictions of the overall behaviour of the composite, which can be unreasonably stiff.[4,30] The origin of this excessive stiffness is to be seeked in the intrinsic non-uniformity (in space) of the actual plastic strain field which can be highly heterogeneous even within a single material phase, a feature which is disregarded by the TFA. Dvorak *et al.*[6] have obtained better results by subdividing each phase into several subdomains. Unfortunately, although the refinement does improve the predictions, a rather fine subdivision of the phases is often necessary to achieve a satisfactory agreement,[16] resulting in a prohibitive increase of the number of internal variables entering the effective constitutive relations. These observations have motivated the development of alternative approximate schemes.[18]

3.2. *Non-uniform transformation fields*

The aim of the NTFA is to account for the non-uniformity of the plastic strain field. The field of anelastic strains is decomposed on a set of fields, called *plastic modes*, $\boldsymbol{\mu}^{(k)}$:

$$\boldsymbol{\varepsilon}_{\mathsf{vp}}(\boldsymbol{x},t) = \sum_{k=1}^{M} \varepsilon_{\mathsf{vp}}^{(k)}(t)\boldsymbol{\mu}^{(k)}(\boldsymbol{x}). \tag{24}$$

Unlike in the classical Transformation Field Analysis, the modes $\boldsymbol{\mu}^{(k)}$ are *non-uniform* (not even piecewise uniform) and depend on the spatial variable $\boldsymbol{x}$. The idea is that their spatial variations capture the salient features of the plastic flow in the unit-cell. They can be determined either analytically or numerically. Their total number, M, can be different (larger or smaller) from the number N of phases. The $\boldsymbol{\mu}^{(k)}$ are tensorial fields, whereas the corresponding variables $\varepsilon_{\mathsf{vp}}^{(k)}$ are scalar variables.

Further assumptions will be made to simplify the theory:

H1: *The support of each mode is entirely contained in a single material phase.* It follows from this assumption that one can attach to each mode a characteristic function $\chi^{(k)}$, elastic moduli $\boldsymbol{L}^{(k)}$, and a dissipation potential $\psi^{(k)}$ which are those of the phase supporting this mode. $M(r)$ will denote the number of modes with support in a given phase $V^{(r)}$.

H2: *The modes are incompressible:*

$$\mathrm{tr}(\boldsymbol{\mu}^{(k)}) = 0. \tag{25}$$

This assumption stems from the fact that the $\boldsymbol{\mu}^{(k)}$ are meant to represent (visco)plastic strain fields. As a consequence of this assumption, the field $\boldsymbol{\varepsilon}_{\mathsf{vp}}$ given by the decomposition (24) is incompressible, expected, with no restriction on the components $\varepsilon_{\mathsf{vp}}^{(k)}$.

H3: *The modes are orthogonal:*

$$\langle \boldsymbol{\mu}^{(k)} : \boldsymbol{\mu}^{(\ell)} \rangle = 0 \quad \text{when } k \neq \ell. \tag{26}$$

This condition is obviously met when the modes have their support in different material phases but has to be imposed to the modes when their support is in the same material phase.

H4: *The modes are normalised:*

$$\left\langle \mu_{\mathsf{eq}}^{(k)} \right\rangle = 1. \tag{27}$$

3.3. *Reduced variables and influence factors*

Using the decomposition (24) into (17), one obtains that

$$\boldsymbol{\varepsilon}(\boldsymbol{x}) = \boldsymbol{A}(\boldsymbol{x}) : \bar{\boldsymbol{\varepsilon}} + \sum_{\ell=1}^{M} \boldsymbol{\eta}^{(\ell)}(\boldsymbol{x})\varepsilon_{\mathsf{vp}}^{(\ell)}, \tag{28}$$

where $\boldsymbol{\eta}^{(\ell)}(\boldsymbol{x}) = \boldsymbol{D} * \boldsymbol{\mu}^{(\ell)}(\boldsymbol{x})$ is the strain at point $\boldsymbol{x}$ due to the presence of an eigenstrain $\boldsymbol{\mu}^{(\ell)}(\boldsymbol{x}')$ at point $\boldsymbol{x}'$, the average strain $\bar{\varepsilon}$ being zero.

Upon multiplication of Eq. (28) by $\boldsymbol{\mu}^{(k)}$ and averaging over V, one obtains

$$e^{(k)} = \boldsymbol{a}^{(k)} : \bar{\varepsilon} + \sum_{\ell=1}^{M} D_N^{(k\ell)} \varepsilon_{\mathsf{vp}}^{(\ell)}, \tag{29}$$

where the *reduced strains* $e^{(k)}$, *the reduced localisation tensors* $\boldsymbol{a}^{(k)}$, and the influence factors $D_N^{(k\ell)}$ (N stands for NTFA) are defined as

$$e^{(k)} = \langle \boldsymbol{\mu}^{(k)} : \varepsilon \rangle, \quad \boldsymbol{a}^{(k)} = \langle \boldsymbol{\mu}^{(k)} : \boldsymbol{A} \rangle, \quad D_N^{(k\ell)} = \langle \boldsymbol{\mu}^{(k)} : \boldsymbol{\eta}^{(\ell)} \rangle. \tag{30}$$

By analogy with the equation defining the reduced strain $e^{(k)}$ in (30), one can define

$$e_{\mathsf{vp}}^{(k)} = \langle \boldsymbol{\mu}^{(k)} : \varepsilon_{\mathsf{vp}} \rangle = \langle \boldsymbol{\mu}^{(k)} : \boldsymbol{\mu}^{(k)} \rangle \varepsilon_{\mathsf{vp}}^{(k)} \quad \text{(no summation over } k). \tag{31}$$

Reduced stresses can be associated by duality to the generalised viscoplastic strains $\varepsilon_{\mathsf{vp}}^{(k)}$ (the notations are chosen so as to highlight the analogy between the reduced stress $\tau^{(k)}$ and the resolved shear stress on the kth system in crystal plasticity):

$$\tau^{(k)} = \langle \boldsymbol{\mu}^{(k)} : \boldsymbol{\sigma} \rangle, \quad x^{(k)} = \langle \boldsymbol{\mu}^{(k)} : \boldsymbol{\mathfrak{X}} \rangle. \tag{32}$$

3.4. *Constitutive relations for the reduced variables*

It remains to specify the *reduced constitutive relations* relating the reduced strains and stresses.

A first set of equations is obtained upon substitution of the stress–strain relation $(12)_1$ into the definition $(32)_1$ of the reduced stresses $\tau^{(k)}$:

$$\tau^{(k)} = \langle \boldsymbol{\mu}^{(k)} : \boldsymbol{L} : (\varepsilon - \varepsilon_{\mathsf{vp}}) \rangle .$$

Elastic isotropy of the phases and assumptions H1 and H2 for the modes $\boldsymbol{\mu}^{(k)}$ lead to

$$\tau^{(k)} = 2G^{(k)}(e^{(k)} - e_{\mathsf{vp}}^{(k)}), \tag{33}$$

where $G^{(k)}$ denotes the shear modulus of phase r containing the support of mode k.

The second set of equations concerns the evolution of the generalised variables $e_{\mathsf{vp}}^{(k)}$ and $x^{(k)}$. Using definition (31) of $e_{\mathsf{vp}}^{(k)}$ and Eqs. (9)

and (10) for the evolution of the viscoplastic strain field $\varepsilon_{\mathsf{vp}}(\boldsymbol{x})$, one obtains that

$$\dot{e}^{(k)}_{\mathsf{vp}} = \langle \boldsymbol{\mu}^{(k)} : \dot{\varepsilon}_{\mathsf{vp}} \rangle = \frac{3}{2}\left\langle \dot{p}\frac{\boldsymbol{\mu}^{(k)} : \boldsymbol{\mathcal{A}}}{\mathcal{A}_{\mathsf{eq}}} \right\rangle,$$
$$\boldsymbol{\mathcal{A}} = \boldsymbol{\sigma} - \mathfrak{X}, \dot{p} = -\frac{\partial \psi}{\partial R}(\mathcal{A}_{\mathsf{eq}}, R). \tag{34}$$

At this stage, an additional approximation must be introduced to derive a relation between the $\dot{e}^{(k)}_{\mathsf{vp}}$, the $\tau^{(\ell)}$, and $x^{(\ell)}$. Different approximations are discussed by Michel and Suquet[18] (uncoupled and coupled models) to which the reader is referred for further details. It follows from this work that the most accurate model is the so-called *coupled model* where the force acting on a mode is the quadratic average of all the generalised forces acting on all modes contained in the same phase. For a given phase r, the generalised force $\mathcal{A}^{(r)}$ is defined as

$$\mathcal{A}^{(r)} = \left(\sum_{k=1}^{M(r)} |\tau^{(k)} - x^{(k)}|^2 \right)^{1/2}. \tag{35}$$

In this relation $M(r)$ denotes the number of modes having their support in phase r. Then, the relation (34) is modified by replacing $\mathcal{A}_{\mathsf{eq}}$ by $\mathcal{A}^{(r)}$ and R by $\mathcal{R}^{(r)}$:

$$\dot{e}^{(k)}_{\mathsf{vp}} = \frac{3}{2}\dot{p}^{(r)}\frac{\tau^{(k)} - x^{(k)}}{\mathcal{A}^{(r)}}, \quad \dot{p}^{(r)} = -\frac{\partial \psi^{(r)}}{\partial R}(\mathcal{A}^{(r)}, \mathcal{R}^{(r)}),$$
$$\mathcal{R}^{(r)} = R^{(r)}(p^{(r)}), \tag{36}$$

where, again, r is the phase containing the support of $\boldsymbol{\mu}^{(k)}$. The plastic multiplier $\dot{p}^{(r)}$ is the same for all modes having their support in the same phase r.

Finally, in order to obtain an evolution equation for the $x^{(k)}$ the last equation in (8) is multiplied by $\boldsymbol{\mu}^{(k)}$ and averaged over V:

$$\dot{x}^{(k)} = \langle \boldsymbol{\mu}^{(k)} : \dot{\mathfrak{X}} \rangle = \frac{2}{3}H^{(k)}\dot{e}^{(k)}_{\mathsf{vp}} - \langle \dot{p}\xi\boldsymbol{\mu}^{(k)} : \mathfrak{X} \rangle. \tag{37}$$

Then, replacing as previously $\mathcal{A}_{\mathsf{eq}}$ by $\mathcal{A}^{(r)}$ and R by $\mathcal{R}^{(r)}$ in the expression of the plastic multiplier $\dot{p}$, one obtains

$$\dot{x}^{(k)} = \frac{2}{3}H^{(k)}\dot{e}^{(k)}_{\mathsf{vp}} - \dot{p}^{(r)}\xi^{(k)}x^{(k)}. \tag{38}$$

In summary, the constitutive relations for the model are

$$\left.\begin{array}{c} e^{(k)} = \boldsymbol{a}^{(k)} : \bar{\varepsilon} + \sum_{\ell=1}^{M} D_N^{(k\ell)} \varepsilon_{\mathsf{vp}}^{(\ell)}, \\ \tau^{(k)} = 2G^{(k)}(e^{(k)} - e_{\mathsf{vp}}^{(k)}), \\ \mathcal{A}^{(r)} = \left(\sum_{k=1}^{M(r)} |\tau^{(k)} - x^{(k)}|^2 \right)^{1/2}, \quad \mathcal{R}^{(r)} = R^{(r)}(p^{(r)}), \\ \dot{e}_{\mathsf{vp}}^{(k)} = \frac{3}{2}\dot{p}^{(r)} \frac{\tau^{(k)} - x^{(k)}}{\mathcal{A}^{(r)}}, \quad \dot{p}^{(r)} = \frac{\partial \psi^{(r)}}{\partial \mathcal{A}_{\mathsf{eq}}} \left(\mathcal{A}^{(r)}, \mathcal{R}^{(r)} \right), \\ \dot{x}^{(k)} = \frac{2}{3} H^{(k)} \dot{e}_{\mathsf{vp}}^{(k)} - \dot{p}^{(r)} \xi^{(k)} x^{(k)}. \end{array}\right\} \quad (39)$$

The system of differential equations (39) is to be solved at each integration point of the structure (macroscopic level). At each time increment, knowing the increment in macroscopic strain, the resolution of the system yields the $e_{\mathsf{vp}}^{(k)}$ from which the $\varepsilon_{\mathsf{vp}}^{(k)}$ can be obtained by inversion of (31).

Once the internal variables $\varepsilon_{\mathsf{vp}}^{(k)}$ are determined, the local stress field in the composite resulting from (13) and (28) reads as

$$\left.\begin{array}{l} \boldsymbol{\sigma}(\boldsymbol{x}, t) = \boldsymbol{L}(\boldsymbol{x}) : \boldsymbol{A}(\boldsymbol{x}) : \bar{\varepsilon}(t) + \sum_{k=1}^{M} \boldsymbol{\rho}^{(k)}(\boldsymbol{x}) \varepsilon_{\mathsf{vp}}^{(k)}(t), \\ \text{where} \quad \boldsymbol{\rho}^{(k)}(\boldsymbol{x}) = \boldsymbol{L}(\boldsymbol{x}) : \left(\boldsymbol{\eta}^{(k)}(\boldsymbol{x}) - \boldsymbol{\mu}^{(k)}(\boldsymbol{x}) \right). \end{array}\right\} \quad (40)$$

The effective constitutive relations for the composite are obtained by averaging this stress field:

$$\bar{\boldsymbol{\sigma}}(t) = \widetilde{\boldsymbol{L}} : \bar{\varepsilon}(t) + \sum_{k=1}^{M} \langle \boldsymbol{\rho}^{(k)} \rangle \varepsilon_{\mathsf{vp}}^{(k)}(t). \quad (41)$$

The localisation tensors $\boldsymbol{a}^{(k)}$, the influence factors $D_N^{(k\ell)}$, the effective stiffness $\widetilde{\boldsymbol{L}}$, and the tensors $\langle \boldsymbol{\rho}^{(k)} \rangle$ are computed once for all.

3.5. *Choice of the plastic modes*

The plastic modes are essential for the accuracy of the method. However, there is *no universal choice* for these modes and they should rather be chosen according to the type of loading which the structure is likely to be subjected to. This implies that the user has an *a priori* idea of the triaxiality of the macroscopic stress field, as well as of its intensity and its time history. For instance, when the structure schematically depicted in Fig. 1 is subjected to pure bending, the macroscopic stress is expected to have a strong uniaxial component. Therefore, the plastic modes should incorporate information about the response of the unit-cell under uniaxial tension (and compression if the response is not symmetric in tension/compression). But, close to points where the plate is supported, the macroscopic stress will likely exhibit a non-negligible amount of transverse shear and transverse normal stress so that plastic modes accounting for the unit-cell response under transverse shear and transverse tension-compression should also be present in the set of modes. Similarly, if one is interested in the response of the structure under monotone loading with limited amplitude, the information about the response of the unit-cell will be limited to certain monotone loading paths in stress space up to a limited amount of deformation.

Given the complexity of the microstructures under consideration, the plastic modes are not determined analytically but numerically from *actual viscoplastic strain fields* in the unit-cell. Different unit-cell responses along the different loading paths of macroscopic stresses stemming from the above qualitative analysis are determined numerically. Second, the plastic modes are extracted from the microscopic viscoplastic strain fields at a given macroscopic strain, which depends on the range of macroscopic strains which is expected in the structural computation. Different or additional loadings can be considered, depending on the problem and keeping in mind that it is desirable to approach as closely as possible the macroscopic loading paths expected at the different integration points of the composite structure.

One of the building assumption of the NTFA is the mode orthogonality (hypothesis H3). If this prerequisite is obviously met when the modes have their support in different material phases, it has to be imposed to the modes which have their support in the same material phase. Let $\boldsymbol{\theta}^{(k)}(\boldsymbol{x})$, $k = 1, \ldots, M_T(r)$ be potential candidates to be plastic modes in phase r. The procedure used to obtain these fields will be detailed in due time but they will not satisfy assumption H3 in general. The Karhunen–Loève decomposition (also known as the *proper orthogonal decomposition* or as the *principal component analysis*) is used to construct a set of (visco)plastic modes $\boldsymbol{\mu}^{(k)}(\boldsymbol{x})$, $k = 1, \ldots, M_T(r)$ from these fields $\boldsymbol{\theta}^{(k)}(\boldsymbol{x})$:

$$\boldsymbol{\mu}^{(k)}(\boldsymbol{x}) = \sum_{\ell=1}^{M_T(r)} v_\ell^{(k)} \boldsymbol{\theta}^{(\ell)}(\boldsymbol{x}), \tag{42}$$

where $\boldsymbol{v}^{(k)}$ and $\lambda^{(k)}$ are the eigenvectors and eigenvalues of the correlation matrix

$$\sum_{j=1}^{M_T(r)} g_{ij} v_j^{(k)} = \lambda^{(k)} v_i^{(k)}, \quad g_{ij} = \langle \boldsymbol{\theta}^{(i)} : \boldsymbol{\theta}^{(j)} \rangle. \tag{43}$$

It is straightforward to check that the resulting modes are orthogonal (as any set of eigenvectors of symmetric matrices):

$$\langle \boldsymbol{\mu}^{(k)} : \boldsymbol{\mu}^{(\ell)} \rangle = \lambda^{(k)} \quad \text{if } k = \ell, \ \text{otherwise } 0. \tag{44}$$

Another advantage of the Karhunen–Loève decomposition is that the NTFA model is almost insensitive to modes with small intensity, or in other words to modes $\boldsymbol{\mu}^{(k)}$ corresponding to small eigenvalues $\lambda^{(k)}$. Therefore, in practice, among the total $M_T(r)$ modes, it is sufficient to consider in the model the first $M(r)$ modes corresponding to the largest eigenvalues (see Ref. 22 for more details).

3.6. *Reduced localisation tensors and influence factors*

Once the plastic modes are chosen, the localisation and influence tensors can be determined by solving only the linear problems. The strain localisation tensor $\boldsymbol{A}$ is obtained by solving successively six linear elasticity problems[a]:

$$\boldsymbol{\sigma}(\boldsymbol{x}) = \boldsymbol{L}(\boldsymbol{x}) : \boldsymbol{\varepsilon}(\boldsymbol{u}(\boldsymbol{x})), \quad \operatorname{div}(\boldsymbol{\sigma}(\boldsymbol{x})) = 0, \quad \langle \boldsymbol{\varepsilon} \rangle = \bar{\boldsymbol{\varepsilon}}, \tag{45}$$

where $\bar{\boldsymbol{\varepsilon}}$ is taken to be equal successively to one of the second-order tensors $\boldsymbol{i}^{(ij)}$ with components

$$i_{mn}^{(ij)} = \frac{1}{2}(\delta_{im}\delta_{jn} + \delta_{in}\delta_{jm}).$$

Let $\boldsymbol{u}^{(ij)}$ and $\boldsymbol{\sigma}^{(ij)}$ denote the displacement field and the stress field solution of (45) with $\bar{\boldsymbol{\varepsilon}} = \boldsymbol{i}^{(ij)}$. The components of the fourth-order strain-localisation tensor $\boldsymbol{A}$, of the fourth-order effective stiffness tensor $\widetilde{\boldsymbol{L}}$ and of the second-order reduced strain-localisation tensor $\boldsymbol{a}^{(k)}$ read as

$$\begin{gathered} A_{ijmn}(\boldsymbol{x}) = \varepsilon_{ij}(u^{(mn)}(\boldsymbol{x})), \\ \widetilde{L}_{ijmn} = \langle \sigma_{ij}^{(mn)} \rangle, \quad a_{ij}^{(k)} = \langle \boldsymbol{\mu}^{(k)} : \boldsymbol{\varepsilon}(\boldsymbol{u}^{(ij)}) \rangle. \end{gathered} \tag{46}$$

[a]Six problems in dimension 3, but only three problems in plane strain and four problems in generalised plane strain, see Ref. 15 for further details.

To obtain the influence factors $D_N^{(k\ell)}$ and the second-order tensors $\boldsymbol{\rho}^{(k)}$, M linear elasticity problems have to be solved:

$$\boldsymbol{\sigma}(\boldsymbol{x}) = \boldsymbol{L}(\boldsymbol{x}) : (\varepsilon(\boldsymbol{u}(\boldsymbol{x})) - \boldsymbol{\mu}(\boldsymbol{x}))), \quad \operatorname{div}(\boldsymbol{\sigma}(\boldsymbol{x})) = 0, \quad \langle \varepsilon \rangle = \mathbf{0}, \tag{47}$$

with $\boldsymbol{\mu} = \boldsymbol{\mu}^{(k)}$. Let $\boldsymbol{u}^{(\ell)}$ denote the displacement field solution of (47) with $\boldsymbol{\mu} = \boldsymbol{\mu}^{(\ell)}$. Note that $\boldsymbol{\rho}^{(\ell)}$ is the stress field solution of (47). Then,

$$D_N^{(k\ell)} = \langle \boldsymbol{\mu}^{(k)} : \varepsilon(\boldsymbol{u}^{(\ell)}) \rangle. \tag{48}$$

The Finite Element Method (FEM) was used in the two examples presented in Secs. 3.9 and 4.4 to solve the linear elasticity problems (45) and (47).

3.7. *Time-integration of the NTFA model: Strain control*

This section is devoted to the time-integration of the NTFA model at the level of a single macroscopic material point when the individual constituents are elasto-viscoplastic (the reader is referred to Ref. 19 for rate-independent elastoplasticity). The history of macroscopic strain $\bar{\varepsilon}(t)$ is prescribed on the time interval $[0, T]$.

Equations (39) to be solved form a system of nonlinear differential equations. Its time-integration requires a time-discretization and an iterative procedure within each time-step. The time interval $[0, T]$ is decomposed into a finite number of time-steps $[t, t + \Delta t]$. All reduced variables at time t are assumed to be known. The reduced variables and the macroscopic stress at time $t + \Delta t$ are obtained as follows.

Time step $t + \Delta t$, iterate $i + 1$:

The reduced strains $(e^{(k)})^i_{t+\Delta t}$, $k = 1, \ldots, M$ being known,

- **Step 1:** Compute the plastic multipliers $(p^{(r)})^i_{t+\Delta t}$, $r = 1, \ldots, N$, the reduced stresses $(\tau^{(k)})^i_{t+\Delta t}$, and the reduced back-stresses $(x^{(k)})^i_{t+\Delta t}$, $k = 1, \ldots, M$ (see following paragraph).
- **Step 2:** Compute the reduced viscoplastic strains $(e^{(k)}_{\mathsf{vp}})^i_{t+\Delta t}$ and $(\varepsilon^{(k)}_{\mathsf{vp}})^i_{t+\Delta t}$. For $k = 1, \ldots, M$:

$$(e^{(k)}_{\mathsf{vp}})^i_{t+\Delta t} = (e^{(k)})^i_{t+\Delta t} - \frac{(\tau^{(k)})^i_{t+\Delta t}}{2G^{(k)}}, \quad (\varepsilon^{(k)}_{\mathsf{vp}})^i_{t+\Delta t} = \frac{(e^{(k)}_{\mathsf{vp}})^i_{t+\Delta t}}{\langle \boldsymbol{\mu}^{(k)} : \boldsymbol{\mu}^{(k)} \rangle}.$$

- **Step 3:** Compute the macroscopic stress $\bar{\boldsymbol{\sigma}}^i_{t+\Delta t}$:

$$\bar{\boldsymbol{\sigma}}^i_{t+\Delta t} = \widetilde{\boldsymbol{L}} : \bar{\varepsilon}_{t+\Delta t} + \sum_{k=1}^{M} \langle \boldsymbol{\rho}^{(k)} \rangle (\varepsilon^{(k)}_{\mathsf{vp}})^i_{t+\Delta t}.$$

- **Step 4:** Update the reduced strains $(e^{(k)})^i_{t+\Delta t}$. For $k = 1, \ldots, M$:

$$(e^{(k)})^{i+1}_{t+\Delta t} = \boldsymbol{a}^{(k)} : \bar{\varepsilon}_{t+\Delta t} + \sum_{\ell=1}^{M} D_N^{(k\ell)} \, (\varepsilon^{(\ell)}_{\mathsf{vp}})^i_{t+\Delta t}.$$

  Go to 1.

The convergence test reads:

$$\operatorname*{Max}_{k=1,\ldots,M} |(e^{(k)})^{i+1}_{t+\Delta t} - (e^{(k)})^i_{t+\Delta t}| < \delta \|\bar{\varepsilon}_{t+\Delta t} - \bar{\varepsilon}_t\|.$$

A typical value for δ is $\delta = 10^{-6}$. The norm for second-order tensors used in the right-hand side of the convergence test is $\|\boldsymbol{a}\| = \max_{i,j} |a_{ij}|$.

Step 1 in details

In order to determine the plastic multipliers $(p^{(r)})^i_{t+\Delta t}$, the reduced stresses $(\tau^{(k)})^i_{t+\Delta t}$ and the reduced back-stresses $(x^{(k)})^i_{t+\Delta t}$ at step 1 of the above described procedure, the last four equations (39) are rewritten in the form of a first-order differential equation for these three unknowns. This is done for each phase separately. For a given phase r, the differential system to be solved in the time interval $[t, t + \Delta t]$ can be written as

$$\dot{\boldsymbol{y}} = \boldsymbol{f}(\boldsymbol{y}), \tag{49}$$

where the initial data at time t (beginning of the time interval) is known from the previous time step, and where

$$\left.\begin{array}{c}
\boldsymbol{y} = \{y_s\}_{s=1,2M(r)+1}, \quad \boldsymbol{f} = \{f_s\}_{s=1,2M(r)+1}, \\[2mm]
\{y_s\}_{s=1,M(r)} = \{\tau^{(k)}\}_{k=1,M(r)}, \\[2mm]
\{f_s\}_{s=1,M(r)} = \left\{ 2G^{(k)} \left(\dot{e}^{(k)} - \dfrac{3}{2}\, \dot{p}^{(r)} \, \dfrac{\tau^{(k)} - x^{(k)}}{\mathcal{A}^{(r)}} \right) \right\}_{k=1,M(r)}, \\[2mm]
\{y_s\}_{s=M(r)+1,2M(r)} = \{x^{(k)}\}_{k=1,M(r)}, \\[2mm]
\{f_s\}_{s=M(r)+1,2M(r)} = \left\{ \left(H^{(k)} \dfrac{\tau^{(k)} - x^{(k)}}{\mathcal{A}^{(r)}} - \xi^{(k)} x^{(k)} \right) \dot{p}^{(r)} \right\}_{k=1,M(r)}, \\[2mm]
\{y_s\}_{s=2M(r)+1} = \{p^{(r)}\}, \quad \{f_s\}_{s=2M(r)+1} = \{\dot{p}^{(r)}\}.
\end{array}\right\} \tag{50}$$

In (50), the generalised force $\mathcal{A}^{(r)}$ is known according to the $\tau^{(k)}$ and the $x^{(k)}$, $k = 1, \ldots, M(r)$, by (35), the plastic multiplier $\dot{p}^{(r)}$ according to $\mathcal{A}^{(r)}$, and $p^{(r)}$ by (36), and the strain-rates $\dot{e}^{(k)}$, $k = 1, \ldots, M(r)$, are given by

$$\dot{e}^{(k)} = \frac{(e^{(k)})^i_{t+\Delta t} - (e^{(k)})_t}{\Delta t}. \tag{51}$$

A Runge–Kutta scheme of order 4 with step control is used to solve the system (49) and (50). The solution in a sub-interval $[t_0, t_1]$ contained in $[t, t + \Delta t]$ is determined by a trial and error procedure. A first trial solution $\boldsymbol{y}(t_1)$ is computed with the time-step $t_1 - t_0$. Then a second solution $\boldsymbol{y}'(t_1)$ is computed with two time-steps of equal size $(t_1 - t_0)/2$. The difference $d = \max_s(|y'_s(t_1) - y_s(t_1)| \,/\, |y'_s(t_1)|)$ is evaluated. If $d > \delta$, the solution is discarded and the time-step is reduced by a factor which depends on the ratio d/δ. If $d \leq \delta$, the solution $\boldsymbol{y}'(t_1)$ is retained and the next time-step is multiplied by a factor which depends on the ratio δ/d. The sub-interval $[t_0, t_1]$ is initialised as $[t, t + \Delta t]$. A typical value for δ is $\delta = 10^{-4}$.

3.8. *Time-integration of the NTFA model: Stress control*

It is often convenient (or necessary) to impose *the direction* of the macroscopic stress $\bar{\boldsymbol{\sigma}}$:

$$\bar{\boldsymbol{\sigma}}_t = \lambda(t)\boldsymbol{\Sigma}^0, \tag{52}$$

where $\boldsymbol{\Sigma}^0$ is the imposed direction of stress. This is typically the situation which is met in the simulation of the response to uniaxial tension.

In rate-independent plasticity, especially in ideal plasticity, or in viscoplasticity with power-law materials such as those considered in Sec. 3.9, it is not appropriate to control directly the level of stress $\lambda(t)$. An arc-length method is preferable[15,17] and the loading is applied by imposing

$$\bar{\varepsilon}_t : \boldsymbol{\Sigma}^0 = \dot{E}_0 t,$$

where $\dot{E}_0$ is the imposed strain-rate (in the direction of the applied stress). As in the strain-controlled method, all reduced variables at time t are assumed to be known. At time $t + \Delta t$, the condition $\bar{\varepsilon}_{t+\Delta t} : \boldsymbol{\Sigma}^0 = \dot{E}_0(t + \Delta t)$ is imposed. The macroscopic stress $\lambda(t + \Delta t)$ is to be determined in addition to the reduced variables. An iterative procedure is used to impose the direction of stress (52) as follows:

Time step $t + \Delta t$, iterate $i + 1$:

The reduced strain $(e^{(k)})^i_{t+\Delta t}$, $k = 1, \ldots, M$ being known and a macroscopic strain $\bar{\varepsilon}^i_{t+\Delta t}$ meeting the condition $\bar{\varepsilon}^i_{t+\Delta t} : \boldsymbol{\Sigma}^0 = \dot{E}_0\,(t + \Delta t)$

being known:

- **1, 2:** Perform steps 1, 2 of the procedure described in Sec. 3.7.
- **3, 4:** Perform steps 3 and 4 of the same algorithm using $\bar{\varepsilon}^i_{t+\Delta t}$ in place of $\bar{\varepsilon}_{t+\Delta t}$.
- **5:** Compute the level of macroscopic stress:

$$\lambda^i_{t+\Delta t} = \frac{\boldsymbol{\Sigma}^0 : \widetilde{\boldsymbol{L}}^{-1} : \bar{\boldsymbol{\sigma}}^i_{t+\Delta t}}{\boldsymbol{\Sigma}^0 : \widetilde{\boldsymbol{L}}^{-1} : \boldsymbol{\Sigma}^0},$$

and update $\bar{\varepsilon}^i_{t+\Delta t}$:

$$\bar{\boldsymbol{\varepsilon}}^{i+1}_{t+\Delta t} = \bar{\boldsymbol{\varepsilon}}^i_{t+\Delta t} + \widetilde{\boldsymbol{L}}^{-1} : \left(\lambda^i_{t+\Delta t}\boldsymbol{\Sigma}^0 - \bar{\boldsymbol{\sigma}}^i_{t+\Delta t}\right).$$

Go to 1.

The test used to check convergence now reads:

$$\operatorname*{Max}_{k=1,\ldots,M} |(e^{(k)})^{i+1}_{t+\Delta t} - (e^{(k)})^i_{t+\Delta t}| < \delta \|\bar{\varepsilon}^{i+1}_{t+\Delta t} - \bar{\varepsilon}_t\|,$$

$$\|\bar{\varepsilon}^{i+1}_{t+\Delta t} - \bar{\varepsilon}^i_{t+\Delta t}\| < \delta \|\bar{\varepsilon}^{i+1}_{t+\Delta t}\|.$$

A typical value for δ is $\delta = 10^{-6}$.

3.9. *Example 1: Effective response of a dual-phase inelastic composite*

The composite materials under consideration in this section are two-phase composites where the two phases play similar (interchangeable) roles in the microstructure.

3.9.1. *Material data*

Both phases are elastoviscoplastic with linear elasticity and a power-law viscous behaviour (corresponding to the dissipation potential (9) with $R = 0$). The material characteristics of phases 1 and 2 read respectively:

$$E^{(1)} = 100\,\text{GPa}, \quad \nu^{(1)} = 0.3, \quad \sigma_0^{(1)} = 250\,\text{MPa},$$

$$\dot{\varepsilon}_0 = 10^{-5}\,\text{s}^{-1}, \quad n_1 = 1,$$

and

$$E^{(2)} = 180\,\text{GPa}, \quad \nu^{(2)} = 0.3, \quad \sigma_0^{(2)} = 50\,\text{MPa}, \quad \dot{\varepsilon}_0 = 10^{-5}\,\text{s}^{-1}.$$

The rate-sensitivity exponent n_2 of phase 2 is varied from 1 to 8. This variation corresponds to the fact that the rate-sensitivity exponent varies significantly with temperature, and our objective here is to assess the accuracy of the NTFA used with a single set of plastic modes determined independently for an intermediate value of the rate-sensitivity exponent.

3.9.2. *Microstructure*

The two-dimensional unit-cell consists of 80 "grains" in the form of regular and identical hexagons. The material properties of these hexagons are prescribed randomly to be either that of phase 1 or that of phase 2 under the constraint that both phases have equal volume fraction ($c_1 = c_2 = 0.5$). 25 different configurations have been generated (same configurations as in Ref. 15). One configuration has been selected among these 25 realisations, namely the one which gives, when $n_1 = n_2 = 1$ and when the phases are incompressible, an effective response which is the closest to the exact result for interchangeable microstructures (given by the self-consistent scheme). This configuration is shown in Fig. 2 (phase 1 is the darkest phase). Each hexagon is discretised into 64 eight-node quadratic FEs with four Gauss points (5120 quadratic elements and 15,649 nodes in total). The unit-cell is subjected to an in-plane simple shear loading with a uniform strain-rate $\dot{\gamma} = \sqrt{3}\dot{\varepsilon}_0$:

$$\bar{\varepsilon}(t) = \frac{\bar{\gamma}(t)}{2}\,(\boldsymbol{e}_1 \otimes \boldsymbol{e}_2 + \boldsymbol{e}_2 \otimes \boldsymbol{e}_1), \quad \bar{\gamma}(t) = \dot{\gamma} t. \tag{53}$$

Fig. 2. Covering of the unit-cell by regular hexagons of phase 1 (dark) and phase 2. Realisation used for the implementation of the NTFA.

Periodic boundary conditions are applied on the boundary of the unit-cell. All computations are performed with the same global time-step $\Delta t = 2/\sqrt{3}\,\text{s}$.

3.9.3. *Plastic modes*

The plastic modes retained for application of the NTFA are given by the Karhunen–Loève procedure from two initial fields $\boldsymbol{\theta}^{(k)}(\boldsymbol{x})$ in each phase corresponding respectively to viscoplastic strain fields determined numerically at small strains ($\bar{\gamma} = 0.03\%$) and at large strains ($\bar{\gamma} = 12\%$). The procedure delivers four orthogonal modes, two modes with support in phase 1 and two modes with support in phase 2. Snapshots of the equivalent strain $\mu_{\text{eq}}^{(k)}$ of the four modes are shown in Figs. 3 and 4.

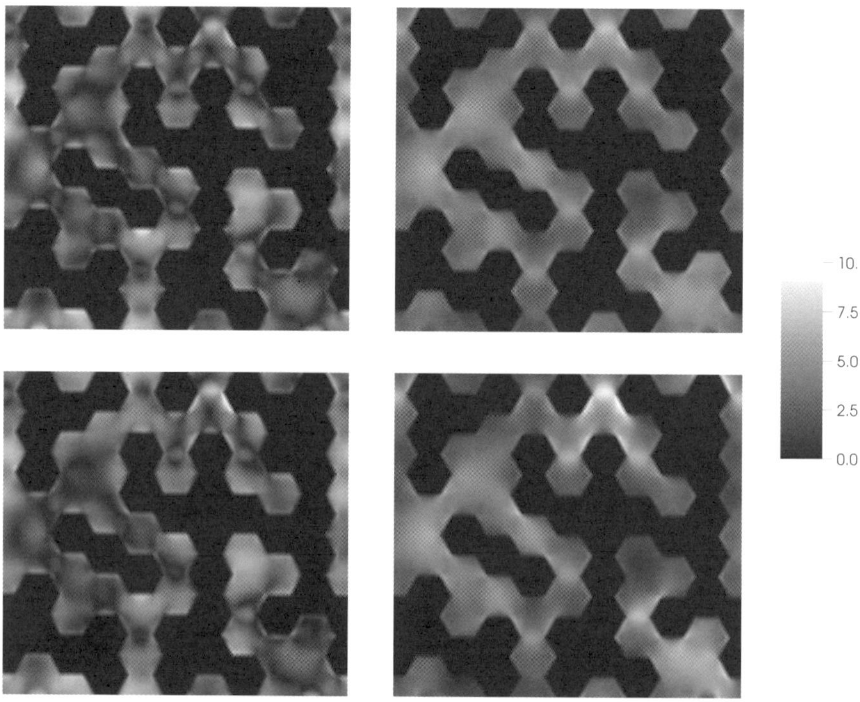

Fig. 3. Dual-phase material. Plastic modes for phase 1. Snapshot of the equivalent strain $\mu_{\text{eq}}^{(k)}$, $k = 1, 2$. At top: $n_2 = 1$. At bottom: $n_2 = 8$. From left to right: modes for small and large strains. The look-up table is the same for all four snapshots.

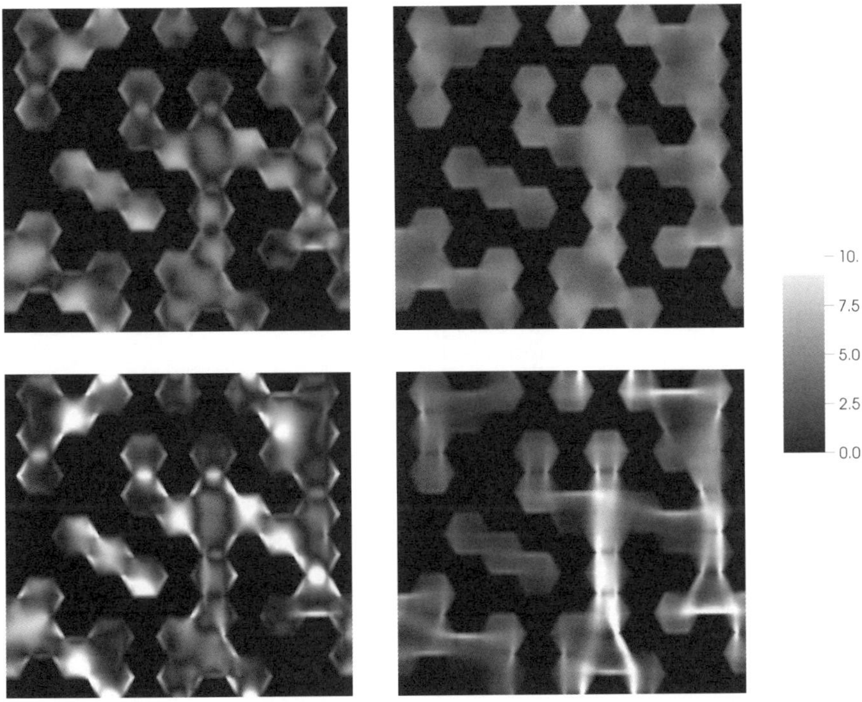

Fig. 4. Dual-phase material. Plastic modes for phase 2. Snapshot of the equivalent strain $\mu_{\text{eq}}^{(k)}$, $k = 3, 4$. At top: $n_2 = 1$. At bottom: $n_2 = 8$. From left to right: modes for small and large strains. The look-up table is the same for all four snapshots.

3.9.4. *Discussion of the results*

The macroscopic stress–strain response ($\bar{\sigma}_{12}$ versus $\bar{\gamma}$) is shown in Fig. 5 when $n_2 = 1$ and $n_2 = 8$. The full-field computation which serves as the reference is shown as a solid line. NTFA(1) refers to the NTFA model with *a single mode* in each phase (the viscoplastic strain field at large strains), whereas NTFA(2) refers to the NTFA model with two modes per phase. If the model NTFA(1) predicts accurately the asymptotic stress response at large strains, the model NTFA(2) is required for a better agreement in the transient regime where elastic and viscous effects are of comparable order, since, as can be seen in Figs. 3 and 4, the features of the modes for small and large strains are rather different.

The variation of the macroscopic asymptotic stress (creep stress at constant strain-rate) is shown in Fig. 6 as a function of the rate-sensitivity

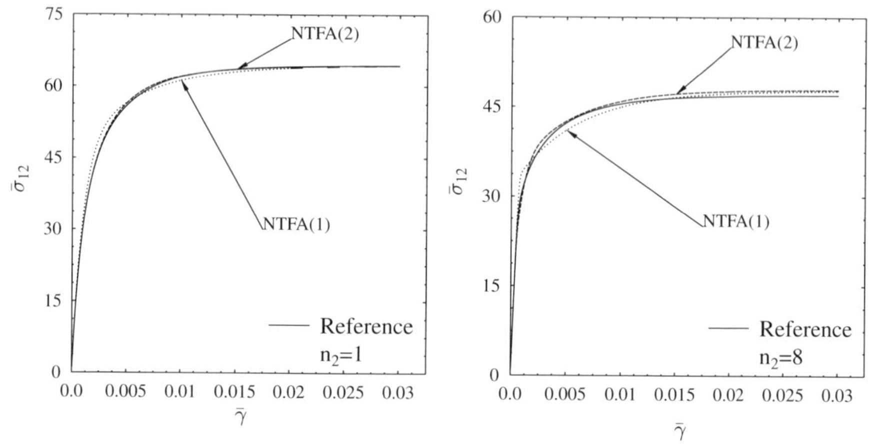

Fig. 5. Dual-phase material. Response of the unit-cell under macroscopic shear deformation (53). At left: $n_2 = 1$. At right: $n_2 = 8$.

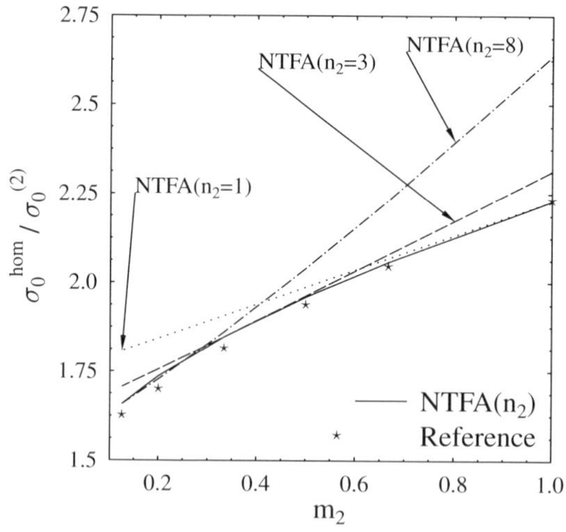

Fig. 6. Dual-phase material. Dependence of the creep stress on the rate-sensitivity exponent $m_2 = 1/n_2$ of the second phase.

exponent $m_2 = 1/n_2$ of phase 2. The full-field computations are shown as stars. The solid line corresponds to the NTFA model implemented with plastic modes which vary with n_2 (viscoplastic strain fields are computed for each value of n_2 and the corresponding plastic modes are deduced by means of the Karhunen–Loève procedure). The results shown as NTFA($n_2 = n$) were obtained by considering a single set of modes identified once for all with $n_2 = n$. NTFA($n_2 = 1$) overestimates the macroscopic creep stress of the composite for large values of n_2. The snapshot of the modes for $n_2 = 8$ shows a rather significant amount of strain localisation in phase 2. NTFA($n_2 = 8$) overestimates the creep stress for small non-linearity (which is consistent with the property of minimisation of the dissipation potential). NTFA($n_2 = 3$) is a reasonable compromise.

4. Application of the NTFA to Structural Problems

4.1. *Implementation of the NTFA method*

The implementation of the NTFA method consists of four different steps. The first two steps are "material steps" in the sense that they are concerned only with computations performed at the unit-cell level, independently of any macroscopic structural problem, except for the choice of the modes which is influenced by the type of macroscopic stress that the material is likely to sustain (as explained in Sec. 3.5). These first two steps can be performed once for all. The last two steps are the structural computation itself and a localisation step which is essential in the prediction of more local phenomena (such as the lifetime of the structure in fatigue). The four steps are as follows:

Step A: Prior to the resolution of any structural problem, choices and preliminary computations have to be made following Secs. 3.5 and 3.6:

(a) Choose the plastic modes $\boldsymbol{\mu}^{(k)}$.
(b) Compute the local fields $\boldsymbol{\eta}^{(k)}$ and the strain localisation tensor $\boldsymbol{A}$ defined in (28)–(46) and used in the localisation step D below. Then compute the reduced localisation tensors $\boldsymbol{a}^{(k)}$, the influence factors $D_N^{(k\ell)}$ entering the constitutive relations (39), the effective stiffness $\widetilde{\boldsymbol{L}}$ and the tensors $\langle \boldsymbol{\rho}^{(k)} \rangle$ entering the expression of the macroscopic stress (41). This is done once for all by solving linear elasticity problems on the unit-cell (see Sec. 3.6).

Step B: Set up a time-integration scheme to integrate the constitutive relations (39) along a prescribed path of macroscopic strain $\bar{\varepsilon}(t)$ or macroscopic stress $\bar{\boldsymbol{\sigma}}(t)$. This can be done using the schemes proposed in Secs. 3.7 and 3.8.

Step C: Incorporate the NTFA model (or more specifically the time-integration scheme of step B) into a FE code. Find the history of macroscopic stresses $\bar{\boldsymbol{\sigma}}(\boldsymbol{X}, t)$ and strains $\bar{\varepsilon}(\boldsymbol{X}, t)$ at every macroscopic material point $\boldsymbol{X}$ in the structure.

Step D: It is often useful to determine the local strains and stresses $\varepsilon(\boldsymbol{X}, \boldsymbol{x})$ and $\boldsymbol{\sigma}(\boldsymbol{X}, \boldsymbol{x})$ in the actual composite structure and not only the macroscopic strain and stress $\bar{\varepsilon}(\boldsymbol{X})$ and $\bar{\boldsymbol{\sigma}}(\boldsymbol{X})$ (which are smoother fields, being averages of the corresponding local fields over a volume element). This *localisation step* is greatly simplified by the NTFA.

Unlike in the exact homogenised problem where the microscopic and macroscopic variables are closely coupled, all steps can be performed independently in the present approach. Steps A and B have already been discussed in Sec. 3 and we shall concentrate the discussion on steps C and D.

4.2. *Implementation of the NTFA in a Finite Element code (step C)*

This section deals with the incorporation of NTFA in a Finite Element code to solve a structural problem. After discretization of the structure into macroscopic finite elements, the unknowns pertaining to the structural (i.e. macroscopic) problem are denoted by overlined letters, e.g. $\bar{\boldsymbol{u}}(\boldsymbol{X})$, $\bar{\boldsymbol{\sigma}}(\boldsymbol{X})$, etc. Arrays of discrete unknowns are denoted with braces, e.g. $\{\bar{\boldsymbol{u}}\}$ denotes the array of discrete unknowns associated with the displacement field $\bar{\boldsymbol{u}}$, and matrices are denoted with brackets, e.g. $[\boldsymbol{K}]$ denotes the assembled stiffness matrix associated with the effective stiffness of the composite $\widetilde{\boldsymbol{L}}$.

The structural problem is solved incrementally after the time discretization of the interval of study. All significant variables (displacement, stresses) being known at time t, the unknowns at time $t + \Delta t$ are determined by the equilibrium equations and the macroscopic (or homogenised) constitutive relations.

Equilibrium of the structure implies

$$
\begin{gathered}
{}^{\mathsf{T}}\{\bar{\boldsymbol{v}} - \bar{\boldsymbol{u}}_{t+\Delta t}\}\,\{\boldsymbol{R}\}_{t+\Delta t} = 0, \\
\{\boldsymbol{R}\}_{t+\Delta t} = -\left\{ \sum_e \int_e {}^{\mathsf{T}}[\boldsymbol{B}]\,\{\bar{\boldsymbol{\sigma}}\}_{t+\Delta t}\, d\boldsymbol{X} \right\},
\end{gathered}
\tag{54}
$$

where $\bar{\boldsymbol{v}}$ is an arbitrary kinematically admissible displacement field, $[\boldsymbol{B}]$ is the classical FE matrix relating displacements and strains, i.e. $\{\bar{\varepsilon}^e\} = [\boldsymbol{B}]\{\bar{\boldsymbol{u}}^e\}$, and e denotes a finite element.

Equation (54) is a non-linear equation which can be solved by an iterative Newton scheme as follows.

Time step $t + \Delta t$, iterate $I + 1$:

$\{\boldsymbol{\Delta}\bar{\boldsymbol{u}}\}^I_{t+\Delta t}$ being known at each nodal point of the structure,

- **Step a:** Compute the stresses $\{\bar{\boldsymbol{\sigma}}\}^I_{t+\Delta t}$ at each integration point of each FE of the structure (see paragraph below).
- **Step b:** Check convergence. If convergence is not reached, solve the linear system:

$$[\boldsymbol{K}]^I_{t+\Delta t}\{\boldsymbol{\delta}\bar{\boldsymbol{u}}\}^I_{t+\Delta t} = \{\boldsymbol{R}\}^I_{t+\Delta t}.$$

- **Step c:** Update $\{\boldsymbol{\Delta}\bar{\boldsymbol{u}}\}^I_{t+\Delta t}$:

$$\{\boldsymbol{\Delta}\bar{\boldsymbol{u}}\}^{I+1}_{t+\Delta t} = \{\boldsymbol{\Delta}\bar{\boldsymbol{u}}\}^I_{t+\Delta t} + \{\boldsymbol{\delta}\bar{\boldsymbol{u}}\}^I_{t+\Delta t}.$$

  Go to step a.

The global stiffness matrix $[\boldsymbol{K}]^I_{t+\Delta t}$ can be chosen among many different possibilities. One of the simplest one, which was used in the example presented in Sec. 4.4, is the initial elastic stiffness:

$$[\boldsymbol{K}]^I_{t+\Delta t} = [\boldsymbol{K}] = \sum_e [\boldsymbol{k}^e], \quad \text{where} \quad [\boldsymbol{k}^e] = \int_e {}^{\mathsf{T}}[\boldsymbol{B}][\widetilde{\boldsymbol{L}}][\boldsymbol{B}]d\boldsymbol{X}. \tag{55}$$

No particular convergence problem was observed with this elementary method.

In the convergence test used in step b, the norm of the equilibrium residues is checked:

$$\max_j\{|R_j|\}^I_{t+\Delta t} < \delta \max_{j'}\{|\mathcal{R}_{j'}|\}^I_{t+\Delta t}, \tag{56}$$

where $\{\boldsymbol{\mathcal{R}}\}^I_{t+\Delta t}$ denotes the array of reactions at nodal points on the boundary of the structure. A typical value for δ is $\delta = 10^{-6}$.

Computation of $\{\bar{\boldsymbol{\sigma}}\}^I_{t+\Delta t}$

Consider an integration point $\boldsymbol{X}$ of a finite element e in the structure. The strain at $\boldsymbol{X}$ reads

$$\{\bar{\varepsilon}(\boldsymbol{X})\}^I_{t+\Delta t} = [\boldsymbol{B}(\boldsymbol{X})]\{\bar{\boldsymbol{u}}^e\}^I_{t+\Delta t}, \quad \{\bar{\boldsymbol{u}}^e\}^I_{t+\Delta t} = \{\bar{\boldsymbol{u}}^e\}_t + \{\boldsymbol{\Delta}\bar{\boldsymbol{u}}^e\}^I_{t+\delta t}.$$

Then the iterative procedure of Sec. 3.7, applied with $\bar{\varepsilon}_{t+\Delta t} = \{\bar{\varepsilon}(\boldsymbol{X})\}^I_{t+\Delta t}$, delivers the stress $\{\bar{\boldsymbol{\sigma}}(\boldsymbol{X})\}^I_{t+\Delta t}$ at point $\boldsymbol{X}$.

Note that the procedure of Sec. 3.7 requires the knowledge of the reduced variables at time t. These variables are $(p_r)_t$, $r = 1, \ldots, N$ and $(e^{(k)})_t$, $(\tau^{(k)})_t$, $(x^{(k)})_t$, $k = 1, \ldots, M$. It is therefore necessary to store these scalar variables at each integration point of each finite element of the structure.

4.3. *Localisation rules*

The strain and stress fields $\bar{\varepsilon}(\boldsymbol{X})$ and $\bar{\boldsymbol{\sigma}}(\boldsymbol{X})$ delivered by the structural analysis are averaged fields. Their value at a macroscopic point $\boldsymbol{X}$ is the average over the microscopic variable $\boldsymbol{x}$ of the zeroth order terms $\varepsilon^0(\boldsymbol{X}, \boldsymbol{x})$ and $\boldsymbol{\sigma}^0(\boldsymbol{X}, \boldsymbol{x})$ in the expansion of the strain and stress fields, when $\boldsymbol{x}$ varies in the unit-cell. The averaged fields do not capture the rapid oscillations (and most importantly the peaks) of the actual strain and stress fields at the microscopic scale.

Mathematical analysis shows that these zeroth order terms in the asymptotic expansion (3) provide, after rescaling, a better approximation of $\varepsilon^\eta(\boldsymbol{X})$ and $\boldsymbol{\sigma}^\eta(\boldsymbol{X})$ than $\bar{\varepsilon}(\boldsymbol{X})$ and $\bar{\boldsymbol{\sigma}}(\boldsymbol{X})$ by setting

$$\tilde{\varepsilon}^\eta(\boldsymbol{X}) = \varepsilon^0\left(\boldsymbol{X}, \frac{\boldsymbol{X}}{\eta}\right), \quad \tilde{\boldsymbol{\sigma}}^\eta(\boldsymbol{X}) = \boldsymbol{\sigma}^0\left(\boldsymbol{X}, \frac{\boldsymbol{X}}{\eta}\right). \tag{57}$$

In *linear elasticity* it has been shown theoretically[27] and observed numerically[7] that $\tilde{\varepsilon}^\eta$ and $\tilde{\boldsymbol{\sigma}}^\eta$ are pointwise approximations of ε^η and $\boldsymbol{\sigma}^\eta$ and not only weak approximations (as are $\bar{\varepsilon}$ and $\bar{\boldsymbol{\sigma}}$), except in a boundary layer close to the boundary of the structure where the periodicity conditions can be in contradiction with the actual boundary conditions (boundary layer terms must be added to have a good approximation up to the boundary).

In *linear elasticity* the zeroth order terms ε^0 and $\boldsymbol{\sigma}^0$ in the expansion of the strain and stress fields are nothing else than the local fields ε and $\boldsymbol{\sigma}$ solution of the local problem (12) and are therefore related to their average by means of the localisation tensors $\boldsymbol{A}$ and $\boldsymbol{B}$:

$$\varepsilon^0(\boldsymbol{X}, \boldsymbol{x}) = \boldsymbol{A}(\boldsymbol{x}) : \bar{\varepsilon}(\boldsymbol{X}), \quad \boldsymbol{\sigma}^0(\boldsymbol{X}, \boldsymbol{x}) = \boldsymbol{B}(\boldsymbol{x}) : \bar{\boldsymbol{\sigma}}(\boldsymbol{X}). \tag{58}$$

Therefore, a good approximation of the actual strain and stress fields can be obtained by solving *independently* the structural problem to find the macroscopic fields $\bar{\varepsilon}(\boldsymbol{X})$ and $\bar{\boldsymbol{\sigma}}(\boldsymbol{X})$ and six unit-cell problems to find the stress-localisation tensors $\boldsymbol{A}$ and $\boldsymbol{B}$. Then the two results are combined by means of (58) to give a good approximation of the actual strain and stress fields in the composite structure (with a possible exception at the boundary, as discussed above). In other words, the local fields $\boldsymbol{\varepsilon}(\boldsymbol{X}, \boldsymbol{x})$ and $\boldsymbol{\sigma}(\boldsymbol{X}, \boldsymbol{x})$ (or good approximations of them) can be obtained by *post-processing* the macroscopic strain and stress fields $\bar{\boldsymbol{\varepsilon}}(\boldsymbol{X})$ and $\bar{\boldsymbol{\sigma}}(\boldsymbol{X})$. A full decoupling of scales can be achieved.

In *non-linear problems*, and in particular in the presence of viscoplasticity or plasticity, no simple relation such as (58) exists. Rigorously speaking, there is no explicit decoupling of scales. If no approximation is made, the microscopic fields $\boldsymbol{\varepsilon}^0(\boldsymbol{X}, \boldsymbol{x})$ and $\boldsymbol{\sigma}^0(\boldsymbol{X}, \boldsymbol{x})$ are intimately coupled to the macroscopic fields $\bar{\boldsymbol{\varepsilon}}(\boldsymbol{X})$ and $\bar{\boldsymbol{\sigma}}(\boldsymbol{X})$, and all microscopic and macroscopic fields must be determined in the course of a coupled computation. The field localisation is not performed as a post-processing step but as a part of the structural analysis. As already underlined, the cost of this computational procedure can be formidable.

The NTFA avoids this complication, thanks to the relations (28) and (40), admittedly at the expense of the approximation (24). First, as shown in Sec. 4.2, the structural problem is solved independently of the unit-cell calculations (performed once for all). Second, the microscopic fields are deduced from their macroscopic counterpart by means of the *explicit and linear relations* (28) and (40):

$$\left.\begin{aligned} \boldsymbol{\varepsilon}(\boldsymbol{X}, \boldsymbol{x}, t) &= \boldsymbol{A}(\boldsymbol{x}) : \bar{\boldsymbol{\varepsilon}}(\boldsymbol{X}, t) + \sum_{k=1}^{M} \boldsymbol{\eta}^{(k)}(\boldsymbol{x}) \varepsilon_{\mathsf{vp}}^{(k)}(\boldsymbol{X}, t), \\ \boldsymbol{\sigma}(\boldsymbol{X}, \boldsymbol{x}, t) &= \boldsymbol{L}(\boldsymbol{x}) : \boldsymbol{A}(\boldsymbol{x}) : \bar{\boldsymbol{\varepsilon}}(\boldsymbol{X}, t) + \sum_{k=1}^{M} \boldsymbol{\rho}^{(k)}(\boldsymbol{x}) \varepsilon_{\mathsf{vp}}^{(k)}(\boldsymbol{X}, t). \end{aligned}\right\} \tag{59}$$

The macroscopic state variables $(\bar{\boldsymbol{\varepsilon}}(\boldsymbol{X}), \varepsilon_{\mathsf{vp}}^{(k)}(\boldsymbol{X}))$ are outputs of the structural computation performed with the homogenised NTFA model. The relation (59) can be used to post-process these fields and obtain an accurate approximation of the actual strain and stress fields $\boldsymbol{\varepsilon}^\eta$ and $\boldsymbol{\sigma}^\eta$ by setting

$$\left.\begin{aligned} \tilde{\boldsymbol{\varepsilon}}^\eta(\boldsymbol{X}, t) &= \boldsymbol{A}\left(\frac{\boldsymbol{X}}{\eta}\right) : \bar{\boldsymbol{\varepsilon}}(\boldsymbol{X}, t) + \sum_{k=1}^{M} \boldsymbol{\eta}^{(k)}\left(\frac{\boldsymbol{X}}{\eta}\right) \varepsilon_{\mathsf{vp}}^{(k)}(\boldsymbol{X}, t), \\ \tilde{\boldsymbol{\sigma}}^\eta(\boldsymbol{X}, t) &= \boldsymbol{L}\left(\frac{\boldsymbol{X}}{\eta}\right) : \boldsymbol{A}\left(\frac{\boldsymbol{X}}{\eta}\right) : \bar{\boldsymbol{\varepsilon}}(\boldsymbol{X}, t) + \sum_{k=1}^{M} \boldsymbol{\rho}^{(k)}\left(\frac{\boldsymbol{X}}{\eta}\right) \varepsilon_{\mathsf{vp}}^{(k)}(\boldsymbol{X}, t). \end{aligned}\right\} \tag{60}$$

4.4. *Example 2: Fatigue of a metal-matrix composite structure*

In this section the NTFA model is applied to a structural problem. A plate composed of a inner core (thickness 4 mm), made of a metal-matrix composite, surrounded by two outer layers of pure matrix (thickness of 0.5 mm each) is subjected to a cyclic four-point bending test. By symmetry, only half of the plate is considered as shown in Fig. 1 where the unit-cell generating the core of the plate by periodicity is also shown.

The matrix is elastoviscoplastic with purely non-linear kinematic hardening (the isotropic hardening is negligible $R(p) = \sigma_y$):

$$\left.\begin{aligned} &E_m = 60\,\mathrm{GPa}, \quad \nu_m = 0.3, \quad \sigma_y = 20\,\mathrm{MPa}, \quad n = 5, \\ &\eta = \sigma_0 \dot{\varepsilon}_0^{-\frac{1}{n}} = 100\,\mathrm{MPa\,s^{\frac{1}{n}}}, \quad H = 10\,\mathrm{GPa}, \quad \xi = 1000\,\mathrm{MPa}. \end{aligned}\right\} \tag{61}$$

The metal matrix is reinforced by long circular fibres arranged at the nodes of a square array. The fibre volume fraction is 25%. The fibres are linear elastic with Young's modulus and Poisson's ratio:

$$E_f = 300\,\mathrm{GPa}, \quad \nu_f = 0.25. \tag{62}$$

The plate is simply supported at points B and B', and periodic (in time) displacements at points A and A' are prescribed. Depending on the amplitude of the displacement, the structure is likely to undergo viscoplastic deformations leading to fatigue failure. There are three possible failure mechanisms at the microscopic scale: fibre breakage, fibre–matrix debonding, and matrix failure. At high temperature, when the matrix is viscoplastic as considered in this study, matrix damage is the dominant mechanism.[13] Therefore, a first modeling assumption is that the failure of the composite occurs by matrix failure. To predict matrix failure, a model due to Skelton[26] for low-cycle fatigue is used (a comparison of different lifetime prediction methods including Skelton's model can be found in Ref. 1). The model is based on the energy dissipated by viscoplasticity during the stabilised cycle:

$$w = \int_{\text{cycle}} \boldsymbol{\sigma} : \dot{\boldsymbol{\varepsilon}}_{\mathsf{vp}}\, dt. \tag{63}$$

Skelton's model is based on the assumption (confirmed experimentally) that the number of cycles to failure N_f for a material under cyclic thermomechanical fatigue tests in the low-cycle regime is related to the energy dissipated w by

$$wN_f^\beta = C, \tag{64}$$

where C and β are material constants independent of the thermomechanical loading.

In the framework of these two working assumptions (failure of the composite governed by matrix failure, and matrix failure governed by the criterion (64)), one can predict the lifetime of the composite structure subjected to a cyclic thermomechanical loading at the expense of resolving the stress and strain fields at the smallest scale in order to apply the criterion (64). This procedure is extremely heavy and the aim of this section is to demonstrate that an accurate prediction can be obtained by means of the NTFA at a much reduced cost, involving only a purely macroscopic computation, followed by a proper postprocessing of the macroscopic fields.

4.4.1. *Meshes*

The fine mesh accounting for all microstructural details of the heterogeneous structure is shown in Fig. 7(a). The mesh of the inner core is obtained by repeating the mesh of the unit-cell shown in Fig. 7(c) which consists of 80 six-node triangular elements (three Hammer points) in the fibre and 128 eight-node quadrilateral elements (four Gauss points) in the matrix, for a total of 208 elements and 577 nodes. The same unit-cell mesh was

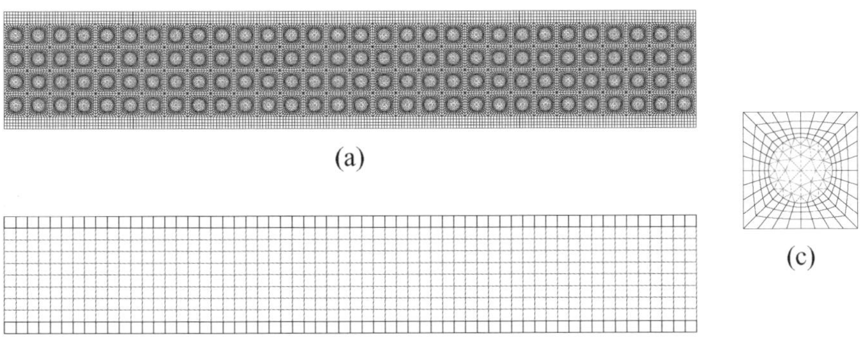

(a) (b) (c)

Fig. 7. Meshes used in the analysis of the composite plate shown in Fig. 1. (a) Fine mesh of the heterogeneous structure. (b) Coarse mesh used for the analysis of the homogenised structure by means of the NTFA model. (c) Mesh of the unit-cell generating, by periodicity, the mesh of the inner core of the plate as shown in (a).

used for the unit-cell preliminary computations (effective properties, plastic modes, influence factors, localisation fields $\boldsymbol{A}$ and $\boldsymbol{\eta}^{(k)}$). The resulting mesh for the heterogeneous structure consists of 26,880 quadratic elements (6 or 8 nodes) and 71,601 nodes in total. The mesh used in the homogenised computations is shown in Fig. 7(b) and consists of only 600 eight-node quadrilateral elements and 1941 nodes.

4.4.2. *Loading*

The boundary conditions applied to the right half of the cross section of the plate (refer to Fig. 1 for the location of points A, A', B, and B') are

$$\left.\begin{array}{lll} X_1 = 0\text{: } \bar{u}_1(0, X_2) = 0, & \bar{t}_2(0, X_2) = 0, & -\dfrac{h}{2} \leq X_2 \leq \dfrac{h}{2}, \\ \text{Point } A\text{: } \bar{t}_1\left(X_1^A, \dfrac{h}{2}\right) = 0, & \bar{u}_2\left(X_1^A, \dfrac{h}{2}\right) = \bar{u}, & \\ \text{Point } A'\text{: } \bar{t}_1\left(X_1^A, -\dfrac{h}{2}\right) = 0, & \bar{u}_2\left(X_1^A, -\dfrac{h}{2}\right) = \bar{u}, & \\ \text{Point } B\text{: } \bar{t}_1\left(X_1^B, -\dfrac{h}{2}\right) = 0, & \bar{u}_2\left(X_1^B, -\dfrac{h}{2}\right) = 0, & \\ \text{Point } B'\text{: } \bar{t}_1\left(X_1^B, \dfrac{h}{2}\right) = 0, & \bar{u}_2\left(X_1^B, \dfrac{h}{2}\right) = 0, & \\ X_1 = L\text{: } \bar{t}_1(L, X_2) = 0, & \bar{t}_2(L, X_2) = 0, & -\dfrac{h}{2} \leq X_2 \leq \dfrac{h}{2}, \end{array}\right\} \tag{65}$$

with $h = 5\,\text{mm}$, $L = 30\,\text{mm}$, $X_1^A = 10\,\text{mm}$, and $X_1^B = 25\,\text{mm}$. The traction on the boundary of the structure is denoted by $\bar{\boldsymbol{t}} = \bar{\boldsymbol{\sigma}} \cdot \boldsymbol{N}$. The vertical displacement $\bar{u}$ imposed at points A and A' is periodic in time with period $\mathcal{T}$. It is a piecewise linear function of time, varying linearly between $\bar{u}_{\max}$ and $-\bar{u}_{\max}$, as shown in Fig. 8. The loading frequency $f = 1/\mathcal{T}$ is prescribed as $f = 0.1\,\text{Hz}$, whereas the maximal displacement at points A and A' is varied: $\bar{u}_{\max} = 0.15$, 0.2, 0.25, 0.35, and $0.5\,\text{mm}$. The loading frequency being kept constant in the different loading cases, varying the maximal displacement prescribed to A and A' results in different velocities for these points and therefore in different strain-rates in the structure. All computations were performed with a global time-step $\Delta t = (\frac{0.25}{\bar{u}_{\max}})10^{-2}\,\text{s}$.

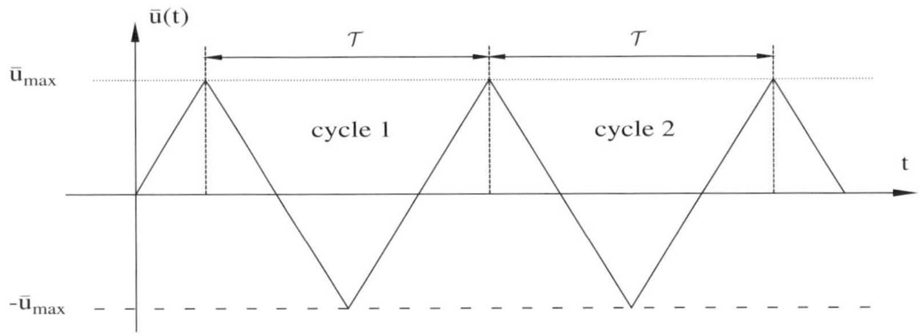

Fig. 8. History of the prescribed displacements at points A and A'.

4.4.3. *Plastic modes*

The choice of the modes depend in general on the type of loading that the structure is likely to undergo. Although it is expected that the dominant stress will be uniaxial tension–compression in the horizontal direction, transverse shear and even transverse normal stress cannot be excluded. So, the three types of stress (horizontal, vertical, and shear) will be considered in the analysis leading to the choice of the modes.

The NTFA model is implemented with five plastic modes in the matrix, and the macroscopic model has therefore five internal variables. These modes were obtained by subjecting the unit-cell to cyclic loading along three different directions of macroscopic stress:

$$\left.\begin{aligned}
\boldsymbol{\Sigma}^{(1)} &= \boldsymbol{e}_1 \otimes \boldsymbol{e}_1 + \Sigma_{33}^{(1)} \boldsymbol{e}_3 \otimes \boldsymbol{e}_3, \quad \bar{\varepsilon}_{33} = 0,\\
\boldsymbol{\Sigma}^{(2)} &= \boldsymbol{e}_1 \otimes \boldsymbol{e}_2 + \boldsymbol{e}_2 \otimes \boldsymbol{e}_1 + \Sigma_{33}^{(2)} \boldsymbol{e}_3 \otimes \boldsymbol{e}_3, \quad \bar{\varepsilon}_{33} = 0,\\
\boldsymbol{\Sigma}^{(3)} &= \boldsymbol{e}_2 \otimes \boldsymbol{e}_2 + \Sigma_{33}^{(3)} \boldsymbol{e}_3 \otimes \boldsymbol{e}_3, \quad \bar{\varepsilon}_{33} = 0.
\end{aligned}\right\} \tag{66}$$

The components $\Sigma_{33}^{(i)}$ are *a priori* left free and determined *a posteriori* as the reactions to the constraint $\bar{\varepsilon}_{33} = 0$. The computations at the unit-cell level are performed in plane strains, in concordance with the plane strain conditions which prevail at the structural level.

The unit-cell is subjected to a cyclic loading along each of the three stress directions (66). The problem is strain-controlled (as described in Sec. 3.8). The macroscopic strain in the imposed stress direction varies between $\bar{\varepsilon}_{\max} : \boldsymbol{\Sigma}^{(i)}$ and $-\bar{\varepsilon}_{\max} : \boldsymbol{\Sigma}^{(i)}$, with $\bar{\varepsilon}_{\max} : \boldsymbol{\Sigma}^{(i)} = 0.0025$, $i = 1, 2, 3$. The variation of the macroscopic strain in time is a triangular profile similar to that shown in Fig. 8 where the prescribed strain-rate is $\dot{\bar{\varepsilon}} : \mathbf{\Sigma}^{(i)} = 10^{-3}\,\mathrm{s}^{-1}$, $i = 1, 2, 3$. All computations are performed with the same global time-step $\Delta t = 10^{-2}\,\mathrm{s}$ until the response of the material point undergoing the largest viscoplastic dissipation (as defined through the scalar quantity (63)) reaches a stabilised cycle.

For each of the three loading cases (66) the viscoplastic strain fields at each quarter of all cycles are stored. In other words, for a given cycle c beginning at time t_c and with period $\mathcal{T}$, the viscoplastic strain fields at time t_c, $t_c + \mathcal{T}/4$, $t_c + \mathcal{T}/2$, and $t_c + 3\mathcal{T}/4$ are stored. This is done for all cycles until the "hottest" point in the unit-cell reaches a stabilised cycle.

Let $\boldsymbol{\theta}_i^{(k)}(\boldsymbol{x})$, $k = 1, \ldots, M_T^{(i)}$, $i = 1, 2, 3$ denote the whole set of fields stored according to this procedure. $M_T^{(i)}$ denotes the total number of fields stored along the ith loading direction $\mathbf{\Sigma}^{(i)}$. The number of modes is first reduced for each loading direction by applying the Karhunen–Loève procedure described in Sec. 3.5 separately to the three family of fields $\boldsymbol{\theta}_i^{(k)}(\boldsymbol{x})$, $i = 1, 2, 3$. The modes with the highest intensity (corresponding to the highest eigenvalue of the correlation matrix) are extracted for each loading case. The five modes finally retained for further use in the NTFA are the shear mode (macroscopic stress being a pure shear) with the highest intensity and the two modes with highest intensity for the two other loading cases (tension–compression in the horizontal and vertical direction, respectively). Taken separately, these modes are sufficient for the NTFA to reproduce accurately the response of the unit-cell along the loading direction from which they were extracted. Lastly, since these five modes were selected independently, they do not necessarily meet the orthogonality condition (44). Another application of the Karhunen–Loève procedure is made, leading finally to five modes satisfying all the desirable requirements. Snapshots of the equivalent strain of the five modes are given in Fig. 9.

4.4.4. *Accuracy of the NTFA model at the level of a material point*

A first check of the accuracy of the NTFA model with these five modes can be performed at the level of a macroscopic material point by comparing the overall response of the unit-cell as predicted by the NTFA with full-field FEM computations. The comparison for uniaxial tension–compression and pure shear is shown in Fig. 10 and the agreement between the model and the reference results is seen to be excellent.

A more local comparison can be performed by examining the stress–strain response, not of the whole unit-cell as was done in Fig. 10, but *at the*

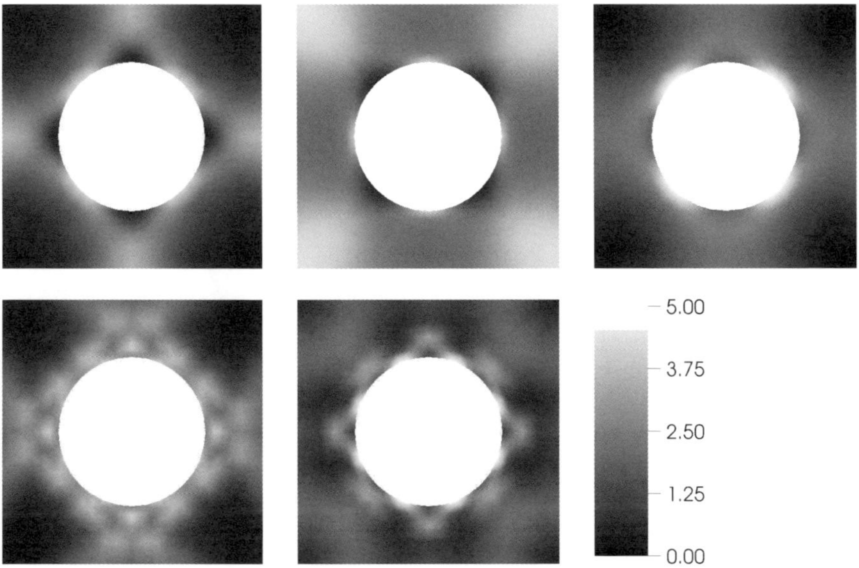

Fig. 9. Snapshots of the equivalent strain $\mu_{\mathrm{eq}}^{(k)}$, $k = 1, \ldots, 5$ for the five orthogonal plastic modes in the matrix. The look-up table is the same for all five snapshots.

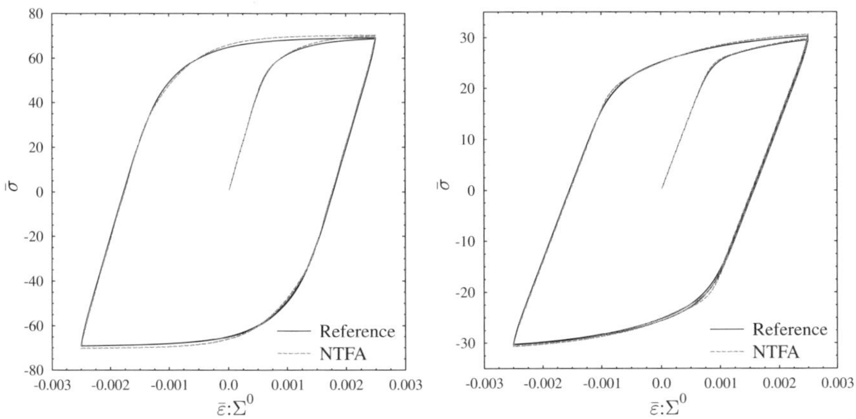

Fig. 10. Unit-cell response. Comparison between full-field FEM computations (black solid line) and the NTFA model with the five modes shown in Fig. 9 (grey dashed line). Overall stress–strain response of the unit-cell. At left: Traction–compression in the horizontal direction (loading case $i = 1$ in (66)). At right: Pure shear (loading case $i = 2$ in (66)).

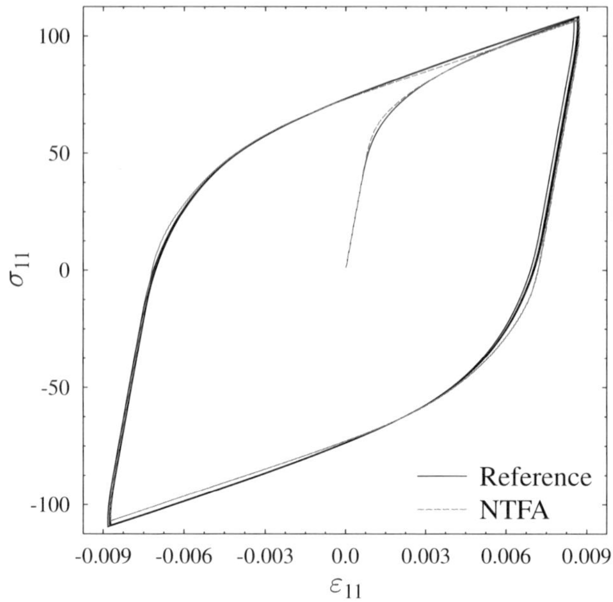

Fig. 11. Unit-cell response. Comparison between full-field FEM computations (black solid line) and the NTFA model with the five modes shown in Fig. 9 (grey dashed line). Stress–strain response at the hottest point in the unit-cell. Tension–compression in the horizontal direction (loading case $i = 1$ in (66)).

material point in the unit-cell undergoing the largest dissipated energy (63). This is done for uniaxial tension–compression in the horizontal direction in Fig. 11. Again, the agreement is seen to be excellent.

Finally, it is also of interest to compare the prediction of the model for the energy dissipated along the stabilised cycle with full-field simulations. This is done in Fig. 12. The model makes use of the localisation rules (60) to estimate the energy (63). The NTFA model captures well the local distribution of the energy dissipated in the unit-cell. The energy turns out to be maximal at the fibre-matrix interface. The reference FEM simulation gives $w_{\max} = 2.134\,\text{MPa}$, whereas the NTFA model predicts $w_{\max} = 2.231\,\text{MPa}$.

4.4.5. *Accuracy of the NTFA model at the structure level*

The accuracy of the NTFA at the structure level is assessed first by comparing the force–displacement response and second by comparing the

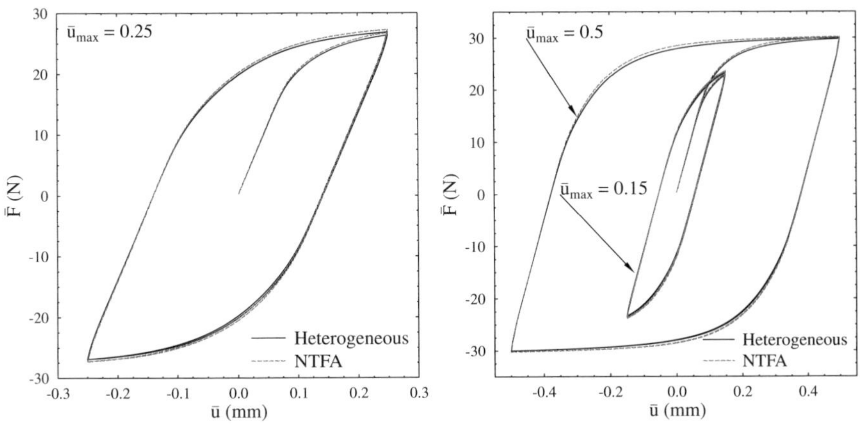

Fig. 12. Unit-cell response. Snapshot of the energy w dissipated in the unit-cell by viscoplasticity along the stabilised cycle. Uniaxial horizontal tension–compression. At left: Reference full-field FEM simulation. At right: Prediction of the NTFA model. The look-up table is the same for both snapshots.

distribution of the energy dissipated along the stabilised cycle, for two different structural simulations:

(a) The first simulation is performed with a very fine mesh of the heterogeneous structure (Fig. (7a)) and accounts for all detailed heterogeneities. It is considered as the *exact* response of the composite structure with a small but non-vanishing value of η.
(b) The second simulation is performed on a coarse mesh, using at each integration point of the mesh the homogenised NTFA model.

A first element of comparison is provided in Fig. 13 where the force–displacement (the force is the sum of the reactions at points A and A') response of the structure predicted by the homogenised NTFA model (dashed line) is compared to the detailed simulation with full account of the heterogeneities (solid line). The two graphs correspond to three different values of the maximal displacement $\bar{u}_{\max} = 0.25\,\text{mm}$ (at left) and $\bar{u}_{\max} = 0.15$ and $0.5\,\text{mm}$ (at right). The agreement is good in all cases.

A more local comparison can be made by examining the response of the most severely loaded unit-cell in the structure (where the energy dissipated is maximal). The stress and strain fields for the NTFA model are obtained by means of relations (60). The quantities used for comparison in Fig. 14 are the stress and strain averaged on this particular unit-cell. The

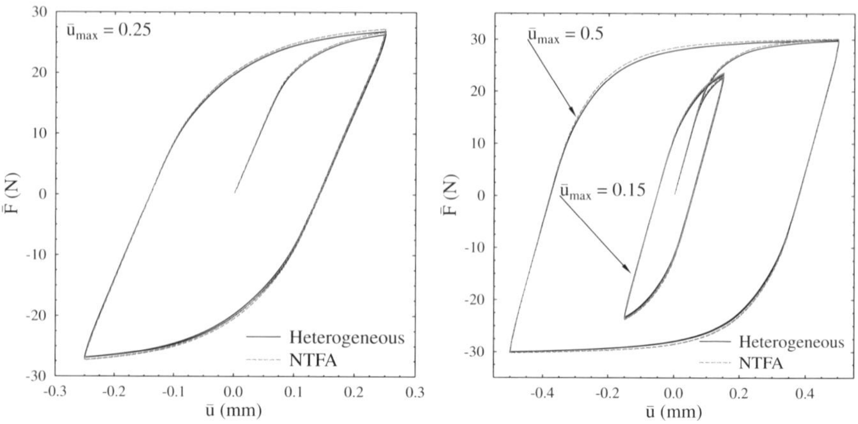

Fig. 13. Four-point bending. Comparison between the heterogeneous FE analysis (solid line) and the NTFA homogenised model (dashed line). Global force–displacement response. Left: $\bar{u}_{\max} = 0.25$ mm. Right: $\bar{u}_{\max} = 0.15$ and 0.5 mm.

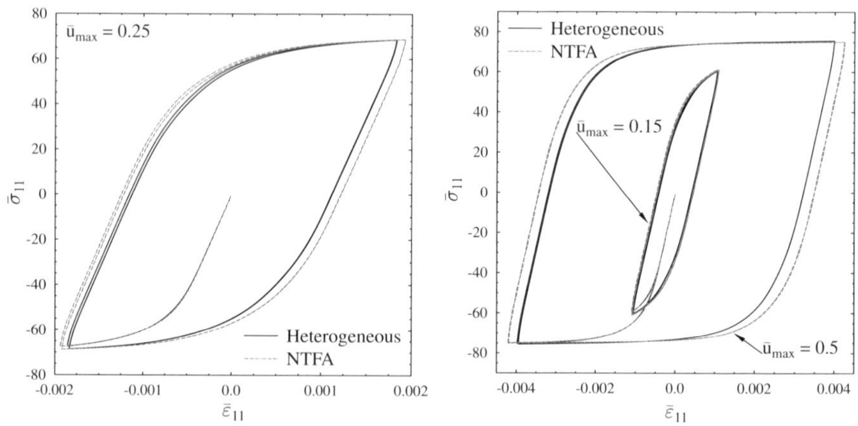

Fig. 14. Four-point bending. Comparison between the heterogeneous FE analysis (black solid line) and the NTFA homogenised model (grey dashed line). Average-stress/average-strain response of the most solicited unit-cell in the structure. At left: $\bar{u}_{\max} = 0.25$ mm. At right: $\bar{u}_{\max} = 0.15$ and 0.5 mm.

agreement in the stress level is rather good, but the NTFA seems to slightly overestimate the amount of local strain.

Finally, as exposed in the introduction of this section, the quantity of interest here is the lifetime of the structure which is directly related to the energy dissipated at the "hottest" point in the structure through the relation (64). The use of the NTFA model raises two questions:

(1) Is the location of the hottest point correctly predicted by the model?
(2) Is the amount of energy dissipated correctly estimated by the model?

To answer these questions, the heterogeneous FE analysis and the macroscopic structural simulation using the homogenised NTFA model are run until the structure reaches a stabilised cycle. The energy dissipated along this stabilised cycle is directly available in the heterogeneous simulation. In the NTFA model it can be directly deduced from the macroscopic results by means of the localisation rules (60). To answer the first question, the two snapshots (full-field computation and NTFA model) of the energy w over the whole structure are shown in Fig. 15 ($\bar{u}_{\max} = 0.25\,\text{mm}$). This very local quantity is reasonably well predicted by the NTFA model. A close-up of the same energy distribution in the region where w is maximal is shown in Fig. 16. As can be seen from these figures, the location of the hottest point is well predicted by the NTFA model. To answer the second question, the stabilised cycles at the hottest point in the structure are shown in Fig. 17. Given the very local character of this information, the agreement of the model's prediction with the detailed computation can be considered as good, the model overestimating the amount of strain at this hottest point. A further pointwise comparison of the maximum $w_{\max}$ of the energy is shown in Fig. 18. Independent of maximal displacement prescribed to the structure, the NTFA overestimates by about 25% of the maximum of the dissipated energy (this estimation is related to the overestimation of the strain at the hottest point). Therefore, the lifetime of the structure will be underestimated

Fig. 15. Four-point bending. Comparison between the heterogeneous Finite Element analysis and the NTFA homogenised model snapshot of the energy w dissipated in the structure along the stabilized cycle (normalized by its maximum). $\bar{u}_{\max} = 0.25\,\text{mm}$. At top: Full heterogeneous simulation (reference). At bottom: Prediction of the NTFA model using the localization rules. The look-up table is the same for both snapshots.

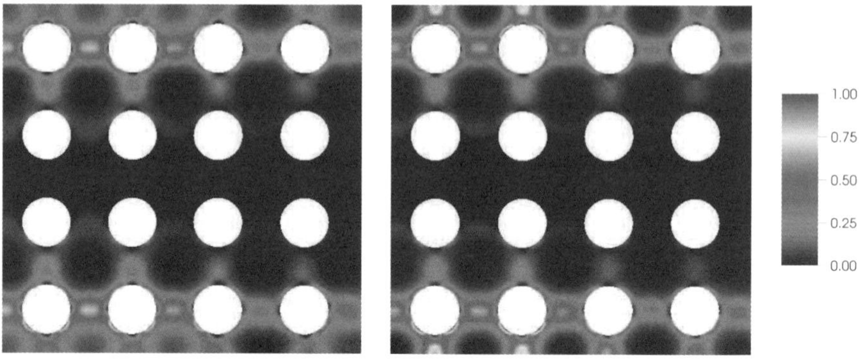

Fig. 16. Four-point bending. Distribution of the dissipated energy w (normalized by its maximum). Stabilized cycle. $\bar{u}_{\max} = 0.25$ mm. Close-up in the most severely loaded region. At left: Full heterogeneous simulation (reference). At right: Prediction of the NTFA model using the localization rules. The look-up table is the same for both snapshots.

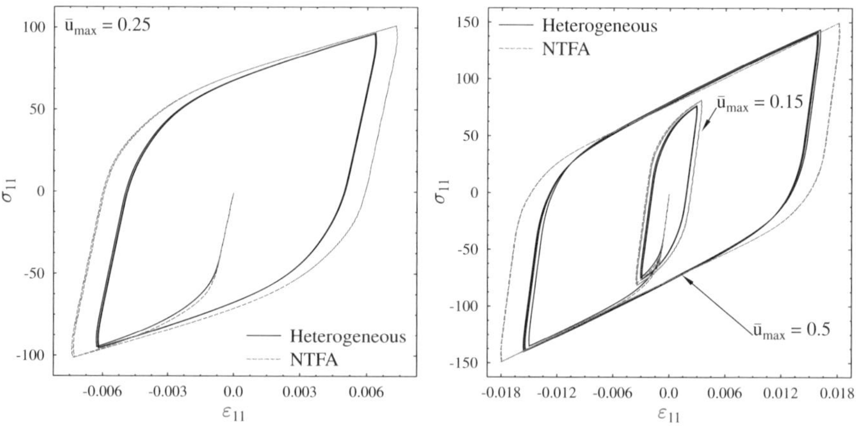

Fig. 17. Four-point bending. Stress/strain response at the hottest point in the structure. Comparison between the heterogeneous FE analysis (black solid line) and the NTFA homogenized model (grey dashed line). At left: $\bar{u}_{\max} = 0.25$ mm. At right: $\bar{u}_{\max} = 0.15$ and 0.5 mm.

by a similar amount, which is a quite reasonable error (on the safe side), given the fact that no coupled multiscale computation is required by the NTFA model but only a postprocessing of a purely macroscopic simulation.

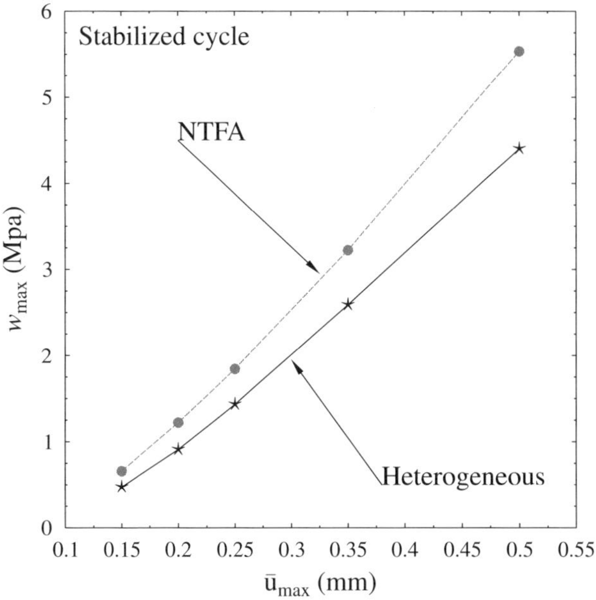

Fig. 18. Influence of the maximal displacement $\bar{u}_{max}$ on the maximum of the dissipated energy. Stabilised cycle. Reference heterogeneous simulation (black solid line) and prediction of the NTFA model (grey dashed line).

5. Conclusion

The *Non-uniform Transformation Field Analysis* is a newly proposed micromechanical scheme for multiscale problems with non-linear phases. This model is based on a drastic reduction of the number of variables describing the microscopic (visco)plastic strain field performed by means of the Karhunen–Loève procedure (proper orthogonal decomposition). It delivers effective constitutive relations for non-linear composites expressed in terms of a small number of internal variables which are the components of the microscopic plastic field over a finite set of plastic modes.

This reduced model can be easily incorporated in a structural computation. A numerical scheme is proposed to integrate in time the homogenised constitutive relations at each integration point of the structure. The predictions of the model compare well to the results of large-scale computations over the whole composite structure, accounting for all detailed information. The agreement is good not only in terms of global quantities (force/displacement) but also in terms of local quantities. For instance, the lifetime of a structure subjected to cyclic loading has been predicted with a fatigue criterion based on the energy dissipated along a cycle in the matrix. The agreement between the model and the large-scale heterogeneous computation is very good.

References

1. S. Amiable, S. Chapuliot, A. Constantinescu and A. Fissolo, A comparison of lifetime prediction methods for a thermal fatigue experiment, *Int. J. Fatigue* **28**, 692–706 (2006).
2. A. Bensoussan, J. L. Lions and G. Papanicolaou, *Asymptotic Analysis for Periodic Structures* (North-Holland, Amsterdam, 1978).
3. J. L. Chaboche, Cyclic viscoplastic constitutive equations, Part I: A thermodynamically consistent formulation, *J. Appl. Mech.* **60**, 813–821 (1993).
4. J. L. Chaboche, S. Kruch, J. F. Maire and T. Pottier, Towards a micromechanics based inelastic and damage modeling of composites, *Int. J. Plasticity* **17**, 411–439 (2001).
5. G. Dvorak and Y. Benveniste, On transformation strains and uniform fields in multiphase elastic media, *Proc. Roy. Soc. Lond. A* **437**, 291–310 (1992).
6. G. Dvorak, Y. A. Bahei-El-Din and A. M. Wafa, The modeling of inelastic composite materials with the transformation field analysis, *Model. Simul. Mater. Sci. Eng.* **2**, 571–586 (1994).
7. F. Feyel and J. L. Chaboche, FE2 multiscale approach for modelling the elastoviscoplastic behaviour of long fibre SiC/Ti composite materials, *Comput. Meth. Appl. Mech. Eng.* **183**, 309–330 (2000).
8. J. Fish, K. Shek, M. Pandheeradi and M. S. Shepard, Computational plasticity for composite structures based on mathematical homogenization: Theory and practice, *Comput. Meth. Appl. Mech. Eng.* **148**, 53–73 (1997).
9. J. Fish and Q. Yu, Computational mechanics of fatigue and life predictions for composite materials and structures, *Comput. Meth. Appl. Mech. Eng.* **191**, 4827–4849 (2002).
10. J. M. Guedes and N. Kikuchi, Preprocessing and postprocessing for materials based on the homogenization method with adaptative finite element methods, *Comput. Meth. Appl. Mech. Eng.* **83**, 143–198 (1990).
11. P. Kattan and G. Voyiadjis, Overall damage and elastoplastic deformation in fibrous metal matrix composites, *Int. J. Plasticity* **9**, 931–949 (1993).
12. J. Lemaitre and J. L. Chaboche, *Mechanics of Solid Materials* (Cambridge University Press, Cambridge, 1994).
13. J. Llorca, Fatigue of particle- and whisker-reinforced metal-matrix composites, *Prog. Mater. Sci.* **47**, 283–353 (2002).

14. J. Mandel, *Plasticité Classique et Viscoplasticité*, Vol. 97, CISM Lecture Notes (Springer-Verlag, Wien, 1972).
15. J. C. Michel, H. Moulinec and P. Suquet, Effective properties of composite materials with periodic microstructure: A computational approach, *Comp. Meth. Appl. Mech. Eng.* **172**, 109–143 (1999).
16. J. C. Michel, U. Galvanetto and P. Suquet, Constitutive relations involving internal variables based on a micromechanical analysis, in *Continuum Thermomechanics: The Art and Science of Modelling Material Behaviour*, eds. G. A. Maugin, R. Drouot and F. Sidoroff (Klüwer Academic Publishers, 2000), pp. 301–312.
17. J. C. Michel, H. Moulinec and P. Suquet, A computational method for linear and non-linear composites with arbitrary phase contrast, *Int. J. Numer. Meth. Eng.* **52**, 139–160 (2001).
18. J. C. Michel and P. Suquet, Nonuniform transformation field analysis, *Int. J. Solids Struct.* **40**, 6937–6955 (2003).
19. J. C. Michel and P. Suquet, Computational analysis of nonlinear composite structures using the nonuniform transformation field analysis, *Comp. Meth. Appl. Mech. Eng.* **193**, 5477–5502 (2004).
20. G. W. Milton, *The Theory of Composites* (Cambridge University Press, Cambridge, 2002).
21. J. R. Rice, On the structure of stress-strain relations for time-dependent plastic deformation in metals, *J. Appl. Mech.* **37**, 728–737 (1970).
22. S. Roussette, J. C. Michel and P. Suquet, Nonuniform transformation field analysis of elastic-viscoplastic composites, *Composites Sci. Technol.* doi: 10.1016/j.compscitech.2007.10.032 (2007).
23. H. Samrout, R. El Abdi and J. L. Chaboche, Model for 28CrMoV5-8 steel undergoing thermomechanical cyclic loadings, *Int. J. Solids Struct.* **34**, 4547–4556 (1997).
24. E. Sanchez-Palencia, Comportement local et macroscopique d'un type de milieux physiques hétérogènes, *Int. J. Eng. Sci.* **12**, 331–351 (1974).
25. E. Sanchez-Palencia, *Nonhomogeneous Media and Vibration Theory*, Vol. 127, Lecture Notes in Physics (Springer-Verlag, Heidelberg, 1980).
26. R. P. Skelton, Energy criterion for high temperature low cycle fatigue failure, *Mater. Sci. Technol.* **7**, 427–439 (1991).
27. P. Suquet, Une méthode duale en homogénéisation: Application aux milieux élastiques, *J. Méca. Th. Appl.* (Special issue) 79–98 (1982).
28. P. Suquet, Local and global aspects in the mathematical theory of plasticity, in *Plasticity Today: Modelling, Methods and Applications*, eds. A. Sawczuk and G. Bianchi (Elsevier, London, 1985), pp. 279–310.
29. P. Suquet, Elements of homogenization for inelastic solid mechanics, in *Homogenization Techniques for Composite Media*, eds. E. Sanchez-Palencia and A. Zaoui, Vol. 272, Lecture Notes in Physics (Springer Verlag, New York, 1987), pp. 193–278.
30. P. Suquet, Effective properties of nonlinear composites, in *Continuum Micromechanics*, ed. P. Suquet, Vol. 377, CISM Lecture Notes (Springer Verlag, New York, 1997), pp. 197–264.

31. L. Tartar, Estimations de coefficients homogénéisés, in *Computing Methods in Applied Sciences and Engineering*, eds. R. Glowinski and J. L. Lions, Vol. 704, Lecture Notes in Mathematics (Springer Verlag, Berlin, 1977), pp. 364–373.
32. K. Terada and N. Kikuchi, A class of general algorithms for multi-scale analyses of heterogeneous media, *Comp. Meth. Appl. Mech. Eng.* **190**, 5427–5464 (2001).

MULTISCALE APPROACH FOR THE THERMOMECHANICAL ANALYSIS OF HIERARCHICAL STRUCTURES

Marek J. Lefik*,‡, Daniela P. Boso†,§ and Bernhard A. Schrefler†,¶

*Chair of Geotechnical Engineering and Engineering Structures
Technical University of Łódz, Al.Politechniki 6, 90 924 Łódz, Poland

†Dipartimento di Costruzioni e Trasporti, Università di Padova
Via Marzolo, 9, 35131 Padova, Italy
‡emlefik@p.lodz.pl
§boso@dic.unipd.it,
¶bas@dic.unipd.it

In this chapter we briefly review the most common methods to obtain equivalent properties and then consider full multiscale modelling. Both linear and non-linear material behaviours are considered. The case of composites with periodic microstructure is dealt with in detail and an example shows the capability of the method. Particular importance is also given to non-conventional methods which make use of Artificial Neural Networks (ANN). It is shown how ANN can be used either to substitute the overall material relationship (ANN routines can be easily incorporated in a Finite Element code) or to identify the parameters of the constitutive relation between averages (i.e. relating volume-averaged field variables).

1. Introduction

Composite materials are commonly applied in engineering practice. They allow to take advantage of the different properties of the component materials, of the geometric structure and of the interaction between the constituents to obtain a tailored behaviour as a final result.

Composite materials are usually multiscale in nature, i.e. the scale of the constituents is of lower order than the scale of the resulting material and structure. To fix the ideas, we speak of *macroscopic scale* as the particular scale in which we are interested in (e.g. at structural level) while the lower scales are referred to as *microscopic scales* (sometimes an intermediate scale is called *mesoscopic scale*). We exclude here scales at atomic level, which would require a separate study.

For most of the analyses of composite structures, *effective* or *homogenised* material properties are used, instead of taking into account the individual component properties and their geometrical arrangements. A lot of effort went into the development of mathematical and numerical models to derive homogenised material properties directly from those of the constituents and from their microstructure. Many engineering problems are solved at the macroscopic scale with such homogenised properties. However, in many instances such analyses are not accurate enough.

In principle it would be possible to refer directly to the microscopic scale, but such microscopic models are often far too complex to handle for the analysis of a large structure. Further, the obtained data would be often redundant and complicated procedures would be required to extract information of interest.

A way out is what is now commonly known as *multiscale modelling*, where macroscopic and microscopic models are coupled to take advantage of the efficiency of macroscopic models and the accuracy of the microscopic ones. The scope of such multiscale modelling is to design combined macroscopic–microscopic computational methods that are more efficient than solving the full microscopic model and at the same time give the information that we need to the desired accuracy.[1]

In the case of material and structural multiscale modelling and in homogenisation in general, one usually proceeds from the lower scales upward in order to obtain equivalent material properties. However, it is also important to be able to step down through the scales until the desired scale of the real, not homogenised, material is reached. This technique is often known as *unsmearing*, *localisation*, or *recovering method.* Usually, in a global analysis both aspects need to be pursued, think for instance of a damage or fracture analysis. The procedure may be either of a *serial coupling*, which represents some sort of data passing up and down the scales, or *concurrent coupling* where both microscale and macroscale models are strongly interwoven and have to be addressed continuously as the computation goes on. This last case is particularly the case in non linear situations.

1.1. *Bounds and other estimates*

Over the last decades, a large body of literature was developed, which deals with the micromechanical modelling techniques for heterogeneous materials. As far as the effective properties are concerned, the various approaches may be divided into two main categories depending upon the microstructure characteristics.

In case of composites with a microstructure sufficiently regular to be considered periodic, if the constitutive behaviour of the individual materials is linear, the effective properties may be determined in terms of unit cell problems with appropriate boundary conditions. If the composite materials have non-linear constitutive behaviours, the problem is still deterministic and effective properties may be determined again in terms of unit-cell problems with appropriate boundary conditions, at least provided that the underlying potentials are convex. The case of periodic composites will be dealt with in Sec. 2.

If the microstructure is not regular the effective properties cannot be determined exactly. However, it is possible to define the range of the possible effective behaviour in terms of bounds, depending on some parameters characterising the microstructure. Various homogenisation techniques have been developed in this sense. These methods go back to the works of Voigt[2] and Reuss,[3] which provide two microfield extremes for the effective moduli of N-phase composites with prescribed volume fraction. Voigt's strain field is one where the heterogeneities and the matrix are perfectly bonded, that is kinematically admissible, while Reuss' strains are such that the tractions at the phase boundaries are in equilibrium, that is statically admissible.

Later on Hill[4] and Paul[5] formulated bounds for polycrystals with given orientation distribution functions. More refined bounds are presented in the works of Hashin and Shtrikman[6–8] and Beran.[9] Alternatively *ad hoc* procedures have been proposed to estimate the effective behaviour of composites with special classes of microstructures. Perhaps the best known among these is the self-consistent method.[10–14] The standard self-consistent method is based on the assumption that the particle is embedded in the effective medium instead of the matrix for calculations. In other words, the "matrix material" of the Eshelby[15] formalism is replaced by the effective medium. Unfortunately, the self-consistent method can give unreliable results in case of voids, high volume fractions, or rigid inclusions (see Ref. 16). To improve this approach, the generalised self-consistent method encases the particles in a shell of matrix material surrounded by the effective medium.[17] However, this method also exhibits problems, primarily due to mixing scales of information in a phenomenological manner.

For composites with random microstructures and non-linear material behaviour, the first information is given by the approximation of Taylor[18] followed by the bounds of Bishop and Hill[19–21] for rigid perfectly plastic polycrystals. The Taylor–Bishop–Hill estimates may be seen as a generalisation for non-linear composites of the Voigt–Reuss–Hill bounds. Various generalisations of the self-consistent method are also available in literature (e.g. Hill,[21] Hutchinson,[22] and Berveiller and Zaoui[23]).

The works of Wills[24,25] and Talbot[26] provide extensions of the Hashin–Shtrikman variational principles for non-linear composites. Their work is followed by the introduction of several new variational principles making use of appropriately chosen "linear comparison composites", which allow the determination of Hashin–Shtrikman and more general bounds and estimates, directly from the corresponding estimates for linear composites. These include the works by Ponte Castaneda,[27,28] Talbot and Willis,[29] Suquet,[30] and Olson.[31]

There exist a lot of other approaches that seek to estimate or bound the aggregate responses of micro-heterogeneous materials. A complete survey is outside the scope of the present work, and we refer the reader to the works of Hashin,[32] Mura,[33] Aboudi,[16] Nemat-Nasser and Hori,[34] and recently Torquato[35] for such reviews or to the extensive works of Llorca and coworkers.[36–44]

2. Asymptotic Theory of Homogenisation

2.1. *Asymptotic analysis*

Asymptotic analysis does not only permit to obtain equivalent material properties, but also allows to solve the full structural problem down to stresses in the constituent materials at the micro- (or local) scale. It is mostly applied to linear two-scale problems, but it can be extended to non-linear analysis and to several scales as will be shown further on. We do not intend to give here a full account of the underlying theory. The interested reader will find in the works of Bensoussan *et al.*[45] and Sanchez-Palencia[46] the rigorous formulation of the method, its application in many fields and further references.

We will however show in detail its finite element (FE) solution, because it is a basic ingredient of many multi-scale analyses. For the moment we consider just two levels, the micro- (or local) and the macro- (or global) level. These levels are shown in Fig. 1, where the structure is periodic and asymptotic analysis can be successfully applied.[47]

Periodicity means that if we consider a body Ω with periodic structure and a generic mechanical or geometric property a (for example, the

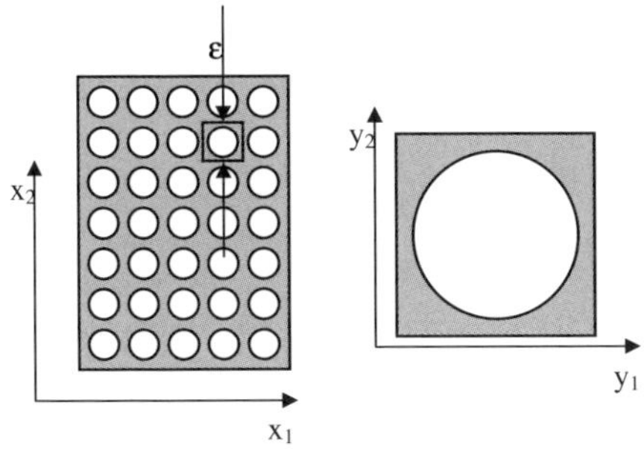

Fig. 1. Example of a periodic structure with two levels: global on the left-hand side and local on the right-hand side.

constitutive tensor), we have

$$\text{if } \mathbf{x} \in \Omega \quad \text{and} \quad (\mathbf{x} + \mathbf{Y}) \in \Omega \Rightarrow a(\mathbf{x} + \mathbf{Y}) = a(\mathbf{x}), \tag{1}$$

where $\mathbf{Y}$ is the (geometric) period of the structure. Hence, the elements of a are $\mathbf{Y}$-periodic functions of the position vector $\mathbf{x}$.

2.2. *Statement of the problem and assumptions*

The first important assumption for asymptotic analysis is that it must be possible to distinguish two length scales associated with the macroscopic and microscopic phenomena. The characteristic size of the single cell of periodicity is assumed to be much smaller than the geometric dimensions of the structure under analysis which means that a clear scale separation is possible. This means that the ratio of these scales defines the small parameter ε (Fig. 1):

$$\mathbf{y} = \frac{\mathbf{x}}{\varepsilon}. \tag{2}$$

Two sets of coordinates related by (2) formally express this separation of scales between macro- and microphenomena: the global coordinate vector $\mathbf{x}$ refers to the whole body Ω, and the stretched local coordinate vector $\mathbf{y}$ is related to the single, repetitive cell of periodicity.

In this way the single cell is mapped into the unitary domain Y (here and in the remainder Y indicates the unitary domain occupied by the cell of periodicity and not the period of the composite material like in (1)).

In the asymptotic analysis the normalised cell of periodicity is mapped onto a sequence of finer and finer structures as ε tends to 0. If the equivalent material properties as defined below are employed, the considered fields (e.g. displacement, temperature, etc.) converge toward the homogeneous macroscopic solution, as the microstructural parameter ε tends to 0. In this sense problems for a heterogeneous body and a homogenised one are equivalent. (For more details concerning the mathematical meaning, see Refs. 45 and 46.)

We now consider a problem of thermo-elasticity defined in a heterogeneous body such as that depicted in Fig. 1, defined by the usual equations (3) to (9):

Balance equations

$$\sigma^{\varepsilon}_{ij,j}(\mathbf{x}) + f_i(\mathbf{x}) = 0, \tag{3}$$

$$q^{\varepsilon}_{i,i} - r = 0. \tag{4}$$

Constitutive equations

$$\sigma^{\varepsilon}_{ij}(\mathbf{x}) = a^{\varepsilon}_{ijkl}(\mathbf{x})e_{kl}(\mathbf{u}^{\varepsilon}(\mathbf{x})) - \alpha^{\varepsilon}_{ij}\theta, \tag{5}$$

$$q^{\varepsilon}_i = -K^{\varepsilon}_{ij}\theta_j. \tag{6}$$

Small strain definition

$$e_{ij}(\mathbf{u}^{\varepsilon}(\mathbf{x})) = 0.5(u^{e}_{i,j}(\mathbf{x}) + u^{e}_{j,i}(\mathbf{x})). \tag{7}$$

Boundary conditions and continuity conditions on the interfaces between different materials S_J

$$\sigma^{\varepsilon}_{ij}(\mathbf{x})n_j = 0 \text{ on } \partial\Omega_1 \quad \text{and} \quad u^{\varepsilon}_i(\mathbf{x}) = 0 \text{ on } \partial\Omega_2, \tag{8a}$$

$$q^{\varepsilon}_i(\mathbf{x})n_i = 0 \text{ on } \partial\Omega_q \quad \text{and} \quad \theta^{\varepsilon}(\mathbf{x}) = 0 \text{ on } \partial\Omega_\theta, \tag{8b}$$

$$[u^{\varepsilon}_i(\mathbf{x})] = 0 \quad [\sigma^{\varepsilon}_{ij}(\mathbf{x})n_j] = 0 \text{ on } S_J, \tag{9a}$$

$$[\theta^{\varepsilon}(\mathbf{x})] = 0 \quad [q^{\varepsilon}_i(\mathbf{x})n_i] = 0 \text{ on } S_J, \tag{9b}$$

where the superscript ε is used to indicate that the variables of the problem depend on the cell dimensions related to the global length. Square parentheses denote the jump of the enclosed value. The other symbols have the usual meaning: $\mathbf{u}$ is the displacement vector, $\mathbf{e}(\mathbf{u}(\mathbf{x}))$ denotes the linearised strain tensor, $\sigma_{ij}(\mathbf{x})$ is the stress tensor, $a_{ijkl}(\mathbf{x})$ is the tensor of elasticity, $K_{ij}(\mathbf{x})$ is the tensor of thermal conductivity, $\alpha_{ij}(\mathbf{x})$ is the tensor of thermal expansion coefficients, $\theta(\mathbf{x})$ and $q_i(\mathbf{x})$ are temperature and heat flux, respectively, and $r(\mathbf{x})$ and $f_i(\mathbf{x})$ stand for thermal sources and mass forces.

Since the components of the elasticity tensor are discontinuous, differentiation (in the above equations and in (16)–(21) below) should be understood in the weak sense. This is the main reason why most of the problems posed in the sequel will be presented in a variational formulation.

We introduce now the second main hypothesis of homogenisation theory: the periodicity of the material characteristics imposes an analogous periodical perturbation on quantities describing the mechanical behaviour of the body. As a consequence we can use the following representation for displacements and temperatures:

$$\mathbf{u}^{\varepsilon}(\mathbf{x}) \equiv \mathbf{u}^0(\mathbf{x}) + \varepsilon\mathbf{u}^1(\mathbf{x},\mathbf{y}) + \varepsilon^2\mathbf{u}^2(\mathbf{x},\mathbf{y}) + \cdots + \varepsilon^k\mathbf{u}^k(\mathbf{x},\mathbf{y}), \tag{10}$$

$$\theta^{\varepsilon}(\mathbf{x}) \equiv \theta^0(\mathbf{x}) + \varepsilon\theta^1(\mathbf{x},\mathbf{y}) + \varepsilon^2\theta^2(\mathbf{x},\mathbf{y}) + \cdots + \varepsilon^k\theta^k(\mathbf{x},\mathbf{y}). \tag{11}$$

Similar expansion with respect to powers of ε results from (10) and (11) for stresses, strains and heat fluxes

$$\sigma^{\varepsilon}(\mathbf{x}) \equiv \sigma^0(\mathbf{x},\mathbf{y}) + \varepsilon\sigma^1(\mathbf{x},\mathbf{y}) + \varepsilon^2\sigma^2(\mathbf{x},\mathbf{y}) + \cdots + \varepsilon^k\sigma^k(\mathbf{x},\mathbf{y}), \tag{12}$$

$$\mathbf{e}^{\varepsilon}(\mathbf{x}) \equiv \mathbf{e}^0(\mathbf{x},\mathbf{y}) + \varepsilon\mathbf{e}^1(\mathbf{x},\mathbf{y}) + \varepsilon^2\mathbf{e}^2(\mathbf{x},\mathbf{y}) + \cdots + \varepsilon^k\mathbf{e}^k(\mathbf{x},\mathbf{y}), \tag{13}$$

$$\mathbf{q}^{\varepsilon}(\mathbf{x}) \equiv \mathbf{q}^0(\mathbf{x},\mathbf{y}) + \varepsilon\mathbf{q}^1(\mathbf{x},\mathbf{y}) + \varepsilon^2\mathbf{q}^2(\mathbf{x},\mathbf{y}) + \cdots + \varepsilon^k\mathbf{q}^k(\mathbf{x},\mathbf{y}), \tag{14}$$

where $\mathbf{u}^k, \boldsymbol{\sigma}^k, \mathbf{e}^k, \theta^k, \mathbf{q}^k$ for $k > 0$ are Y-periodic, i.e. they take the same values on the opposite sides of the cell of periodicity.

The term scaled with the nth power of ε in (10)–(14) is called *term of order n*.

2.3. *Formalism of the homogenisation procedure*

The necessary mathematical tools are the chain rule of differentiation with respect to the microvariable and averaging over a cell of periodicity.

We introduce the assumption (10)–(14) into the equations of the heterogeneous problem (3)–(9) and make use of the rule of differential calculus (see also Ref. 46):

$$\frac{d}{dx_i} f = \left(\frac{\partial}{\partial x_i} + \frac{1}{e}\frac{\partial}{\partial y_i}\right) f = f_{i(x)} + \frac{1}{e} f_{i(y)}. \tag{15}$$

This equation explains also the notation used in the sequel for differentiation with respect to local and global independent variables.

Because of (15) the equilibrium equations split into terms of different orders (the terms of the same power of ε are equated to zero separately, e.g., Eqs. (16) and (19) are of order $1/\varepsilon$).

For the equilibrium equation we have

$$\sigma^0_{ij,j(y)}(\mathbf{x},\mathbf{y}) = 0, \tag{16}$$

$$\sigma^0_{ij,j(x)}(\mathbf{x},\mathbf{y}) + \sigma^1_{ij,j(y)}(\mathbf{x},\mathbf{y}) + f_i(\mathbf{x}) = 0, \tag{17}$$

$$\sigma^1_{ij,j(x)}(\mathbf{x},\mathbf{y}) + \sigma^2_{ij,j(y)}(\mathbf{x},\mathbf{y}) = 0. \tag{18}$$

We have similar expressions for the heat balance equation

$$q^0_{i,i(y)}(\mathbf{x},\mathbf{y}) = 0, \tag{19}$$

$$q^0_{i,i(x)}(\mathbf{x},\mathbf{y}) + q^1_{i,i(y)}(\mathbf{x},\mathbf{y}) - r(\mathbf{x}) = 0, \tag{20}$$

$$q^1_{i,i(x)}(\mathbf{x},\mathbf{y}) + q^2_{i,i(y)}(\mathbf{x},\mathbf{y}) = 0. \tag{21}$$

From Eqs. (7) and (15) it follows that the main term of $\mathbf{e}$ in expansions (13) depends not only on $\mathbf{u}^0$, but also on $\mathbf{u}^1$:

$$e^0_{ij}(x,y) = u^0_{(i,j)(x)} + u^1_{(i,j)(y)} \equiv e_{ij(x)}(\mathbf{u}^0) + e_{ij(y)}(\mathbf{u}^1). \tag{22}$$

The constitutive relationships (5) and (6) assume now the form

$$\sigma^0_{ij}(\mathbf{x},\mathbf{y}) = a_{ijkl}(\mathbf{y})(e_{kl(x)}(\mathbf{u}^0) + e_{kl(y)}(\mathbf{u}^1)) - \alpha_{ij}(\mathbf{y}), \tag{23}$$

$$\sigma^1_{ij}(\mathbf{x},\mathbf{y}) = a_{ijkl}(\mathbf{y})(e_{kl(x)}(\mathbf{u}^1) + e_{kl(y)}(\mathbf{u}^2)) - \alpha_{ij}(\mathbf{y}). \tag{24}$$

$$\vdots$$

$$q^0_k(\mathbf{x},\mathbf{y}) = K_{kl}(\mathbf{y})(\theta^0_{l(x)} + \theta^1_{l(y)}), \tag{25}$$

$$q^1_k(\mathbf{x},\mathbf{y}) = K_{kl}(\mathbf{y})(\theta^1_{l(x)} + \theta^2_{l(y)}). \tag{26}$$

It can be seen that the terms of order n in the asymptotic expansions for stresses (23), (24) and heat flux (25), (26) depend respectively on the displacement and temperature terms of order n and $n+1$. In this way, the influence of the local perturbation on the global quantities is accounted for. This is the reason why for instance we need $\mathbf{u}^1(\mathbf{x},\mathbf{y})$ to define via the constitutive relationship the main term in expansion (12) for stresses (and $\mathbf{u}^2(\mathbf{x},\mathbf{y})$ for the term of order 1, if needed).

2.4. *Global solution*

Referring separately to the terms of the same powers of ε leads to the following variational formulations for the unknowns of successive order of the problem. Starting with the first order, it can be formally shown[46,48] that $\mathbf{u}^1(\mathbf{x}, \mathbf{y})$ and similarly $\theta^1(\mathbf{x}, \mathbf{y})$ may be represented by separate variables

$$u_i^1(\mathbf{x}, \mathbf{y}) = -e_{pq(x)}(\mathbf{u}^0(\mathbf{x}))\chi_i^{pq}(\mathbf{y}) + C_i(\mathbf{x}), \tag{27}$$

$$\theta^1(\mathbf{x}, \mathbf{y}) = \theta^0_{p(x)}(\mathbf{x})\vartheta^p(\mathbf{y}) + C(\mathbf{x}). \tag{28}$$

We will call $\chi^{pq}(\mathbf{y})$ and $\vartheta^p(\mathbf{y})$ the *homogenisation functions* for displacements and temperature, respectively. The *zero order* (sometimes also referred to as *first order*) component of the equation of equilibrium (16) and of heat balance (19) in the light of (27) and (28) yields the following boundary value problems (BVP) for the functions of homogenisation:

$$\text{find } \chi_i^{pq} \in V_Y \text{ such that: } \forall v_i \in V_Y,$$
$$\int_Y a_{ijkl}(\mathbf{y})(\delta_{ip}\delta_{jq} + \chi^{pq}_{i,j(y)}(\mathbf{y}))v_{k,l(y)}(\mathbf{y})\, d\Omega = 0, \tag{29}$$

$$\text{find } \vartheta^p \in V_Y \text{ such that: } \forall \varphi \in V_Y,$$
$$\int_Y K_{ij}(\mathbf{y})(\delta_{ip} + \vartheta^p_{i(y)}(\mathbf{y}))\varphi_{j(y)}(\mathbf{y})\, d\Omega = 0. \tag{30}$$

In the above equations V_Y is the subset of the space of kinematically admissible functions which contains the functions with equal values on the opposite sides of the cell of periodicity Y. The tensor $\boldsymbol{\chi}^{pq}$ and the three scalar functions ϑ^p depend only on the geometry of the cell of periodicity and on the values of the jumps of material coefficients across S_J. Functions $v(\mathbf{y})$ and $\varphi(\mathbf{y})$ are usual test functions having the meaning of Y-periodic displacement and temperature fields, respectively. They are used here to write explicitly the counterparts of the expressions (16) and (19), in which the prescribed differentiations are understood in a weak sense.

The solutions $\boldsymbol{\chi}^{pq}$ and ϑ^p of the *local* (that is defined for a single cell of periodicity) BVPs with periodic boundary conditions (29) and (30) can be interpreted as obtained for the cell subject to a unitary average strain e^{pq} and unitary average temperature gradient $\vartheta_{p(y)}$, respectively. The true values of perturbations are obtained later by scaling $\boldsymbol{\chi}^{pq}$ and ϑ^p with true global strains (gradient of global temperature), as prescribed by (27) and (28).

In the asymptotic expansion for displacements (10) and for temperature (11) the dependence on $\mathbf{x}$ only is marked in the first term. The independence on $\mathbf{y}$ of these functions can be proved (see for example, Ref. 46). The functions depending only on $\mathbf{x}$ define the macrobehaviour of the structure and we will call them *global terms.*

To obtain the global behaviour of stresses and of heat flux the mean values over the cell of periodicity are defined[46]:

$$\tilde{\sigma}^0_{ij}(\mathbf{x}) = |Y|^{-1}\int_Y \sigma^0_{ij}(\mathbf{x},\mathbf{y})\,dY, \quad \tilde{\mathbf{q}}^0(\mathbf{x}) = |Y|^{-1}\int_Y \mathbf{q}^0(\mathbf{x},\mathbf{y})\,dY. \tag{31}$$

Averaging of Eqs. (23) and (25) results in the effective constitutive relationships

$$\tilde{\sigma}^0_{ij}(\mathbf{x}) = a^h_{ijkl}e_{kl}(\mathbf{u}^0), \quad \tilde{q}^0_i = -k^h_{ij}\theta^0_j. \tag{32}$$

In the above equations the effective material coefficients appear. They are computed according to

$$a^h_{ijkl} = |Y|^{-1}\int_Y a_{ijpq}(y)(\delta_{kp}\delta_{lq} + \chi^{pq}_{k,l(y)}(y))\,dY \tag{33}$$

$$k^h_{ij} = |Y|^{-1}\int_Y k_{ip}(\mathbf{y})(\delta_{jp} + \vartheta^j_p(\mathbf{y}))\,dY, \tag{34}$$

$$\alpha^h_{ij} = |Y|^{-1}\int_Y \alpha_{ij}(\mathbf{y})\,dY. \tag{35}$$

The macrobehaviour can be defined now by averaging first-order terms in the equilibrium and flux balance equations (17), (20), and boundary conditions (8a) and (8b), and then substituting the averaged counterparts of stress and heat flux (31) (first-order perturbations vanish in averaging (17) and (20) because of periodicity). Equations (5) and (6) should be replaced by (32), while Eqs. (9a) and (9b) have no more sense since we deal now with homogeneous uncoupled thermo-elasticity.

The heterogeneous structure can now be studied as a homogeneous one with effective material coefficients given by (33)–(35), and global displacements, strains and average stresses, and heat fluxes can be computed. Then we go back to Eq. (23) for the local approximation of stresses. This last step is the above-mentioned *unsmearing* or *re-localisation.*

2.5. *Local approximation of the stress vector*

We note that the homogenisation approach results in two different kinds of stress tensors. The first one is the average stress field defined by (32).

It represents the stress tensor for the homogenised, equivalent but unreal body. Once the effective material coefficients are known, the stress field may be obtained from a standard finite element structural code as explained above.

The other stress field is associated with a family of uniform states of strains $e_{pq(x)}(\mathbf{u}^0)$ over each cell of periodicity Y. This local stress is obtained by introducing Eq. (22) into (23) and results in

$$\sigma_{ij}^0(\mathbf{x},\mathbf{y}) = a_{ijkl}(\mathbf{y})(\delta_{kp}\delta_{lq} - \chi_{k,l(y)}^{pq})e_{pq(x)}(\mathbf{u}^0) - \alpha_{ij}(\mathbf{y})\theta^0. \tag{36}$$

Because of (16) and (29) this tensor fulfils the equations of equilibrium everywhere in Y. If needed, the stress description can be completed with a higher order term in Eq. (12). This approach is presented by Lefik *et al.*[49,50] Finally, the local approximation of heat flux is as follows:

$$q_j^0(\mathbf{y}) = k_{ij}(\mathbf{y})(\delta_{ip} + \vartheta_{i(y)}^p(\mathbf{y}))\theta_{p(x)}^0. \tag{37}$$

2.6. *Finite element analysis applied to the local problem*

For the numerical formulation, it is convenient to use the matrix notation for the above-introduced quantities.

The homogenisation functions are ordered as defined by Eqs. (38) and (39), respectively (the numbers in the superscripts in Eqs. (38), (39) and subscripts in Eqs. (40), (41) refer to the reference axes 1, 2, 3):

$$\mathbf{X}^{\mathrm{T}}(\mathbf{y}) = [\{\boldsymbol{\chi}^{11}(\mathbf{y})\}\{\boldsymbol{\chi}^{22}(\mathbf{y})\}\{\boldsymbol{\chi}^{33}(\mathbf{y})\}\{\boldsymbol{\chi}^{12}(y)\}\{\boldsymbol{\chi}^{23}(\mathbf{y})\}\{\boldsymbol{\chi}^{13}(\mathbf{y})\}]_{3\times 6}, \tag{38}$$

$$\mathbf{T}^{\mathrm{T}}(\mathbf{y}) = [\vartheta^1(\mathbf{y})\vartheta^2(\mathbf{y})\vartheta^3(\mathbf{y})]_{1\times 3}. \tag{39}$$

This is in accordance with the ordering of strains and temperature gradients

$$\mathbf{e} = \{e_{11}\ e_{22}\ e_{33}\ e_{12}\ e_{23}\ e_{13}\}_6^{\mathrm{T}} = \{e_{pq}\}_6^{\mathrm{T}}, \tag{40}$$

$$\theta_p = \{\theta_1\ \theta_2\ \theta_3\}^{\mathrm{T}} = \{\theta_p\}_3^{\mathrm{T}}. \tag{41}$$

In the following, the superscript e denotes the values of a function in the nodes of a FE mesh.

We have the usual representations for each element

$$\mathbf{X}(\mathbf{y}) = \mathbf{N}(\mathbf{y})\mathbf{X}^e, \tag{42}$$

where $\mathbf{N}$ contains the values of standard shape functions.

It is easy to show that the variational formulation (29) can be rewritten as follows:

$$\begin{gathered}\text{find } \mathbf{X} \in V_Y \text{ such that: } \forall \mathbf{v} \in V_Y, \\ \int_Y \mathbf{e}^{\mathrm{T}}(\mathbf{v}(\mathbf{y}))\mathbf{D}(\mathbf{y})(\mathbf{1} + \mathbf{L}\mathbf{X}(\mathbf{y}))\, dY = 0.\end{gathered} \tag{43}$$

In the above, $\mathbf{L}$ denotes the matrix of differential operators and $\mathbf{D}$ contains the material coefficients a_{ijkl} in the repetitive domain. Matrix $\mathbf{X}^{\mathrm{e}}$ which contains the values of homogenisation functions in the nodes of the mesh is obtained as a FE solution of (43). The equation to solve is

$$\begin{gathered}\mathbf{K}\mathbf{X}^{\mathrm{e}} - \mathbf{F} = \mathbf{0}; \quad \mathbf{X} \text{ being } Y\text{-periodic,} \\ \text{with zero mean value over the cell,}\end{gathered} \tag{44}$$

where

$$\mathbf{F} = \int_Y \mathbf{B}^{\mathrm{T}}\mathbf{D}(\mathbf{y}), \quad \mathbf{K} = \int_Y \mathbf{B}^{\mathrm{T}}\mathbf{D}(\mathbf{y})\mathbf{B}, \quad \mathbf{B} = \mathbf{L}\mathbf{N}(\mathbf{y}). \tag{45}$$

It can be shown that $\mathbf{X}$ in (43) (and thus in (44)) is a solution of a BVP, for which the loading consists of unitary average strains over the cell. This is seen in the form of the first equation of (45), which forms a matrix. We thus solve six equations for six functions of homogenisation.

The variational formulation (30) can be represented in a form similar to (44), $\mathbf{T}^{\mathrm{e}}$ being Y-periodic, with a given mean zero value over the cell

$$\mathbf{K}\mathbf{T}^{\mathrm{e}} + \mathbf{F} = \mathbf{0}, \tag{46}$$

where

$$\mathbf{F} = \int_Y \mathbf{B}_\theta^{\mathrm{T}}\mathbf{K}_\theta(\mathbf{y}), \quad \mathbf{K} = \int_Y \mathbf{B}_\theta^{\mathrm{T}}\mathbf{K}_\theta(\mathbf{y})\mathbf{B}_\theta, \quad \mathbf{B} = \mathbf{L}_\theta\mathbf{N}(\mathbf{y}). \tag{47}$$

$\mathbf{K}_\theta$ contains the conductivities k_{ij} of materials in the repetitive domain. Differential operators in $\mathbf{L}_\theta$ are ordered suitably for the thermal problem.

The periodicity conditions can be taken into account using Lagrange multiplier in the construction of a FE code. Also, the requirements of the zero mean value has to be included in the program.

Having computed $\mathbf{X}^{\mathrm{e}}$, $\mathbf{T}^{\mathrm{e}}$ and by consequence $\mathbf{u}^1$ and θ^1, one can derive the effective material coefficients, according to

$$\mathbf{D}^h = |Y|^{-1} \int_Y \mathbf{D}(\mathbf{y})(\mathbf{1} - \mathbf{B}\mathbf{X}^{\mathrm{e}})\, dY, \tag{48}$$

$$\mathbf{K}^h = |Y|^{-1} \int_Y \mathbf{K}_\theta(\mathbf{y})(\mathbf{1} + \mathbf{B}T^{\mathrm{e}})\, dY, \tag{49}$$

$$\boldsymbol{\alpha}^h = |Y|^{-1} \int_Y \boldsymbol{\alpha}(\mathbf{y})\, dY. \tag{50}$$

With the homogenised material coefficients (48)–(50) any thermo-elastic FE code can be used to obtain the global displacements and temperatures. For the unsmearing procedure we need the gradients of temperatures and strains in the regions of interest (see Eqs. (36) and (37)). Strains are directly obtained from standard post-processing, and gradients of temperature can be replaced by their local approximation with finite differences.

To present the graphs of stresses over the single cell, nodal projection can be used. To assure continuity of tangential stresses, this projection should be extended to patches of cells.

2.7. *Asymptotic homogenisation at three levels: Micro, meso, and macro*

Asymptotic theory of homogenisation is applicable also to non-linear situations, if applied iteratively. Further, it can obviously be used to bridge several scales. Here we deal with the case where three scales are bridged by applying in sequential manner the two-scale asymptotic analyses. The behaviour of the components is physically non-linear. Again we refer to thermomechanical behaviour and introduce a micro-, meso-, and macro-level, as shown in Fig. 2.

At the stage of micro- or mesomodelling, some main features of the local structure can be extracted and used then for the macro-analysis. The behaviour of the components, even if elastoplastic, is supposed here to be piecewise linear, so that the homogenisation we perform is piecewise linear. Only monotonic loading and/or temperature variation are considered; otherwise, we should store the whole history and use an incremental analysis.

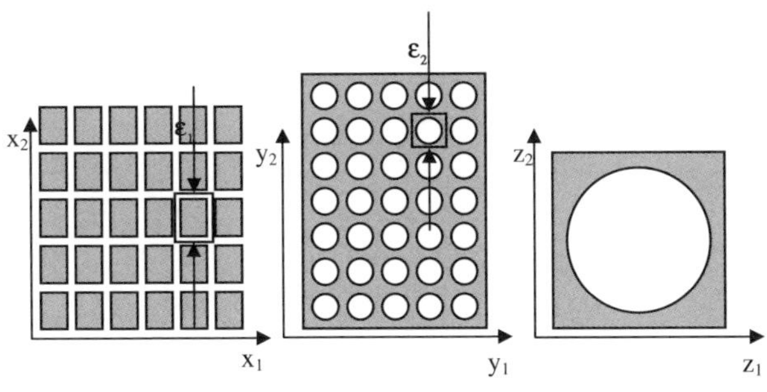

Fig. 2. Example of a periodic structure with three levels: macro (on the left), meso and micro (on the right).

Because of the chosen material properties we deal with a sequence of problems of linear elasticity written for a non-homogeneous material domain and with coefficients that are functions of both temperature and stress level.

At the top level of the hierarchy we consider an elastic body contained in the domain Ω with a smooth boundary $\partial\Omega$. On the part $\partial\Omega_1$ of its boundary, tractions are given. On the remaining part of $\partial\Omega$ (i.e. on $\partial\Omega_2$), displacements are prescribed. The domain Ω as filled with repetitive cells of periodicity Y, shown in Fig. 2, where the material of the body is supposed to be piecewise homogeneous inside Y, as defined in Eq. (1). The governing equations are still (3)–(9).

For the lower level all the formulations are formally the same with one difference: the boundary conditions are those of an infinite body. It is worth to mention that all the macrofields at the microlevel become the microfields at the higher structural level. Effective material coefficients and mean fields obtained with the homogenisation procedure at the lower level enter as local perturbations at the higher step.

Before explaining the application of the homogenisation procedure in sequential form to multilevel non-linear material behaviour, we mention the solution by Terada and Kikuchi,[51] who wrote a two-scale variational statement within the theory of homogenisation. The solution of the microscopic problem at each Gauss point of the finite element mesh for the overall structure, and the deformation histories at time t_{n-1} must be stored until the macroscopic equilibrium state at current time t_n is obtained. This procedure has not been applied to bridging of more than two scales.

A triple scale asymptotic analysis is used by Fish and Yu[52] to analyse damage phenomena occurring at micro-, meso-, and macroscales in brittle composite materials (woven composites). These authors also maintain the second-order term in the displacement expansion (Eq. (10)) and introduce a similar form for the expansion of the damage variable. We recall further that stochastic aspects can also be introduced in the homogenisation procedure.[53]

A three-level homogenisation is now presented, dealing with non-linear, temperature-dependent material characteristics. The two usual tools of homogenisation of the previous section are used, i.e. volume averaging and total differentiation with respect to the global variable $\mathbf{x}$ that involves the local variable $\mathbf{y}$. The homogenisation functions are obtained similar to Eqs. (29) and (30) (only a factor λ is introduced to adapt the solution to the real strain level as explained below):

$$\begin{aligned} &\text{find } \chi_i^{pq} \in V_Y \text{ such that:} \\ &\forall v_i \in V_Y \int_{Y(\lambda)} C_{ijkl}(\mathbf{y}, \lambda, \theta^0)(\delta_{ip}\delta_{jq} + \chi_{i,j(y)}^{pq})v_{k,l(y)}\, d\Omega = 0, \\ &\qquad\qquad \sigma(\lambda, \chi_i^{pq}) \in P. \end{aligned} \tag{51}$$

Material properties depend upon temperature, so that a set of representative temperatures is considered for the material input data and linear interpolation is used between the given values. P is the domain inside the surface of plasticity. The requirement that the stress belongs to the admissible region P (introduced in (51)) is verified via classical unsmearing procedure, described before.

The modification of the algorithm required by the material non-linearity is now explained. We start with the composite cell of periodicity with given elastic components. The uniform strain is increased step-by-step. Effective material coefficients are constant until the stress reaches the yield surface in some points of the cell. The yield surface in the space of stresses is different for each material component, being thus a function of place. The region, where the material yields, is of finite volume at the end of the step; hence, it is easy to replace the material with the yielded one, with the elastic modulus equal to the hardening one, and with Poisson ratio tending to 0.5.

The cell of periodicity is hence transformed in this way: it is made up of one more material and we can start the usual analysis again (uniform strain, new homogenisation function, new stress map over the cell). We identify then the new region where further local yielding occurs, then redefine the cell, and so on. The loop is repeated as many times as needed. In (51) the history of this replacement of materials at the microlevel is marked by λ, the level of the average stress, for which the micro yielding occurs each time. The algorithm is summarised in Box 1.

At the end of each step we can also compute the mean stress over the cell having (generalised) homogenisation functions (see Eqs. (32)) and the effective coefficients can be computed using Eqs. (33)–(35).

As mentioned, an important part of multiscale modelling is the recovery of stress, strain, and displacements at the level of the microstructure. This is obtained from Eqs. (36) and (37) using the following procedure: first global (mean) fields are obtained from the homogeneous analysis where the material is characterised by the effective coefficients (33)–(35); then, we return to the original problem formulation, using homogenisation functions. We recover thus the main parts of the stress and heat flux.

Box 1. Updating yield surface algorithm scheme. It is to note, that for "solving BV problem" mentioned in points (vi) and (viii) it is not always necessary to use the true finite element solution. If the cell of periodicity has not been changed before, this solution can be composed according to (27).

(i) Compute effective coefficients at microlevel;
(ii) Compute effective coefficients at mesolevel;
(iii) Apply increment of forces and/or temperature at the macrolevel, solve global homogeneous problem;
(iv) Compute global strain $E_{ij} : E_{ij} = e_{ij}(\mathbf{u}^0)$ reminding that $E_{ij} = \tilde{e}^\varepsilon(\mathbf{x})$;
(v) Apply E_{ij} to mesolevel cell by equivalent kinematical loading (displacement on the border);
(vi) Solve the kinematical problem at the mesolevel for $\mathbf{w}(\mathbf{y})$, compute stress (unsmearing for mesolevel) and strain E_{ij}; now $E_{ij} = e_{ij}(\mathbf{w}^0)$ and $E_{ij} = \tilde{e}^\varepsilon(\mathbf{y})$;
(vii) Apply E_{ij} from meso- to microlevel cell by equivalent kinematical loading (displacements on the border);
(viii) Solve the kinematical problem at the microlevel for $\mathbf{w}^1(\mathbf{z})$; compute stress (unsmearing for microlevel);
(ix) Verify yielding of the material **in the physically true situation at microlevel**. If yes change mechanical parameter of the material and go to 1, else exit.

Because of the three-level hierarchical structure we are dealing with, the recovery process must be applied twice, and since material characteristics are temperature-dependent and non-linear, the procedure must be applied for each representative temperature and within the context of the correct stress state. We recall that the recovery process starts at the highest structural level while the homogenisation begins at the lowest part of the structural hierarchy.

As an example of application, we consider a superconducting strand used for fusion devices. The structures and the three scales are shown in Figs. 3 and 4, where the single filament (microscale), groups of filaments (mesoscale), and the superconducting strand (macroscale in this case) are shown.

The homogeneous effective properties will be defined for the inner part of the strand, shown on the left of Fig. 4. The diameter of the strand is about 0.80 mm. The application of the theory of homogenisation is justified by the scale separation clearly evidenced in Figs. 3 and 4.

As already indicated, periodic homogenisation is applicable to structures obtained by a multiple translation of a representative volume element (RVE), called in this case the cell of periodicity. The considered strand

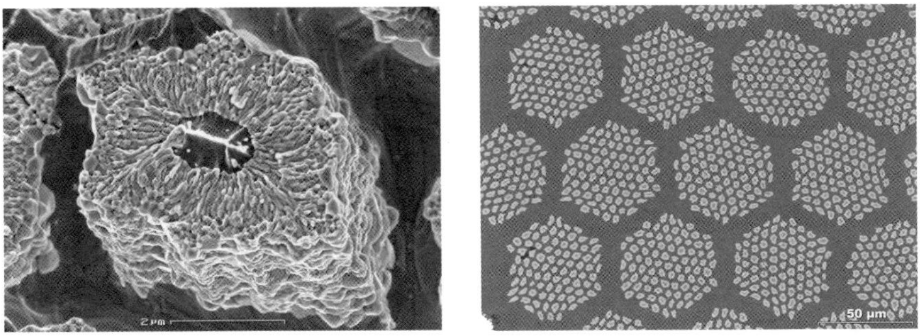

Fig. 3. A single Nb_3Sn filament (left) and Nb_3Sn filaments groups (right); the respective scales are also evidenced. Each filament group is made of 85 filaments. Courtesy of P. J. Lee, University of Wisconsin, Madison Applied Superconductivity Center.

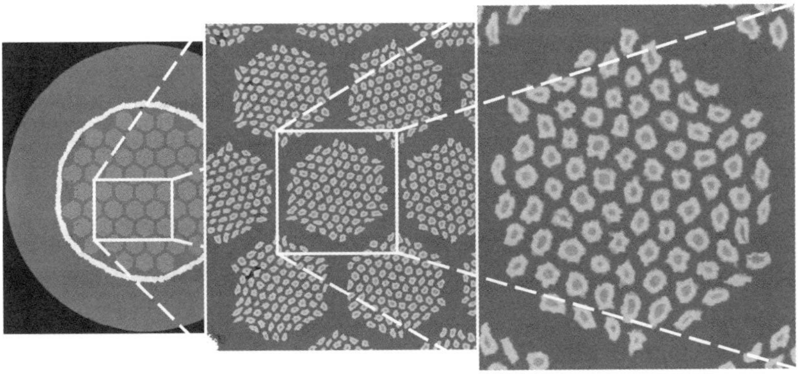

Fig. 4. Three-level hierarchy in the VAC strand. The central part of the strand itself (left) consists of 55 groups of 85 filaments, embedded in tin rich bronze matrix, while the outer region is made of high conductivity copper. Courtesy of P. J. Lee, University of Wisconsin, Madison Applied Superconductivity Center.

shows two different levels of such a translative structure. On the mesolevel we have the repetitive pattern of the superconducting filament in the bronze matrix (Microscale RVE), filling the hexagonal region as illustrated in Fig. 5. The second translative structure is the net of the hexagonal filament groups (Mesoscale RVE) in the body of the single strand shown in Fig. 6. The homogenisation thus splits into two steps, each one dealing with rather similar geometry and a comparable scale separation factor.

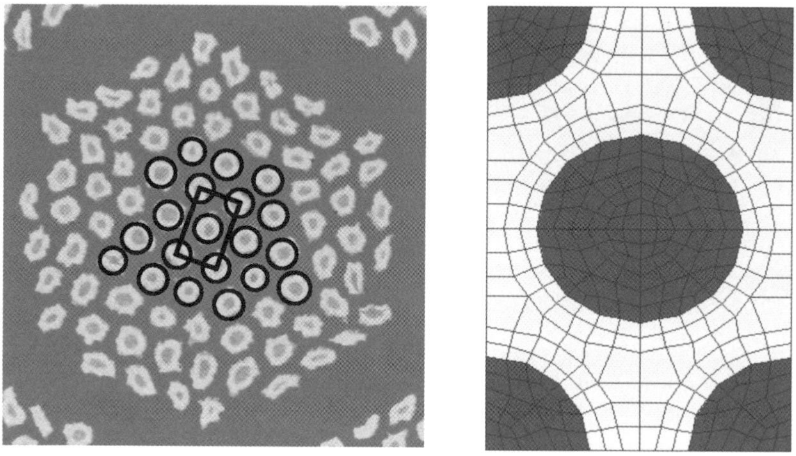

Fig. 5. Microscale unit cell. Light element: bronze material, dark elements: Nb_3Sn alloy. The area of the cell is $9.0 \times 5.6\,\mu$m.

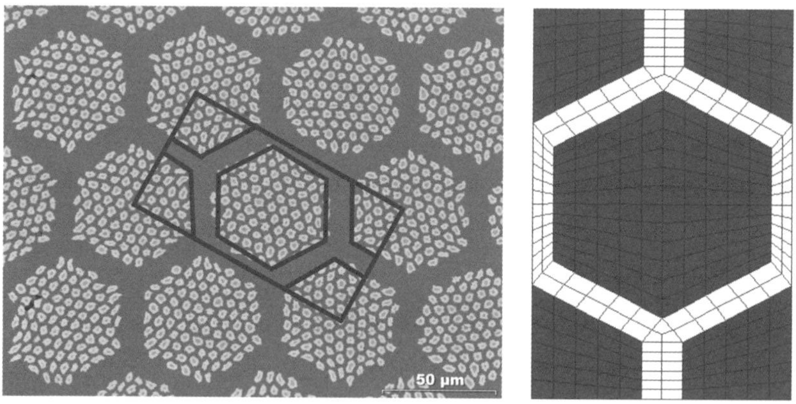

Fig. 6. Mesoscale unit cell. Light element: bronze material, dark elements: homogenised material at microlevel. The area of the cell is $100.0 \times 60.1\,\mu$m.

Boundary conditions for the macro-problem will be given in terms of interaction of the strand with the other strands in the cable,[54] and will be of the type of equations (8a).

To form the Nb_3Sn compound (which is the superconducting material) the strand is kept for 175 h at 923 K. Afterwards, to reach the operating temperature, it is cooled down from 923 K to 4.2 K. In this example, we analyse the effects of such a cool down, using the homogenisation procedure to define the strain state of the strand at 4.2 K due to the different thermal contractions of the materials.[55–57] This strain state is the initial condition for successive operations of the cable. In fact, the helicoidal geometry of the wires inside the cable and of the filaments inside the wire causes an additional strain.[58] Finally, when the magnetic field is applied, electromagnetic forces act as a transversal load on the wires, which behave like continuous beams supported by the contacting wires in their neighbourhood. In this way a bending strain is added to the initial strain.[59,60] It is recalled that the superconductivity of Nb_3Sn filaments is strain-sensitive, and hence a precise knowledge of these strains is of paramount importance. At the end of the cool down in a reacted strand the filaments are in a compressive strain state while the bronze and copper matrices are in a tensile state. We assume that the strand components are in a relaxed state of equilibrium at 923 K without stresses, since the strands remained for several hours at that temperature.

The Nb_3Sn compound has a low thermal contraction but a relatively high elastic modulus and a very high yield strength. The bronze and copper reach their yield limits as soon as the temperature starts decreasing.

Material thermal characteristics are taken from the conductor database.[61] Measurements of elastoplastic properties of the strand components over the whole temperature range 4–923 K are very few.[62–65] Due to their high yield limit the Nb_3Sn filaments can be assumed as elastic over the whole temperature range, with a constant elastic modulus of 160 GPa.[66] Variations of the different material elastic moduli and thermal expansion coefficients vs temperature are shown in Fig. 7 and in Fig. 8, respectively.

After the homogenisation procedure, the equivalent material has an orthotropic behaviour, depending upon the material characteristics and the geometrical configuration of the unit cell.

Thermal expansion is almost linear with temperature, that of bronze being higher than that of Nb_3Sn. The resulting effective coefficients are illustrated in Fig. 8 for the mesolevel (green lines) and for the macrolevel (blue lines): a11, a22, and a33 denote the values of the expansion coefficients referred to as the Cartesian system of coordinates where the third axis is parallel to the longitudinal axis of the strand.

Mechanical characteristics of the single materials and homogeneous results are compared in Fig. 9, showing the diagonal terms D11, D22, D33 of the elasticity tensor as a function of temperature. The peculiar disposition of the superconducting filaments gathered into groups results in an almost

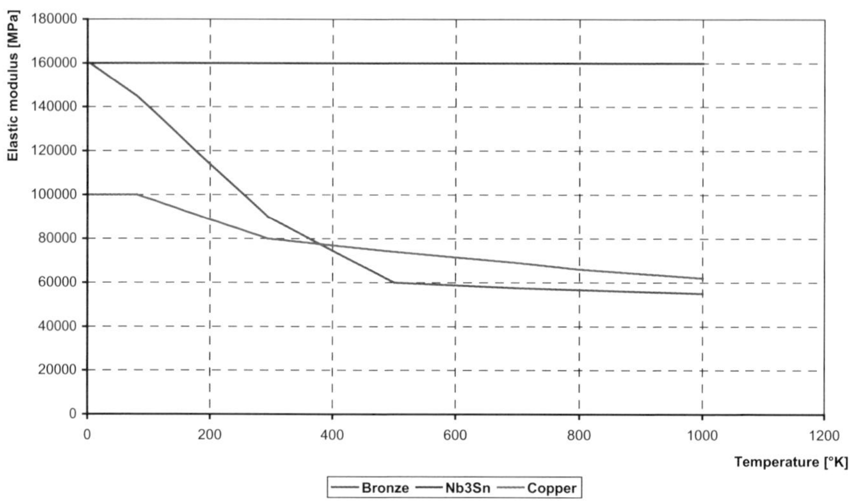

Fig. 7. Variation of bronze (red line), Nb_3Sn (blue line), and copper (green line) elastic modulus vs temperature.

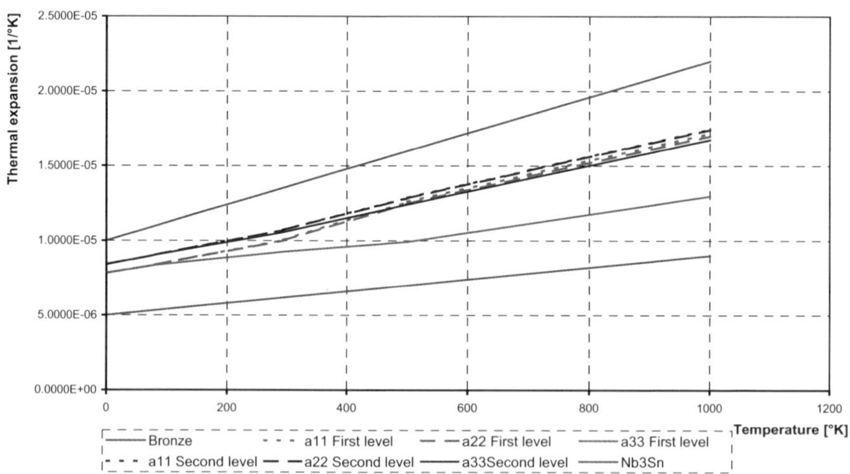

Fig. 8. Thermal expansion [1/K] of bronze (red line), Nb_3Sn (pink line), meso- and macrolevel homogenisation results (green and blue lines, respectively).

isotropic behaviour in the strand cross section, while along the longitudinal direction of the strand the material behaviour is strongly influenced by the superconducting material. The procedure has been validated by comparing results of a homogenised group of filaments and those of a very fine

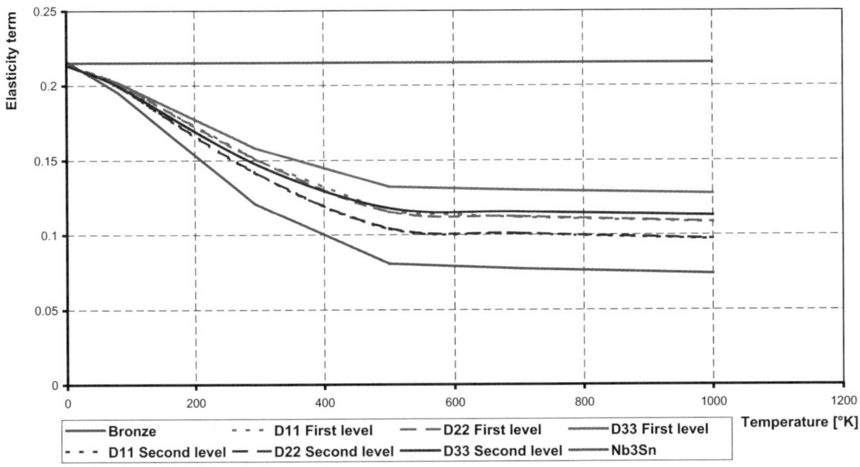

Fig. 9. Main diagonal elasticity terms for bronze (red line), Nb_3Sn (pink line), meso- and macrolevel homogenisation results (green and blue lines, respectively).

discretisation — and then successfully applied to the cool down analysis of a strand.

3. Non-Standard Numerical Techniques in Modelling of Hierarchical Composites

For a non-linear composite or for a complex hierarchical heterogeneity, an adequate description of effective behaviour is usually very difficult to obtain on a purely theoretical way. As presented in the previous sections, the classical, symbolic constitutive law is usually theoretically deduced from known properties of a representative volume, based on a suitable version of homogenisation theory. An alternative to the theoretical development is given by numerical tests of behaviour on a representative volume of the composite. This approach is well known: numerical experiments can be carried out on a representative volume of the composite using, for example, a FE code. Usually the deformation is kinematically imposed and the material properties are deduced from the relation between averaged strain and averaged stress measures, computed from the FE solution. This method is also known as *virtual testing* and is briefly recalled in Sec. 3.2.

In this section we are going to draft an alternative way, tested in some earlier works.[67–72] The possibility to identify the effective material properties from real experiments or numerical simulations or a combination of the two is investigated. In this context the use of an Artificial Neural Network (ANN) is presented, trained with the pairs {mean stress, average strain} or their respective increments, as an approximation of the effective constitutive relationship. ANN is used as a numerical representation of the effective constitutive law and, sometimes, as a numerical tool for the analysis of the constitutive relations between averaged quantities. This method is based on numerical sampling of the mechanical behaviour of a sufficiently large portion of the composite material. ANN approximation replaces a usual symbolic description of the effective constitutive law. We will also mention the so-called *self-learning finite element procedure* introduced first by Ghaboussi[73] and developed by Shin and Pande[74,75] as a very promising form of the ANN training process.

The central point of this method is the description of a constitutive law by means of an ANN incorporated into a FE code. The source of examples that composes the knowledge base of the constitutive data can be given by the FE analysis of a sample of the composite consisting of many repetitive cells. Presentation of this method would exceed the frame of this section; hence, we refer the reader to our preliminary works[69,70] for more details.

3.1. *Definition of effective behaviour based on numerical or real experiment*

Roughly speaking there are two kinds of experiments that can be either performed numerically or executed in the laboratory. We will refer to the first as the *classical* one, while the second is *non-classical*, but more important for our purpose.

By the classical one we mean the test on a sample in which the state of strain or the state of stress are carefully imposed, so that the hypothesis about their homogeneity can be accepted. In experimental tests, it can be problematic because of the friction between the surfaces of the sample and the experimental device, to mention the simplest example. In numerical simulations, the mean values of strain or stress can be easily formulated as

$$\Sigma_{ij} \equiv \frac{1}{|\Omega|}\int_{\Omega} \sigma_{ij}\, d\Omega, \quad \frac{1}{|\Omega|}\int_{\partial\Omega} t_i x_k\, ds = \Sigma_{ik}, \tag{52}$$

$$E_{ij} \equiv \frac{1}{|\Omega|}\int_{\Omega} \varepsilon_{ij}\, d\Omega, \quad E_{ij} = \frac{1}{2|\Omega|}\int_{\partial\Omega} (u_i n_j + u_j n_i)\, d\partial\Omega. \tag{53}$$

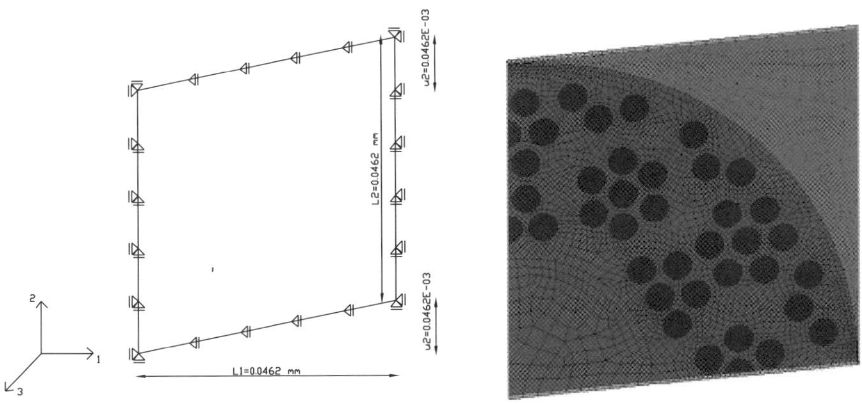

Fig. 10. An example of a numerical experiment. The scheme (on the left) and the deformed FE mesh (on the right) for the numerical shear test of the representative cell are depicted.

In the above expressions the first defines the average value and the second gives the method to achieve these values by using some chosen boundary conditions: imposed stress vector **t** or displacement **u**.

Let us consider a certain number N of different numerical or real experiments (Fig. 10). Their results, written in the form (54a), can always be rearranged to form the system of equations (54b), where $\mathbf{X}$ contains the 21 independent elastic constants (the elements of the effective stiffness matrix) and subscript n refers to the selected data from the nth numerical or real experiment

$$\boldsymbol{\Sigma}_n = \mathbf{D}^{\text{eff}}\mathbf{E}_n, \tag{54a}$$

$$\boldsymbol{\Sigma}_n = \mathbf{X}_{21}^{\text{Deff}}\mathbf{E}_{21\times n}, \quad n = 1\cdots N. \tag{54b}$$

Equation (54b) can be inverted as follows (the analogous information can be collected for the effective compliance matrix):

$$\mathbf{X}^{\text{Deff}} = \boldsymbol{\Sigma}\mathbf{E}^{\text{T}}(\mathbf{E}\mathbf{E}^{\text{T}})^{-1}. \tag{55}$$

Equation (55) represents the least square solution of (54), and it can be obtained without any use of ANN. On the other hand, the training of the ANN can be interpreted as the solution of (55) corresponding to the pattern set obtained from n experimental trials: $\{\boldsymbol{\Sigma}_n, \mathbf{E}_n\}$.

Constitutive equations relate stress with strain, but in laboratory we are able to impose and measure forces and displacements. As mentioned, to identify a constitutive relationship we need a homogeneous distribution of stress and strain. This is even more important when an experiment is performed on the naturalscale of a structure. This represents the second type of experiments, called *non-classical* (in this case the set of measurements of physical and/or geometrical quantities are referred to as *non-classical experimental data*). In this case for our method neither the shape of the sample nor the loading conditions are limited, and we can treat some real structures as a source of experimental knowledge, where measurements concern the quantities that are directly observable. The deduction of the material parameters from the observed behaviour is, in this case, the subject of an inverse analysis. This analysis allows us to determine the material parameters that assure the best fit of the observed feature in the frame of an *a priori* prescribed theoretical or numerical model. This inverse procedure can be carried out particularly well by ANN. The use of ANN to solve the inverse problem is well recognised in literature. Some highly specialised techniques, proposed first by Ghabboussi[73,76] and then by Shin and Pande,[74,75] can be understood as a kind of inverse analysis. We refer here to the so-called *self-learning FE model* introduced by those authors. We underline that the use of ANN is necessary in that last case, while in the frame of the classical inverse analysis a lot of algorithms which work well without ANN have been developed.

The use of the above-presented numerical or experimental tests to the case of composites requires two assumptions:

- The considered volume of the composite is both sufficiently large to exhibit a global, homogeneous-like behaviour and sufficiently small to make the numerical analysis possible.
- It is possible to describe the observed global behaviour in the frame of a homogeneous model, the mechanical nature of which has to be assumed *a priori*. This is a clear disadvantage of this approach with respect to the purely theoretical reasoning, especially asymptotic analysis.

3.2. *Characterisation of the elastic–plastic behaviour of a composite based directly on numerical experiments (virtual testing)*

Let us start with a simple example of the characterisation of the elastic–plastic behaviour of a periodic composite based directly on numerical experiments. A cell of periodicity consisting of a rectangular metal casing with a void inside and a thin epoxy interface with the neighbouring cells is considered. To investigate the global elastic–plastic properties, a global strain tensor $\mathbf{E}^*$ is imposed to the cell, and it is monotonically increased to generate a kinematical loading path. This means that, numerically, a large number of equal kinematical steps are applied to the unit cell. Since the cell is immersed in the whole body, periodic boundary conditions are applied at the opposite sides of the cell. The multiplier α_0 is chosen in such a way that the elastic frontier is reached, and then the displacement is proportionally increased with equal steps (for example, equal to 1/10 of the first one). The homogenised stress tensor $\boldsymbol{\Sigma}$ is then computed for each step of the load history by means of Eq. (56b). The sequence of steps characterised by a fixed $\mathbf{E}^*$, generates a sequence of points in the stress space. Therefore, we have one point in the stress space for each load step. These points are called *interpolation points*: here the behaviour of the homogenised material is known.

$$\begin{cases} \text{microscopic constitutivelaws} \\ \text{div } \sigma = 0 \text{ (micro-equilibrium)}, & \text{(56a)} \\ E_{ij} = \dfrac{1}{|Y|} \displaystyle\int_Y \varepsilon_{ij}\, dY = \left(\alpha_0 + m\dfrac{\alpha_0}{10}\right) E_{ij}^*; m = 0, 1, \ldots, m_{\max}. & \text{(56b)} \end{cases}$$

Repeating the procedure for several different given tensors $\mathbf{E}^*$, we identify the behaviour of the homogenised material in a discrete number of points and for a certain variety of loading situations. At this point we introduce a simplifying hypothesis: we assume that the interpolation points, characterised by the same step number but belonging to different loading paths, lie on the same plastic surface, i.e. they are labelled by the same value of an internal variable. In this manner, by connecting points relating to the corresponding steps of different loading paths, it is possible to construct a series of plastic surfaces generated by the numerical experiments. This surface can be elaborated numerically in order to obtain the information needed for FE procedure such as plastic flow direction, as described by Pellegrino *et al.*[77] and Boso *et al.*[78] The intuitive illustration of this post-processing is given in Fig. 11.

3.3. *ANN in constitutive modelling*

Before introducing the use of ANN for homogenisation a short presentation of this tool of numerical analysis is given.

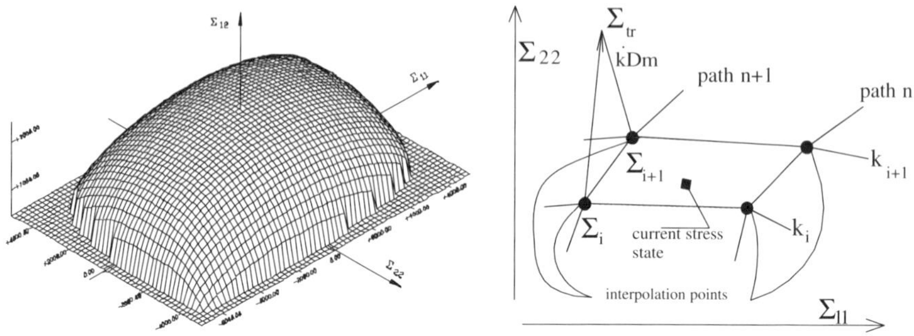

Fig. 11. Numerically defined yield surface (left) and a zoom illustrating the post-processing of the collected data (right).

A Neural Network can be considered as a collection of simple processing units that are mutually interconnected with variable weights. This system of units is organised to transform a given input signal into a given output signal. Both input and output signals are suitably defined to possess a needed physical interpretation. In our case this is a sequence of corresponding values of stresses and strains. The weights of interconnections are shaped to force the desired output signal to be a response to a given input pattern. This is an iterative process called *training* phase of the Network and is based on a set of input data and corresponding known output (*target*). It is stopped when the error between the Neural Network output and the desired one (target) is minimised for a whole set of pairs: {given input, known output}. The interested reader is referred to related textbooks for details concerning the activity of units.

The transfer of input signal i into the output signal o can be prescribed by formula (57) that defines a typical activity of a node (*neurone*). Three actions are executed by each neurone through the network:

- Summation of incoming signals from all connected nodes, weighted by the weights of connexions $\mathbf{w}$.
- Transformation of the sum by a so-called *activation function* of one variable $x \to g(x)$, usually in the form of non-decreasing "cutting off" sigmoid (in (57) parentheses enclose a value of the scalar argument x of the function $g(x)$).
- The computed result (activation of the node i) is weighted by the weight of connection w_{ij} and sent to node j. This is repeated for every connected node.

Expression (57) is written for the jth output from the network containing three layers of neurones (nodes). Weights are labelled with the number of layers by superscript, b are biases. Summation over repeated index is used, except where the index is enclosed by parentheses:

$$o_j = \sum_j w_{js}^{(3)} g_s \left(\sum_r w_{sr}^{(2)} g_r \left(\sum_i w_{ri}^{(1)} i_i + b_r^{(1)} \right) + b_s^{(2)} \right) + b_j^{(3)}. \tag{57}$$

It is worth to point out that the representation of a constitutive law with ANN has the advantage of simplicity. One can observe that in such a representation neither yield surface nor plastic potential is explicitly defined. However, the stress response on any strain input will never fall outside the admissible domain in stress space since the network was trained only with the admissible graphs. We show that the ANN can be used as a tool for the elaboration of experimental data. We use the generated fields as an input for an ANN with hidden layers.

According to our experience, the incremental form of the constitutive law is suitable when one intends to use ANN to approximate it. To this end a special form of ANN has been elaborated. The input of this ANN consists of data defining the current state of stress and strain and the increment of strain along a given path. At the output, the nodes of the last layer are interpreted as increments of stress measure.

The scheme of this network is shown in Fig. 12.

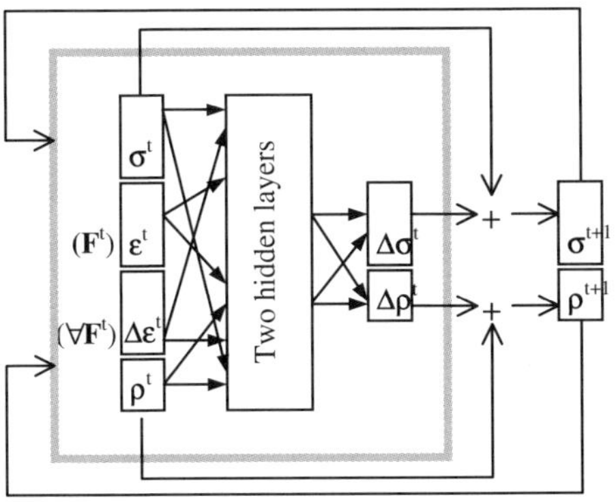

Fig. 12. Scheme of ANN to approximate the incremental constitutive relationship. The trained network is inside of the grey line, external arrows illustrate the use of the network to model an evolution of stress with the temperature or time. F is the deformation gradient that can be used alternatively, ρ is an internal variable, porosity, or apparent density, for example.

In our applications the Neural Network is trained to reflect correctly the set of data from numerical experiments. Each of these experiments requires a solution of a BVP for the kinematical boundary conditions $\mathbf{u}$ for a given average strain $\mathbf{E}$, and results with the average stress Σ over the cell Y. The Network is trained until all the pairs {average strain, average stress} that correspond each other according to the experiment, are successfully associated as the input and relative output of the Network.

The Network automatic generalisation capability enables us to predict the material behaviour, i.e. to produce the graph stress–strain for an arbitrary sequence of stress or strain values. This function of ANN has been described by Chen *et al.*,[79] Hertz *et al.*,[80] and Hu *et al.*[81]

3.4. *Direct use of an ANN to define the effective material behaviour known from laboratory experiments*

The construction of the non-symbolic description of a non-linear constitutive behaviour known from experiments and the use of this representation in a FE code is now considered. This example has been inspired by an engineering analysis of the mechanics of a superconducting cable used for fusion devices. A superconducting cable is made of more than 1000 strands twisted together according to a precise multilevel twisting scheme. The bundle of strands can be considered as a composite body. The stress–strain relation is known from experimental tests. It will be coded with ANN and used then as a part of a FE model.

We analyse the results of compression in the direction perpendicular to the axis of the cable, experiment performed at the University of Twente, The Netherlands.[82] The pairs {displacements, force} or {strain–stress} are collected in Fig. 13 on the right and show some hysteretic loops. This research revealed a very complex irreversible, non-linear behaviour of the cable, due to the complex, interaction between the components of the composite. We train the ANN to simulate these loops correctly. The following input and output pairs will be correctly reproduced by ANN with weights obtained by successive corrections during the training process:

$$\{\text{input nodes values, output nodes}\} = \{(\boldsymbol{\varepsilon}_i, \boldsymbol{\sigma}_i, \boldsymbol{\eta}_i, \Delta\varepsilon_i), \Delta\sigma_{\alpha\beta i}\}.$$

In the above expression we deal not only with increments but also with values of stress measured exactly in the ith point of the "constitutive curve", where η is a scalar parameter, which is very important when we deal with

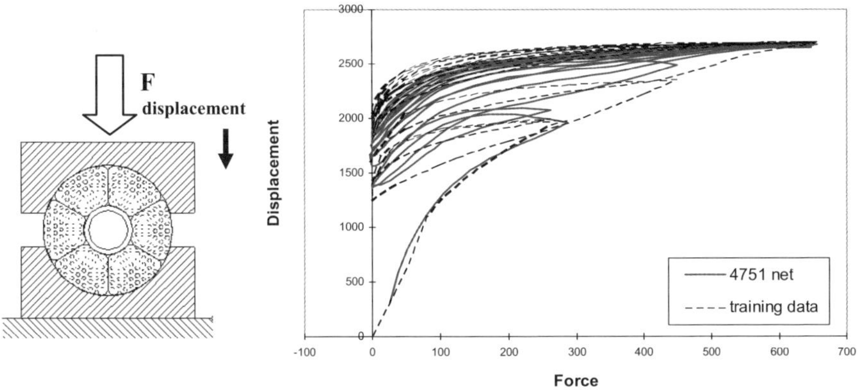

Fig. 13. Scheme of the experiment and hysteretic loops (continuous red line) obtained by the ANN 4751 trained with experimental data (dotted line).

irreversible processes; it is the area under the curve at the current point (see discussion by Lefik *et al.*[69]). The result of the training in the form of autonomous response of the network on the given sequence of strain increments is shown in Fig. 13 on the right (continuous red line).

In this case, the method we propose, which employs the ANN technique, does not require any arbitrary choice of the constitutive model. The numerical description of the observed behaviour is easily and automatically defined. Unfortunately, the weights that it determines have no physical meaning. The description can be incorporated in a very natural manner into any FE code[70] and can be used in the role of a usual constitutive model.

3.5. *Direct use of an ANN to define the effective material behaviour based on numerical experiment*

In this section we present the construction of a non-symbolic description of non-linear constitutive behaviour of a composite, similar to the one presented in Sec. 3.4. The difference is that now we train the ANN using numerical experiments instead of the real ones. We consider a hyperelastic material governed by a Neo-Hookean constitutive relationship in a plane strain condition. Non-homogeneity is caused by the presence of a regular pattern of circular voids. We assume that the material parameters are given (Young's modulus E and Poisson's ratio v).

Application of Eqs. (52) and (53) allowed us to build a constitutive database containing many average stress–average strain graphs obtained according to a scheme:

(i) A given set of kinematical loads (displacements at the borders of the cell of periodicity) has been imposed;
(ii) from the relative FE solution we computed displacement field, mean (over the cell) gradient of deformation $\mathbf{F}_{ik}$ and Cauchy stress Σ_{ik} components.

All boundary conditions have been applied incrementally. For all increments, these data are used as input patterns for the ANN: $\{(\Sigma_{ik}, \Delta\mathbf{F}_{ik}), \Delta\Sigma_{ik}\}$.

Typical examples of imposed boundary conditions and the corresponding deformed mesh are shown in Fig. 14.

The effective constitutive relationship has been approximated with a relatively small ANN with two hidden layers: 7, 20, 7, 3 (or alternatively — three separate networks, one for each component of the stress tensor: $3 \times (7, 15, 8, 1)$). The network has been trained with the computed data

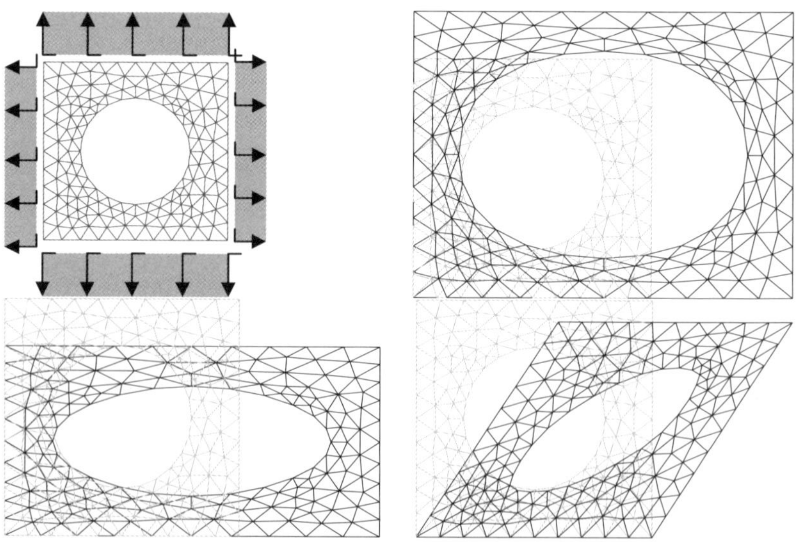

Fig. 14. Cell of periodicity of a composite with circular voids: examples of kinematical loadings and deformed configurations.

and then used inside a FE code (the open source code "Flagshyp"[83] has been adapted to this end).

As a test of the method, an "exact" FE solution of a sample of 7×7 cells of periodicity of the composite has been computed for two simple axial and shear loading–unloading using a fine mesh.

The same geometrical domain filled with homogenised material has been discretized with a coarse mesh and has been solved by a FE code with the trained ANN inserted to generate strain–stress relation. It can be seen in Fig. 15 that the deformed coarse mesh of the homogenised problem almost coincides with the deformation of the fine discretized composite.

3.6. *Approximation of dependence of effective material properties on the microstructural parameters by ANN in multiscale homogenisation*

In the previous sections we have used an ANN to approximate an effective constitutive law of a composite. We were able to define the material behaviour observable at macrolevel, knowing a constitutive description

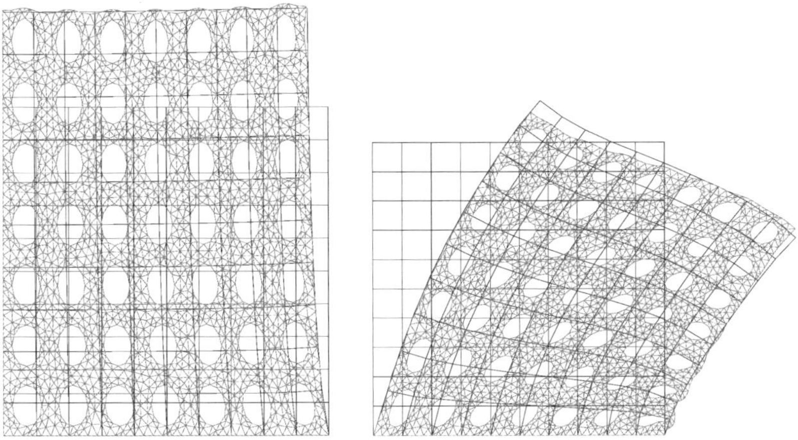

Fig. 15. Tests of quality of the approximation. The coarse meshes, both initial and deformed, are used for the model with ANN as a constitutive subroutine. The fine meshes are used for the heterogeneous case. In the left image the symmetric part of the axial extension test is illustrated (fully constrained along the bottom edge). In the right image the result of the uniform horizontal stress vector applied at the upper edge of the sample is presented (only bottom line fully constrained, plane stress state).

of components and the geometrical characteristics at the microlevel. We used an ANN instead of the functional relationship between an increment of average stress tensor and the relative increment of average strain tensor:

$$\Delta\boldsymbol{\sigma}^{\text{ave}} = \text{ANN@}\{\boldsymbol{\Sigma} \quad \mathbf{E}\}. \tag{58}$$

The symbol @ in (58) denotes an action of an "ANN operator" on the ordered set of values; $\boldsymbol{\Sigma}$, and $\mathbf{E}$ are stress and strain respectively; Δ denotes an increment.

In this section the ANN will be used to approximate, memorise, and even discover the law governing an effective (observable at macroscale) behaviour of composites, the microstructure of which depends on some parameters of mechanical or geometrical nature. In detail, with the ANN, we will identify the functional dependence of the effective constitutive tensor elements D_{ijkl} on the constitutive tensors $\mathbf{D}$ of each of the n materials of the composite and on some scalar parameters c^k characterising the geometry of the microstructure:

$$D^{\text{eff}}_{ijkl} = \text{ANN@}\{\mathbf{D}^{(1)} \cdots \mathbf{D}^{(n)}, c^k, \text{geometry, assembling}\}. \tag{59}$$

The simple consequence of this is the possibility of computation of the effective characteristics of hierarchical composites. In fact if the materials exhibit an internal structure at more than one length scale, this method can be repeated for each structural level or, as well, it allows to bridge one or more structural levels. Some abstract, fractal-like structures can be considered as composites for which the number of structural levels tends to infinity. In such a case the presented method would be particularly useful.[84,85]

Of course, the ANN will be used here as a suitable and powerful tool of approximation of the given knowledge of the macro-behaviour of the composite. The source of knowledge must be found elsewhere (for example, an asymptotic analysis).

In the application we are going to present, the use of ANN is justified by a set of theorems (by various authors, see for example Chen *et al.*[79]) which asserts that ANN is a universal approximator of a function of many variables, of a functional or an operator. Because of this, we are sure that the functional dependencies between effective properties of the composite and the characteristics of the components of the cell of periodicity can be suitably handled with a sufficiently trained ANN.

As stated by Lefik *et al.*,[69] independent variables are the mechanical properties of the components and some parameters describing their geometrical repartition in the cell. The functions to be approximated are known from the direct application of the homogenisation procedure for the unit cell.

The ANN for the approximation of the elements of the effective constitutive tensor is constructed as follows:

- The first group of neurones of the input layer are valued with the given values of the constitutive parameters of the materials of the single microstructural cell. The second group of neurones at the input are interpreted as parameters describing the geometry of the microstructural cell. It has to be possible to describe the geometry of the microstructure by few parameters only. (This second group of neurones is absent in the example presented below since the geometry of the cell is constant in space and does not change with time.) We mention this possibility here, because for Functionally Graded Materials, it is very useful, as shown in an example by Lefik *et al.*[85]
- The output layer contains neurones valued with the values of the effective constitutive parameters of the homogenised material.
- Some hidden layers are constructed to assure the best approximation of the unknown relation between the material properties of components, their geometrical organisation and effective material properties at the output. The best approximation is understood in the usual sense and measured by the test and training errors. It strongly depends on the number and the quality of the data used for the training phase.

Of course, by using any of the homogenisation theories and applying it for each level of the hierarchical structure we are dealing with, we would be able to compute the effective properties of a hierarchical composite without any use of ANN. However, in this case it is necessary to solve at each step of computations (or at each structural level) a BVP for local perturbation using the FE method. The procedure becomes thus time-consuming. The time for single run of computation is usually reasonable; hence, if performed only once for a given composite the procedure is acceptable. Unfortunately, in some of our recent numerical models of hierarchical composites[68,69] the computations of effective coefficients are performed at each step of loading and in many zones of the micro-heterogeneous body. This necessity is due to the fact that locally, at the microlevel, each of the homogeneous components can change its mechanical properties, depending on the stress level they are subject to (fracturing, yielding, damage, etc.). If this is a common feature for many microcells in a zone that can be considered as a macro-domain (being still a subregion of the considered body), new effective properties must be calculated for this region. In practice, this is a region covered by a single element of the global FE mesh. The approach becomes almost impracticable if we repeat FE solution and a suitable post-processing for each load step and for each element of the global mesh in order to obtain the effective constitutive data — an input for a global FE model. A similar numerical scheme and comparable numerical effort is required for the so-called FE^2 procedure.[86–88] In contrast, the same chain of computations can be achieved within a reasonable time when the effective properties are read as an output signal from a well-trained ANN. Because of this, the presented application of the ANN is very important in our numerical practice.

In elasticity, the number of the input parameters varies between two times the number of materials plus the number of geometrical parameters for isotropic components and 21 times the number of materials plus the number of geometrical parameters for anisotropic components. Two material parameters for each material at the input is applicable only when the input data are from the microlevel. The number of the output parameters depends on the type of effective constitutive relationships. This is theoretically known *a priori*. The maximum number is 21, but so far we have tested only effective 3D orthotropy with 9 parameters. It is to note that for a case of porous material, the voids treated as the second material do not require any additional input neuron. If the geometry of all the levels is obtained by a scaling of the same figure, the geometrical parameters in the input layer can be omitted.

The following algorithm is proposed to perform the approximation of the effective characteristics of the composite:

- preparation of the learning data: for chosen random values of the materials data and for each kind (geometry) of cell of periodicity the effective material characteristics are computed by a FE solution of a BVP with periodic conditions suitably post-processed;
- training of the network with the pairs of sets: {given random input and computed (as said above), corresponding output}. Interpretations of input and output data are defined in Eq. (59);

Having the well-trained ANN, starting at the microlevel, for each structural level:

- for each kind of cell of periodicity at the current level:
  - (i) run the Neural Network with input data characterising the current level of the structure to identify the unknown output (*recall* mode of the ANN);
  - (ii) complete the sets of input data for each cell of higher structural level from ANN outputs obtained at the previous level;
- for each cell of periodicity at the next (higher) level of composition:
  - (i) run the same Neural Network in the recall mode with suitably completed input data plus information characterising the geometry of the cell of periodicity of the higher structural level;
- At the macrolevel algorithm stops.

The method is applicable also in the case when the elements of microstructure depend on parameters like temperature, damage parameter, or state of yielding. In such a case the method allows to save a huge amount of computational time, replacing the solution of a BVP by a simple run of ANN in recall mode.

We now show an application[84] of the method to the superconducting strand already presented in Sec. 3.7. We recall that the structure splits into two levels: the microcell is made of a Nb_3Sn inclusion in a bronze matrix, and the mesocell is given by the above composite in the same bronze matrix (Figs. 3–6).

A typical scheme of an ANN with hidden layer for a two-level approximation of effective characteristics when the microcell is made of two materials (in this example: bronze and Nb_3Sn) is illustrated in Fig. 16.

The ANN is used to find the effective stiffness matrix coefficients as a function of temperature. The comparison between the values of the diagonal terms D11, D22, D33 of the elasticity tensor obtained with ANN and by applying the asymptotic homogenisation is shown in Fig. 17. Figure 18 shows the results of the ANN training for the first coefficient of the elasticity tensor of the intermediate level.

This example is carried out using our own FE code for multilevel homogenisation and unsmearing. The efficiency of predictions for two structural levels in the case of bridging across the third, intermediary structural level is clearly observable. Another interesting field of possible

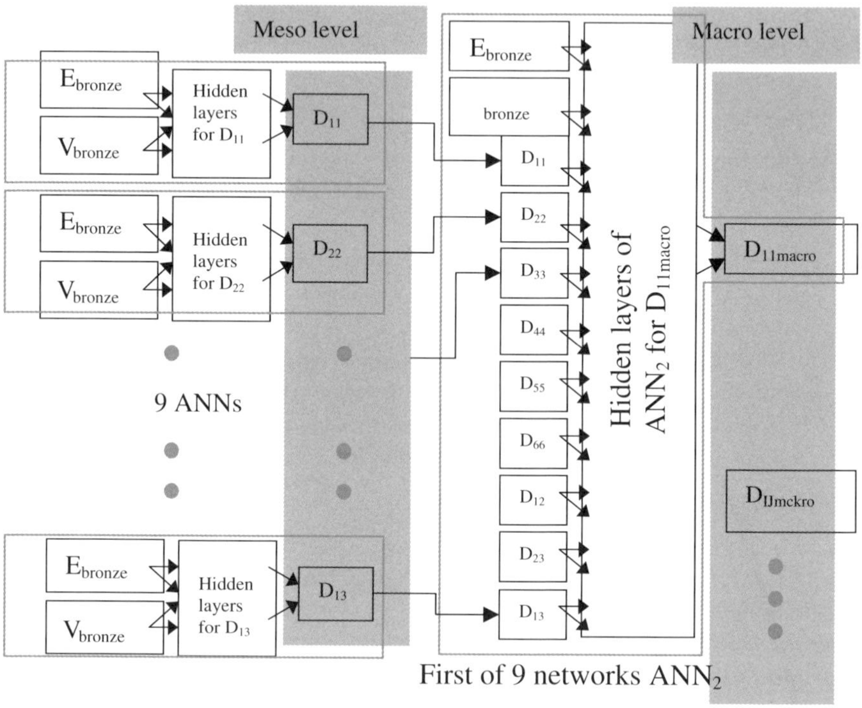

Fig. 16. Scheme of the complex ANN (described in this section) computing terms of the effective stiffness matrix of a two-level composite. A correctly trained ANN_1 that computes effective stiffness at the intermediary level furnishes input data for the ANN_2. This scheme is valid for the case of one material having constant E and v with temperature (Nb_3Sn in our case).

application of the presented numerical techniques is for Functionally Graded Materials (FGM). Such materials can be considered as a generalisation of the usual composite. While for a composite the effective properties are usually constant over the cross section (it can be considered as homogeneous), for FGM the effective properties are functions of the global variable x. This dependence can be obtained by parameterisation of the cell of periodicity. In different regions the cell of periodicity can have different concentrations of inclusions in the matrix or gradually changed shape of inclusion.

Obviously, if the functional dependence of the geometrical parameter of the cell could be associated with a functional dependence of the effective material coefficients, the optimisation and a simple FE analysis

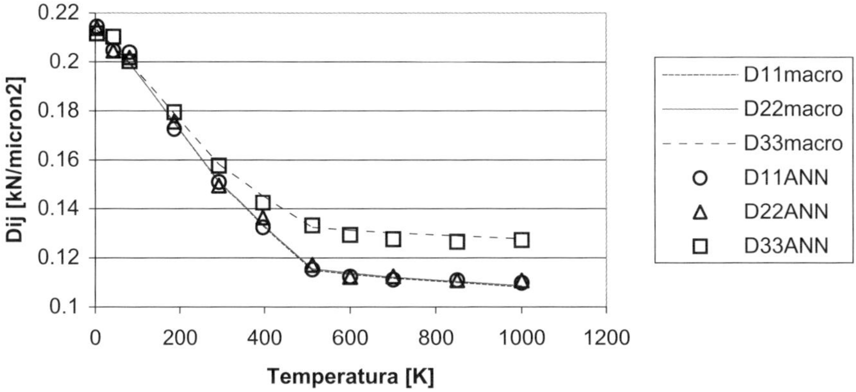

Fig. 17. Evolution of effective stiffness matrix coefficients vs temperature at the macrolevel. Points mark the results of ANN in the recall mode executed twice (micro–macro); lines are obtained from asymptotic homogenisation theory (see previous section, Fig. 9).

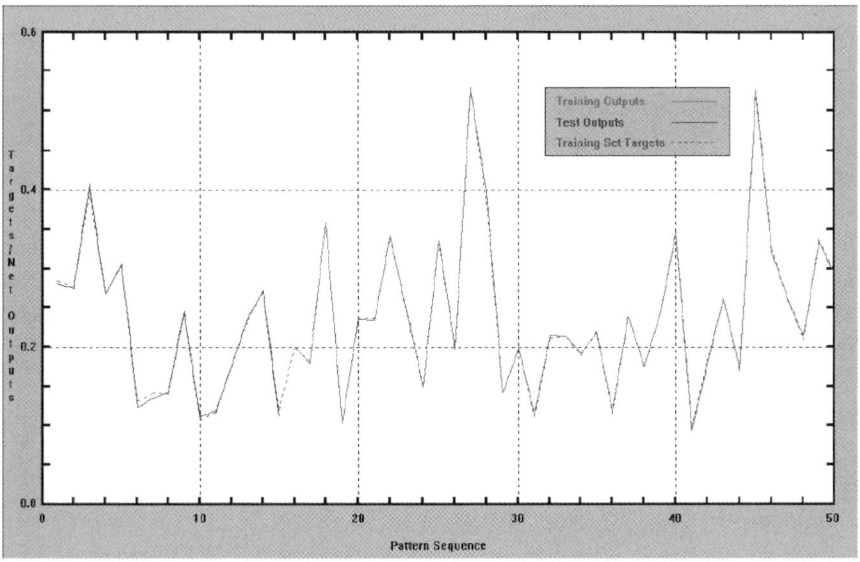

Fig. 18. Results of the training of the ANN {2-3-1} for D_{11}^{meso}. It is seen that testing and learning outputs from the network fit very well with the training target.

of the structure would be numerically practicable. The use of ANN to approximate this functional dependence applied to FGM is analysed by Lefik *et al.*[84,85]

3.7. *ANN as a tool for unsmearing*

A similar use of the ANN can be mentioned for unsmearing in a sequential scheme of homogenisation. The preliminary results of this application have been published in Lefik *et al.*[84,85] It can be described as follows. While the material is elastic we apply the classical asymptotic approach, described in the previous section. According to it, the vector of homogenisation functions allows us to retrieve a field of stress, localised over the cell of periodicity at the lower structural level for each combination of the mean strain gradients. When the material behaviour becomes non-elastic, the homogenisation functions cannot be applied. The localised field can be obtained numerically instead, but this is much more time-consuming. We solve, namely, a BVP for kinematically loaded cell of periodicity. We developed a special purpose FE code to perform homogenisation and unsmearing automatically, through all the structural levels of the composite.

Depending on what we need for the rest of the analysis, two kinds of post-processing of the unsmearing phase are possible. The obvious one is this: is it useful to discover whether the yielding starts inside the cell and if yes — in what material and in how many Gauss points? In practice the local stress is computed always, without looking for this qualitative answer. Local tangent stiffness is changed via return mapping only in the case of yielding. For the cells that remain entirely elastic the stiffness does not vary; most of the computational effort is thus useless through many iterations. The following scheme of the ANN is proposed to reduce useless numerical effort in this problem.

For a given material in the cell a separate ANN is defined:

$$\{\sigma^Y \; n^{ip}\} = \mathrm{ANN@}\{E_{11} \; E_{22} \; E_{33} \; E_{12} \; E_{23} \; E_{13}\}. \tag{60}$$

Within the chosen material domain, σ^Y stands for the value of yield function related to the admissible yield stress, n^{ip} denotes number of integration points (related to the total number of integration points), in which σ^Y is greater than 1.0. The ANN has two output neurones and six (for 3D mechanical problem) input neurones (average strains).

The second type of post-processing is less obvious. Usually, the information concerning the stress state at the lower level appears as the output of the unsmearing procedure requiring the average deformation at the input. Average deformation is directly accessible only at the global level. At each lower level it must be computed via unsmearing. The trained ANN could substitute this procedure, by directly identifying the strain state in the most strained point of the composite (or homogeneous) material, component of the repetitive cell.

These two post-processing procedures, together with the unsmearing itself are numerically costly. The ANN with mean strain state at the input, trained with results of several exemplary runs of computations, can replace both of them. Acting in recall mode during the execution of the homogenisation loops, ANN requires much less numerical effort.

For a given material and for each component of the strain tensor a separate ANN is defined:

$$\varepsilon_{ij} = \mathrm{ANN}@\{E_{11}\ E_{22}\ E_{33}\ E_{12}\ E_{23}\ E_{13}\}. \tag{61}$$

The unsmearing of the strain state is aided by six independent networks (61), each one for different components of the localised strain tensor. The common point for these six networks is that they are trained by six values taken from the same point in the cell. For the VAC-type superconducting strands (Figs. 4–6), when the structure splits into two levels, both (60) and (61) ANNs are successfully trained: first for meso–micro, then for macro–micro unsmearing. The back propagation ANNs of the structure, respectively (6, 5, 2) and (6, 5, 1) work surprisingly well when trained with about 400 and tested with about 200 examples. The examples were prepared using our own FE code for multilevel homogenisation and unsmearing.

4. Concluding Remarks

The non-linear multiscale procedures presented in Secs. 2 and 3 exhibit a good balance between accuracy and computational effort. Obviously, they are not free of limitations.

The asymptotic theory of homogenisation is applicable only to composites exhibiting periodic microstructures, but it gives a comprehensive analysis of the overall constitutive relation (i.e. between volume-averaged field variables).

The ANN representation of any constitutive law requires a certain number of experimental or numerical tests for its training, but it is a flexible tool for representation of the effective behaviour of materials with complex internal microstructure. Usefulness of the hybrid FE–ANN code has been shown, since it opens up new possibilities in comparison with the standard FE codes (the constitutive models can be easily modified) both for sequential and for concurrent multilevel homogenisation. The applications presented in Secs. 3.5 and 3.6 are easy in training and very efficient in recall mode. Representation of the effective constitutive law is very simple (a network of the architecture 3-6-3 is sufficient in the first example). These approximations are good enough to be repeated two times in order to compute effective characteristics and predict some properties of stress and strain fields defined on the microcell, across the intermediary level, making it particularly suitable for hierarchical composite. This representation is "automatic", i.e. it does not require any *a priori* choice or adaptation of the existing constitutive theory for the description of the observed material behaviour. The examples show that the model is possible even in the case of complicated non-linear, inelastic behaviour, and usually convergence is fast. Four steps are enough to obtain a qualitatively good model.

It is worth to underline that continuum-based approaches are not applicable down to the nanoscale as non-continuum behaviour is observed at that scale. Further, nanoscale components are generally used in conjunction with components that are larger and have a mechanical response at different lengths and timescales. As a consequence, single-scale methods such as molecular dynamics or quantum mechanics are generally not applicable in this last case due to the disparity of the scales, and scale bridging is necessary. This is a new and rapidly developing area, and the interested reader is referred, e.g. to the special issue devoted to that topic.[89]

Acknowledgements

Support for this work was partially provided by PRIN 2006091542-003: Thermomechanical multiscale modelling of ITER superconducting magnets, TW7-TMSC-SULMOD: Modelling work to Support ITER Conductor Tests in Sultan, TW6-TMSC-CABLST: Cable Design Effects on Stiffness, KMM-NoE — Knowledge-based multicomponent materials for durable and safe performance — Network of Excellence.

The authors thank P. J. Lee for the permission to reproduce the pictures.

References

1. W. E. B. Engquist, X. Li, W. Ren and E. Vanden-Eijnden, *Commun Computat. Phys.* **2**, 367 (2007).
2. W. Voigt, Über die Bezielung zwischen den beiden Elastizitätskonstanten isotroper Körper, *Wied. Ann.* **38**, 573 (1889).
3. A. Reuss, Berechnung del Fliessgrenze von Mischkristallen auf Grund der Plastizitätbedingung für Einkristalle, *Z. Angew. Math. Mech.* **9**, 49 (1929).
4. R. Hill, *Proc. Phys. Soc. A* **65**, 349 (1952).
5. B. Paul, *Trans. ASME* **218**, 36 (1960).
6. Z. Hashin and S. Shtrikman, *J. Mech. Phys. Solids* **10**, 335 (1962).
7. Z. Hashin and S. Shtrikman, *J. Mech. Phys. Solids* **10**, 343 (1962).
8. Z. Hashin and S. Shtrikman, *J. Mech. Phys. Solids* **11**, 127 (1963).
9. M. Beran, *Nuovo Cimento* **38**, 771 (1965).
10. D. A. G. Bruggeman, *Ann. Phys.* **24**, 636 (1935).
11. A. V. Hershey, *ASME J. Appl. Mech.* **21**, 236 (1954).
12. E. Kröner, *Z. Phys.* **151**, 504 (1958).
13. R. Hill, *J. Mech. Phys. Solids* **13**, 213 (1965).
14. B. Budiansky,*J. Mech. Phys. Solids* **13**, 223 (1965).
15. J. D. Eshelby, *Proc. Roy. Soc. A* **241**, 376 (1957).
16. J. Aboudi, *Mechanics of Composite Materials — A Unified Micromechanical Approach* (Elsevier, Amsterdam, 1992).
17. R. Christensen, *J. Mech. Phys. Solids* **38**(3), 379 (1990).
18. G. L. Taylor, *J. Inst. Metals* **62**, 307 (1938).
19. J. F. W. Bishop and R. Hill, *Phil. Mag.* **42**, 414 (1951).
20. J. F. W. Bishop and R. Hill, *Phil. Mag.* **42**, 1298 (1951).
21. R. Hill, *Mech. Phys. Solids* **13**, 89 (1965).
22. J. W. Hutchinson, *Proc. Roy. Soc. Lond. A* **348**, 101 (1976).
23. B. Berveiller and A. Zaoui, *J. Mech. Phys. Solids* **26**, 325 (1979).
24. J. R. Willis, *ASME J. Appl. Mech.* **50**, 1202 (1983).
25. J. R. Willis, in *Homogeneization and Effective Moduli of Materials and Media*, eds. J. L. Ericksen *et al.* (Springer-Verlag, New York, 1986), p. 247.
26. D. R. S. Talbot and J. R. Willis, *IMA J. Appl. Math.* **35**, 39 (1985).
27. P. Ponte Castañeda, *J. Mech. Phys. Solids* **39**, 45 (1991).
28. P. Ponte Castañeda, *J. Mech. Phys. Solids* **40**, 1757 (1992).
29. D. R. S. Talbot and J. R. Willis, *Int. J. Solids Struct.* **29**, 1981 (1992).
30. P. Suquet, *J. Mech. Phys. Solids* **41**, 981 (1993).
31. T. Olson, *Mater. Sci. Eng. A* **175**, 15 (1994).
32. Z. Hashin, *ASME J. Appl. Mech.* **50**, 481 (1983).
33. T. Mura, *Micromechanics of Defects in Solids,* 2nd edn. (Kluwer Academic Publishers, 1993).
34. S. Nemat-Nasser and M. Hori, *Micromechanics: Overall Properties of Heterogeneous Solids*, 2nd edn. (Elsevier, Amsterdam, 1999).
35. S. Torquato, *Random Heterogeneous Materials: Microstructure and Macroscopic Properties* (Springer-Verlag, New York, 2002).

36. J. Segurado and J. Llorca, *J. Mech. Phys. Solids* **50** (2002).
37. J. Segurado, J. Llorca and C. González, *J. Mech. Phys. Solids* **50** (2002).
38. C. González and J. Llorca, *J. Mech. Phys. Solids* **48**, 675 (2000).
39. C. González and J. Llorca, *Acta Mater.* **49**, 3505 (2001).
40. C. González and J. Llorca, *Mater Sci. Eng. A.* (2002).
41. J. Llorca, *Acta Metall. Mater.* **42**, 151 (1994).
42. J. Llorca, in *Comprehensive Composite Materials, Vol. 3, Metal Matrix Composites*, ed. T. W. Clyne (Pergamon Amsterdam, 2000), p. 91.
43. J. Llorca and C. González, *J. Mech. Phys. Solids* **46**, 1 (1998).
44. P. Poza and J. Llorca, *Metall. Mater. Trans.* **30A**, 869 (1999).
45. A. Bensoussan, J. L. Lions and G. Papanicolau, *Asymptotic Analysis for Periodic Structures* (North-Holland, Amsterdam, 1976).
46. E. Sanchez-Palencia, *Non-Homogeneous Media and Vibration Theory* (Springer Verlag, Berlin, 1980).
47. B. A. Schrefler, in *The Finite Element Method for Solid and Structural Mechanics,* eds. O. C. Zienkiewicz and R. L. Taylor, 6th edn. (Elsevier, 2005), p. 547.
48. G. A. Francfort, in *Numerical Methods for Transient and Coupled Problems*, eds. R. Lewis, E. Hinton, P. Betess and B. A. Schrefler (Pineridge Press, Swansea, 1984), p. 382.
49. M. Lefik and B. A. Schrefler, *Computat. Mech.* **14**(1), 2–15 (1994).
50. M. Lefik and B. A. Schrefler, *Fusion Eng. Design* **24**, 231–255 (1994).
51. K. Terada and N. Kikuchi, *Comput. Methods Appl. Mech. Eng.* **190**, 5427 (2001).
52. J. Fish and Q. Yu, *Int. J. Num. Meth. Eng.* **52**(1–2), 161 (2001).
53. M. Kaminski and B. A. Schrefler, *Comput. Methods Appl. Mech. Eng.* **188**, 1 (2000).
54. H. W. Zhang, D. P. Boso and B. A. Schrefler. *Int. J. Multiscale Computat. Eng.* **1**(04), 359 (2003).
55. D. P. Boso, M. Lefik and B. A. Schrefler, *Cryogenics* **45**(4), 259 (2005).
56. D. P. Boso, M. Lefik and B. A. Schrefler, *Cryogenics* **46**(7/8), 569 (2006).
57. D. P. Boso, M. Lefik and B. A. Schrefler, *IEEE Trans. Appl. Superconductivity* **17**(2), 1362 (2007).
58. D. P. Boso, M. Lefik and B. A. Schrefler, *Cryogenics* **45**(9), 589 (2005).
59. P. L. Ribani, D. P. Boso, M. Lefik, Y. Nunoya, L. Savoldi Richard, B. A. Schrefler and R. Zanino, *IEEE Trans. Appl. Superconductivity* **16**(2), 860 (2006).
60. D. P. Boso, M. Lefik and B. A. Schrefler, *IEEE Trans. Appl. Superconductivity* **16**(2), 1823 (2006).
61. Thermal, electrical and mechanical properties of materials at cryogenic temperatures, *Conductor Database*, Appendix C, Annex II, 24 August (2000).
62. J. Ekin, in *Superconductor Materials Science: Metallurgy, Fabrication and Applications*, eds. S. Foner and B. Schwartz, NATO Advanced Study Institute Series (Plenum Press, 1980).

63. R. P. Reed and A. F. Clark (eds.), *Materials at Low Temperature* (American Society for Metals, Metals Park, OH, 1983).
64. S. Ochiai and K. Osamura, *Acta Metall.* **37**(9), 2539 (1989).
65. G. Rupp, in *Filamentary A15 Superconductors,* eds. M. Suenaga and A. F. Clark (Plenum Press, New York and London), p. 155.
66. N. Mitchell, *Cryogenics* **42**, 311 (2002).
67. D. Gawin, M. Lefik and B. A. Schrefler, *Int. J. Numer. Meth. Eng.* 299 (2001).
68. M. Lefik, in *Proc. XIII Polish Conf. Computer Methods in Mechanics* (Poznań, 1997), p. 723.
69. M. Lefik and B. A. Schrefler, *Fusion Eng. Design* **60**(2), 105 (2002).
70. M. Lefik and B. A. Schrefler, *Comp. Struct.* **80/22**, 1699 (2002).
71. M. Lefik and M. Wojciechowski, in *Proc. AI-METH 2003 — Methods Of Artificial Intelligence*, eds. T. Burczyski, W. Cholewa and W. Moczulski (AI-METH, 2003).
72. M. Lefik, in *Proc. CMM-2003 — Computer Methods in Mech.* (Gliwice, Poland, 2003).
73. J. Ghaboussi, D. A. Pecknold, M. Zhang and R. M. Haj-ali, *Int. J. Numer. Meth. Eng.* **42**, 105 (1998).
74. H. S. Shin and G. N. Pande, *Comp. Geotechnics* **27**, 161 (2000).
75. H. S. Shin and G. N. Pande, in *Intelligent Finite Elements, Computational Mechanics* — New Frontiers for New Millenium, eds. S. Valliapan and N. Khalili (Elsevier, 2001), p. 1301.
76. J. Ghaboussi, J. H. Garrett and X. Wu, *J. Eng. Mech.* **117**, 132–151 (1991).
77. C. Pellegrino, U. Galvanetto and B. A. Schrefler, *Int. J. Numer. Meth. Eng.* **46**, 1609 (1999).
78. D. P. Boso, C. Pellegrino, U. Galvanetto and B. A. Schrefler, *Commun. Num. Meth. Eng.* **16**(9), 615 (2000).
79. T. Chen and H. Chen, *IEEE Trans. on Neural Networks* **6**(4), 911 (1995).
80. J. Hertz, A. Krogh and G. R. Palmer, in *Introduction to the Theory of Neural Computation*, Lecture Notes Vol. I, Santa Fe Institute Studies in the Sciences of Complexity (Addison-Wesley, 1991).
81. Y. H. Hu and J.-N. Hwang (eds.), *Handbook of Neural Network Signal Processing* (CRC Press, 2002).
82. A. Nijuhuis, N. H. W. Noordman and H. H. J. ten Kate, Mechanical and electrical testing of an ITER CS1 model coil conductor under transverse loading in a cryogenic pres, March 10 Preliminary Report, University of Twente (1998).
83. R. D. Wood and J. Bonet, *Nonlinear Continuum Mechanics for Finite Element Analysis* (Cambridge University Press, 1997).
84. M. Lefik and M. Wojciechowski, in *Proc. CMM-2005 — Computer Methods in Mechanics*, 3–6 June, 2005, Czstochowa.
85. M. Lefik and M. Wojciechowski, *Comp. Assisted Mech. Eng. Sci.* **12**, 183 (2005).

86. F. Feyel, *Comp. Mater. Sci.* **16**, 344 (1999).
87. F. Feyel and J. L. Chaboche, *Comp. Methods Appl. Mech. Eng.* **183**, 309 (2000).
88. F. Feyel and J. L. Chaboche, *Comp. Methods Appl. Mech. Eng.* **192**, 3233 (2003).
89. W. K. Liu, D. Qian and M. F. Horstemeyer, *Comp. Methods Appl. Mech. Eng.* **193**, 17 (2004).

RECENT ADVANCES IN MASONRY MODELLING: MICROMODELLING AND HOMOGENISATION

Paulo B. Lourenço

Department of Civil Engineering, University of Minho
Azurém, P-4800-058 Guimarães, Portugal
pbl@civil.uminho.pt

The mechanics of masonry structures have been for long underdeveloped in comparison with other fields of knowledge, presently, non-linear analysis being a very popular field in research. Masonry is a composite material made with units and mortar, which presents a clear microstructure. The issue of mechanical data necessary for advanced non-linear analysis is addressed first, with a set of recommendations. Then, the possibilities of using micromodelling strategies replicating units and joints are addressed, with a focus on an interface finite element model for cyclic loading and a limit analysis model. Finally, homogenisation techniques are addressed, with a focus on a model based on a polynomial expansion of the microstress field. Application examples of the different models are also given.

1. Introduction

Masonry is a building material that has been used for more than 10,000 years, being still widely used today. Masonry partition walls, including rendering, amount typically to ∼15% of the cost of a structural frame building. In different countries, masonry structures still amount to 30%–50% of the new housing developments. Finally, most structures built before the 19th century, still surviving, are built with masonry.

Therefore, research in the field is essential to understand masonry behaviour, to develop new products, to define reliable approaches to assess the safety level, and to design potential retrofitting measures. To achieve these purposes, researchers have been trying to convert the highly indeterminate and non-linear behaviour of masonry buildings into something that can be understood with an acceptable degree of mathematical certainty. The fulfillment of this objective is complex and burdensome, demanding a considerable effort centred on integrated research programmes, able to combine experimental research with the development of consistent constitutive models. In this chapter, some recent approaches regarding masonry modelling and involving the microstructure are reviewed, together with the recommendations for non-linear material data.

2. Masonry Behaviour and Non-Linear Mechanics

Masonry is a heterogeneous material that consists of units and joints. Usually, joints are weak planes and concentrate most damage in tension and shear. Accurate modelling requires a thorough experimental description of the material.[1,2] A basic notion is softening, which is a gradual decrease of mechanical resistance under a continuous increase of deformation forced upon a material specimen or structure (Fig. 1). It is a salient feature of soil, brick, mortar, ceramics, rock or concrete, which fail due to a process of progressive internal crack growth. For tensile failure this phenomenon has been well identified.[3] For shear failure, a softening process is also observed, associated with the degradation of the cohesion in Coulomb friction models.[4] For compressive failure, softening behaviour is highly dependent upon the boundary conditions in the experiments and the size of the specimen.[5] Experimental data seems to indicate that both local and continuum fracturing processes govern the behaviour in uniaxial compression.

2.1. *Non-linear properties of unit and mortar (tension)*

Extensive information on the tensile strength and fracture energy of units exists.[4,6,7] The ductility index d_u, given by the ratio between the fracture energy G_f and the tensile strength f_t, found for brick was between 0.018 and 0.040 mm, as shown in Tables 1 and 2. It is normal that the values are different because different testing procedures and different techniques to calculate the fracture energy have been used. Therefore, the recommended ductility index d_u, in the absence of more information is the average, 0.029 mm.

For stone granites, it is noted that a non-linear relation[7] given by $d_u = 0.239 f_t^{-1.138}$ was found, with d_u in mm and f_t in N/mm^2. For an average granite tensile strength value of 3.5 N/mm^2, the d_u value reads 0.057 mm, which is two times the suggested value for brick.

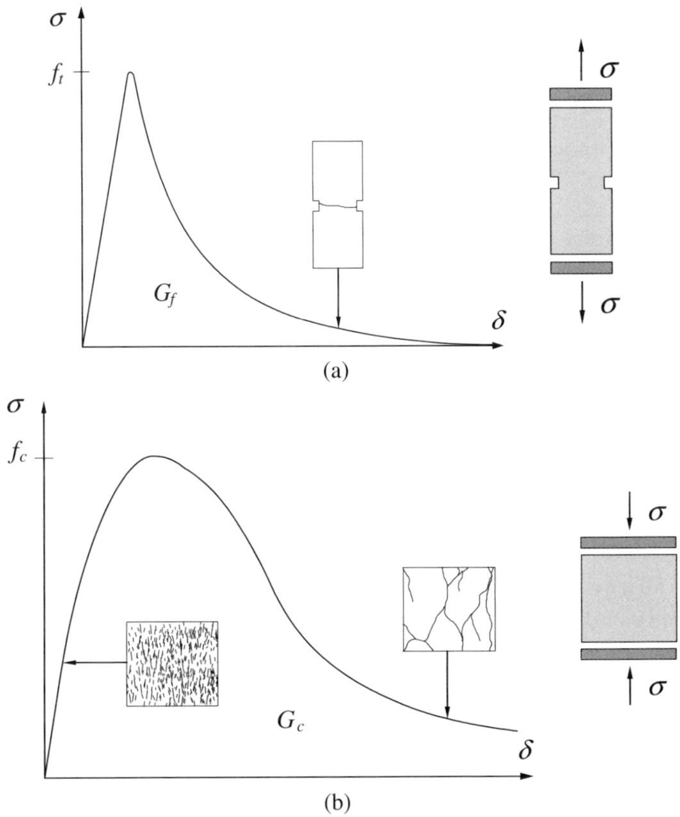

Fig. 1. Softening and the definition of fracture energy: (a) tension; (b) compression. Here, f_t equals the tensile strength, f_c equals the compressive strength, G_f equals the tensile fracture energy and G_c equals the compressive fracture energy. It is noted that the shape of the non-linear response is also considered a parameter controlling the structural response. Nevertheless, for engineering applications, this seems less relevant than the other parameters.

Table 1. Ductility index for different bricks.[6]

Bricks	$f_{t//}/f_{t\perp}$ [-]	$f_{t//}$ [N/mm^2]	d_u [mm]
S	1.18	3.48	0.0169
HP	1.53	4.32	0.0196
HS	1.39	3.82	0.0179
Average	1.4	3.9	0.018

Table 2. Ductility index for different bricks.[4]

Bricks	$f_{t//}/f_{t\perp}$ [-]	$f_{t//}$ [N/mm^2]	d_u [mm]
VE	1.64	2.47	0.0367
JC	1.49	3.51	0.0430
Average	1.6	3.0	0.040

Finally, Model Code 90[8] recommends for concrete (maximum aggregate size 8 mm), the value of $G_f = 0.025\,(f_c/10)^{0.7}$, with G_f in N/mm and f_c in N/mm^2. Assuming that the relation between tensile and compressive strength is 5%,[9] the following expression is obtained: $G_f = 0.025\,(2f_t)^{0.7}$. For an average tensile strength value of 3.5 N/mm^2, G_f is equal to 0.0976 N/mm and d_u reads 0.028 mm, which is similar to the suggested value for brick.

For the mortar, standard test specimens are cast in steel moulds and the water absorption effect of the unit is ignored, being thus the non-representative of the mortar inside the composite. For the tensile fracture energy of mortar, and due to the lack of experimental results, it is recommended to use values similar to brick, as indicated above.

2.2. *Non-linear properties of the interface (tension and shear)*

The research on masonry has been scarce when compared with other structural materials, and experimental data which can be used as input for advanced non-linear models is limited.

The parameters needed for the tensile mode (Mode I) are similar to the previous section, namely the bond tensile strength f_t and the bond fracture energy G_f. The factors that affect the bond between unit and mortar are highly dependent on the units (material, strength, perforation, size, air-dried, pre-wetted, etc.), on the mortar (composition, water contents, etc.) and on workmanship (proper filling of the joints, vertical loading, etc.). A recommendation for the value of the bond tensile strength based on the unit type or mortar type is impossible, but an indication is given in Eurocode 6.[10] It is stressed that the tensile bond strength is very low,[2,4] typically in the range 0.1–0.2 N/mm^2.

Limited information on the non-linear shear behaviour of the interface (Mode II) also exists.[2,4] A recommendation for the value of the bond shear strength (or cohesion) based on the unit type or mortar type is impossible, but an indication is again given in Eurocode 6.[10] The ductility index $d_{u,s}$,

Table 3. Ductility index for different brick/mortar combination.[2]

Combination of unit and mortar	c [N/mm^2]	$d_{u,s}$ [mm]
VE.B	0.65	0.100
VE.C	0.85	0.062
JG.B	0.88	0.147
JG.C	1.85	0.072
KZ.B	0.15	0.087
KZ.C	0.28	0.090
Average	—	0.093

given by the ratio between the fracture energy G_{fs} and the cohesion c, found for different combinations of unit and mortar was between 0.062 and 0.147 mm, as shown in Table 3. The recommended ductility index $d_{u,s}$, in the absence of more information, is the average value of 0.093 mm. It is noted that the Mode II fracture energy is clearly dependent on the normal stress level,[4] and the above values hold for a zero normal stress.

2.3. *Non-linear properties of unit, mortar and masonry (compression)*

The parameters needed for characterising the non-linear compressive behaviour are the peak strain and the post-peak fracture energy. The values proposed for concrete in the Model Code 90[8] are a peak strain of 0.2% and a total compressive fracture energy from Fig. 2. This curve

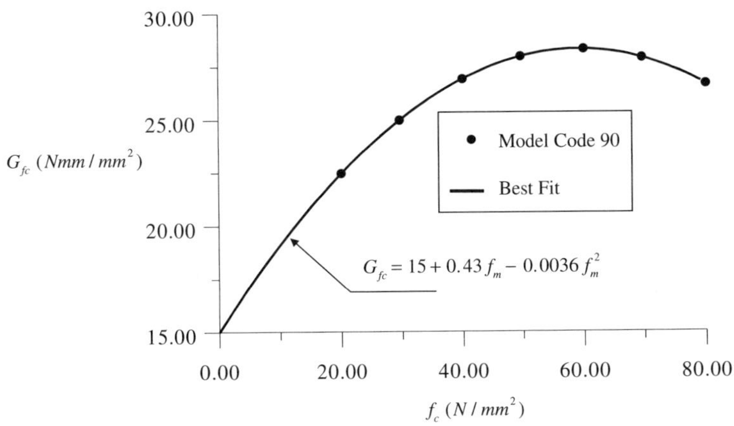

Fig. 2. Compressive fracture energy according to the Model Code 90.[8]

is only applicable for f_c values between 12 and 80 N/mm^2. The average ductility index in compression $d_{u,c}$ resulting from the average value of the graph is 0.68 mm, even if this value changes significantly. Therefore, for compressive strength values between 12 and 80 N/mm^2, the expression for the compressive fracture energy from Fig. 2 is recommended. For f_c values lower than 12 N/mm^2, a $d_{u,c}$ value equal to 1.6 mm is suggested and for f_c values higher than 80 N/mm^2, a $d_{u,c}$ value equal to 0.33 mm is suggested. These are the limits obtained from Model Code 90.

3. Modelling Approaches

In general, the approach towards the numerical representation of masonry can focus on the micromodelling of the individual components, viz unit (brick, block, etc.) and mortar, or the macromodelling of masonry as a composite.[11] Depending on the level of accuracy and the simplicity desired, it is possible to use the following modelling strategies (Fig. 3): (a) Detailed micromodelling, in which unit and mortar in the joints are represented by continuum elements, whereas the unit–mortar interface is represented by discontinuum elements; (b) Simplified micromodelling, in which expanded units are represented by continuum elements, whereas the behaviour of the mortar joints and unit–mortar interface is lumped in discontinuum elements; (c) Macromodelling, in which units, mortar and unit–mortar interface are smeared out in a homogeneous continuum.

In the first approach, Young's modulus, Poisson's ratio and, optionally, inelastic properties of both unit and mortar are taken into account. The interface represents a potential crack/slip plane with initial dummy stiffness to avoid interpenetration of the continuum. This enables the combined action of unit, mortar and interface to be studied under a magnifying

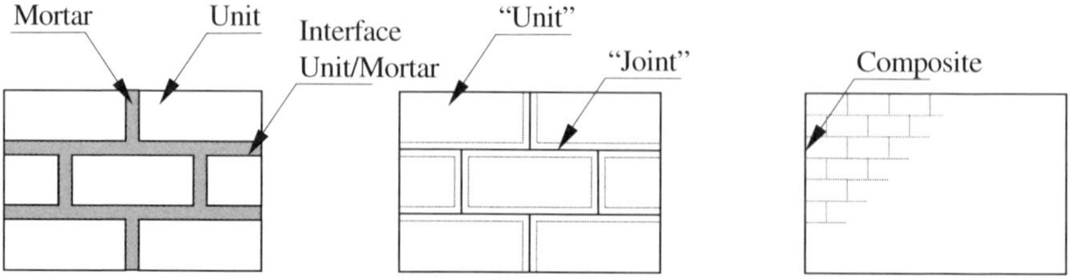

Fig. 3. Modelling strategies for masonry structures: (a) detailed micromodelling; (b) simplified micromodelling; (c) macromodelling.

glass. In the second approach, each joint, consisting of mortar and the two unit–mortar interfaces, is lumped into an average interface while the units are expanded in order to keep the geometry unchanged. Masonry is thus considered as a set of blocks bonded by potential fracture/slip lines at the joints. Some accuracy is lost since Poisson's effect of the mortar is not included. The third approach does not make a distinction between individual units and joints but treats masonry as a homogeneous anisotropic continuum. One modelling strategy cannot be preferred over the other because different application fields exist for micro- and macromodels. In particular, micromodelling studies are necessary to give a better understanding about the local behaviour of masonry structures.

Here, attention will be given to approaches involving some sort of multiscale modelling, using a representation of the geometry of the lower scale and homogenisation approaches.

4. Micromodelling Approaches

Different approaches are possible to represent heterogeneous media, namely, the discrete element method (DEM), the discontinuous finite element method (FEM) and limit analysis (LAn).

The explicit formulation of a discrete (or distinct) element method (DEM) is detailed in an introductory paper.[12] The discontinuous deformation analysis (DDA), an implicit DEM formulation, was originated from a back-analysis algorithm to determine a best fit to a deformed configuration of a block system from measured displacements and deformations.[13] The relative advantages and shortcomings of DDA have been compared with the explicit DEM and FEM,[14] even if significant developments occurred in the last decade, also for masonry structures,[15] particularly with respect to 3D extension, solution techniques, contact representation and detection algorithms. The typical characteristics of DEMs are (a) the consideration of rigid or deformable blocks (in combination with FEM); (b) connection between vertices and sides/faces; (c) interpenetration is usually possible; (d) integration of the equations of motion for the blocks (explicit solution) using the real damping coefficient (dynamic solution) or artificially large (static solution). The main advantages are an adequate formulation for large displacements, including contact update, and an independent mesh for each block, in case of deformable blocks. The main disadvantages are the need for a large number of contact points required for accurate representation of interface stresses and a rather time-consuming analysis, especially for 3D problems.

The FEM remains the most used tool for numerical analysis in solid mechanics, and an extension from standard continuum finite elements (FEs) to represent discrete joints was developed in the early days of non-linear mechanics. Interface elements were initially employed in concrete,[16] in rock mechanics[17] and in masonry,[18] being used since then in a great variety of structural problems. On the contrary, LAn received far less attention from the technical and scientific community for masonry structures.[19] Still, limit analysis has the advantage of being a simple tool, while having the disadvantages that only collapse load and collapse mechanism can be obtained and loading history can hardly be included. Here, recent advances in interface modelling and limit analysis are detailed and applied to illustrative examples.

4.1. *A combined crack–shear–compression interface model*

The application of a micromodelling strategy to the analysis of in-plane masonry structures using FEM requires the use of continuum elements and line interface elements. Usually, continuum elements are assumed to behave elastically, whereas non-linear behaviour is concentrated in the interface elements.

A relation between generalised stress and strain vectors is usually expressed as

$$\boldsymbol{\sigma} = \mathbf{D}\boldsymbol{\varepsilon}, \tag{1}$$

where $\mathbf{D}$ represents the stiffness matrix. For zero-thickness line interface elements, the constitutive relation defined by Eq. (1) expresses a direct relation between the traction vector and the relative displacement vector along the interface, which reads

$$\boldsymbol{\sigma} = \begin{Bmatrix} \sigma \\ \tau \end{Bmatrix} \quad \text{and} \quad \boldsymbol{\varepsilon} = \begin{Bmatrix} \Delta u_n \\ \Delta u_t \end{Bmatrix}. \tag{2}$$

Here, a model capable of representing cracking, shearing and crushing of the interface is addressed.[20] This model is fully based on an incremental formulation of plasticity theory, which includes all the modern concepts used in computational plasticity, such as implicit return mappings and consistent tangent operators.

4.1.1. *Standard plasticity constitutive model*

The constitutive interface model is defined by a convex composite yield criterion, composed by three individual yield functions, where softening behaviour has been included for all modes, reading

$$\begin{aligned} &\text{Tensile criterion}: f_t(\boldsymbol{\sigma}, \boldsymbol{\kappa}_t) = \sigma - \bar{\sigma}_t(\boldsymbol{\kappa}_t), \\ &\text{Shear criterion}: f_s(\boldsymbol{\sigma}, \boldsymbol{\kappa}_s) = |\tau| + \sigma \tan\phi - \bar{\sigma}_s(\boldsymbol{\kappa}_s), \\ &\text{Compressive criterion}: f_c(\boldsymbol{\sigma}, \boldsymbol{\kappa}_c) = (\boldsymbol{\sigma}^T \mathbf{P} \boldsymbol{\sigma})^{1/2} - \bar{\sigma}_c(\boldsymbol{\kappa}_c). \end{aligned} \tag{3}$$

Here, ϕ represents the friction angle, and $\mathbf{P}$ is a projection diagonal matrix, based on material parameters. $\bar{\sigma}_t$, $\bar{\sigma}_s$ and $\bar{\sigma}_c$ are the isotropic effective stresses of each of the adopted yield functions, ruled by the scalar internal variables $\boldsymbol{\kappa}_t$, $\boldsymbol{\kappa}_s$ and $\boldsymbol{\kappa}_c$, respectively. In order to obtain a simple relation between the scalar variable $\boldsymbol{\kappa}_c$ and the plastic multiplier λ_c, the original monotonic compressive criterion (Eq. (3)) was rewritten in square root form. The rate expressions for the evolution of the isotropic hardening variables were assumed to be given by

$$\dot{\boldsymbol{\kappa}}_t = |\Delta\dot{u}_n| = \dot{\lambda}_t, \quad \dot{\boldsymbol{\kappa}}_s = |\Delta\dot{u}_t| = \dot{\lambda}_s \quad \text{and} \quad \dot{\boldsymbol{\kappa}}_c = \frac{\boldsymbol{\sigma}^T \dot{\boldsymbol{\varepsilon}}^p}{\bar{\sigma}_c} = \dot{\lambda}_c. \tag{4}$$

Figure 4 schematically represents the three individual yield surfaces that compose the multisurface interface model in stress space. Associated flow rules were assumed for tensile and compressive modes and a non-associated plastic potential was adopted for the shear mode, with a dilatancy angle ψ, given by

$$g_s = |\tau| + \sigma \tan\psi - \bar{\sigma}_s(\boldsymbol{\kappa}_s). \tag{5}$$

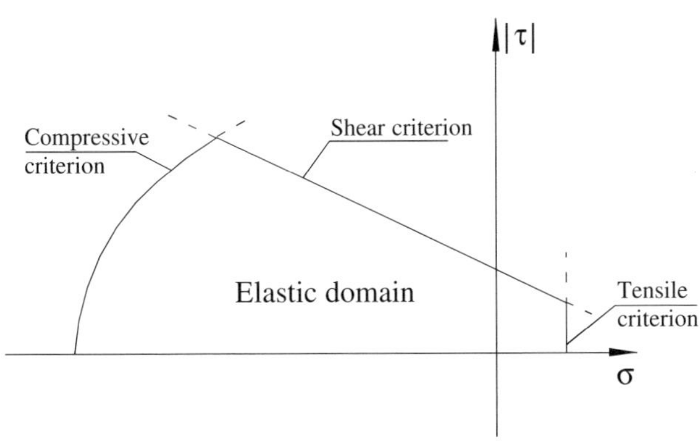

Fig. 4. Multisurface interface model (stress space).

A non-associated flow rule for shear is necessary because friction and dilatancy angles are considerably different.[4]

4.1.2. *Extension for cyclic loading*

In order to include unloading/reloading behaviour in an accurate manner, an extension of the plasticity theory is addressed.[21] Two new auxiliary yield surfaces (termed unloading surfaces) similar to the monotonic ones were introduced in the monotonic model, so that unloading to tension and to compression could be modelled. Each unloading surface moves inside the admissible stress space towards the similar monotonic yield surface. In a given unloading process, when the stress point reaches the monotonic yield surface, the surface used for unloading becomes inactive, and the loading process becomes controlled by the monotonic yield surface. Similarly, if a stress reversal occurs during an unloading process, a new unloading surface is started, subsequently deactivated when it reaches the monotonic envelope or when a new stress reversal occurs. The proposed model comprises six possibilities for unloading/reloading movements.

Both unloading surfaces are ruled by mixed hardening laws, for which a definition of the back-stress vector α is necessary. In this work, the evolution of the back-stress vector is assumed to be given by[22]

$$\dot{\boldsymbol{\alpha}} = (1-\gamma)\dot{\lambda}_U K_t \boldsymbol{u}_\alpha, \tag{6}$$

where K_t is the kinematic tangential hardening modulus, $\dot{\lambda}_U$ is the unloading plastic multiplier rate, and $\boldsymbol{u}_\alpha$ is the unitary vector of $\boldsymbol{\alpha}$. Associated flow rules are assumed during unloading to tension and to compression.

Unloading/reloading to tension can be started from any allowable stress point, except from points on the monotonic tensile surface (Fig. 5(a)) ruled according to the yield function

$$f_{Ut}(\boldsymbol{\sigma}, \boldsymbol{\alpha}, \boldsymbol{\kappa}_{Ut}) = \boldsymbol{\xi}_{(1)} - \bar{\sigma}_{i,Ut}(\gamma\boldsymbol{\kappa}_{Ut}), \tag{7}$$

where $\bar{\sigma}_{i,Ut}$ is the isotropic effective stress and κ_{Ut} is the tensile unloading hardening parameter. The scalar γ provides the proportion of isotropic and kinematic hardening ($0 \leq \gamma \leq 1$). The relative (or reduced) stress vector $\boldsymbol{\xi}$ is given by

$$\boldsymbol{\xi} = \boldsymbol{\sigma} - \boldsymbol{\alpha}. \tag{8}$$

In the same way, unloading/reloading to compression can take place from any acceptable stress point, except from the points on the monotonic

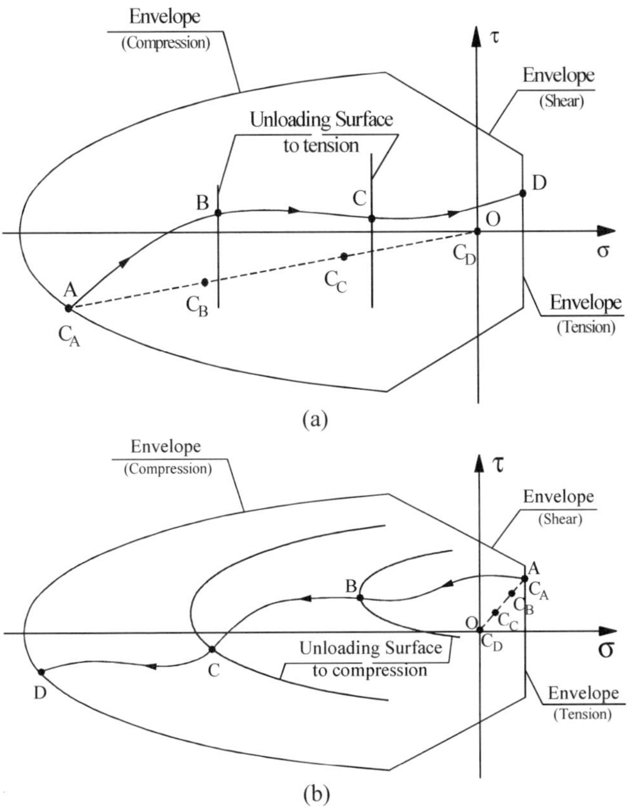

Fig. 5. Hypothetic motion of the unloading surface in stress space to: (a) tension and (b) compression.

compressive surface (see Fig. 5(b)), being controlled by the following yield function:

$$f_{Uc}(\boldsymbol{\sigma}, \boldsymbol{\alpha}, \boldsymbol{\kappa}_{Uc}) = (\boldsymbol{\xi}^{\mathrm{T}}\mathbf{P}\boldsymbol{\xi})^{1/2} - \bar{\sigma}_{i,Uc}(\gamma\boldsymbol{\kappa}_{Uc}), \tag{9}$$

where $\bar{\sigma}_{i,Uc}$ is the isotropic effective stress and $\boldsymbol{\kappa}_{Uc}$ is the compressive unloading hardening parameter.

The evolution of the hardening parameters is given by

$$\dot{\boldsymbol{\kappa}}_{Ut} = |\Delta\dot{u}_n^p| = \dot{\lambda}_{Ut} \quad \text{and} \quad \dot{\boldsymbol{\kappa}}_{Uc} = \frac{\boldsymbol{\xi}^T\dot{\boldsymbol{\varepsilon}}^p}{\bar{\sigma}_{i,Uc}} = \dot{\lambda}_{Uc}. \tag{10}$$

For each of the six hypotheses considered for unloading movements, a curve that relates the unloading hardening parameter $\boldsymbol{\kappa}_U$ and the unloading effective stress $\bar{\sigma}_U$ must be defined. Thus, the adoption of appropriate evolution rules makes possible to reproduce non-linear behaviour during unloading. Physical reasons imply that C^1 continuity must be imposed on all the six $\bar{\sigma}_U - \boldsymbol{\kappa}_U$ curves. Also, all functions must originate positive effective stress values; their derivatives must always be non-negative and its shape must be adequately chosen to fit experimental data, obtained from uniaxial tests. The six different curves adopted in this study are used in the definition of the isotropic and kinematic hardening laws.

The definition of the hardening laws requires four additional material parameters with respect to the monotonic version, which can be obtained from uniaxial cyclic experiments under tensile and compressive loading. These parameters define ratios between the plastic strain expected at some special points of the uniaxial $\sigma - \Delta u_n$ curve and the monotonic plastic strain. Some of these points are schematised in Fig. 6, and are defined as: κ_{1t}, plastic strain at zero stress when unloading from the monotonic tensile envelope (Fig. 6(a)); κ_{1c}, plastic strain at zero stress when unloading from the monotonic compressive envelope (Fig. 6(b)); κ_{2c}, plastic strain at the monotonic tensile envelope when unloading from the monotonic compressive envelope (Fig. 6(b)); $\Delta\kappa_c$, plastic strain increment originated by a reloading from a CT or a CTCT unloading movement (stiffness degradation between cycles).

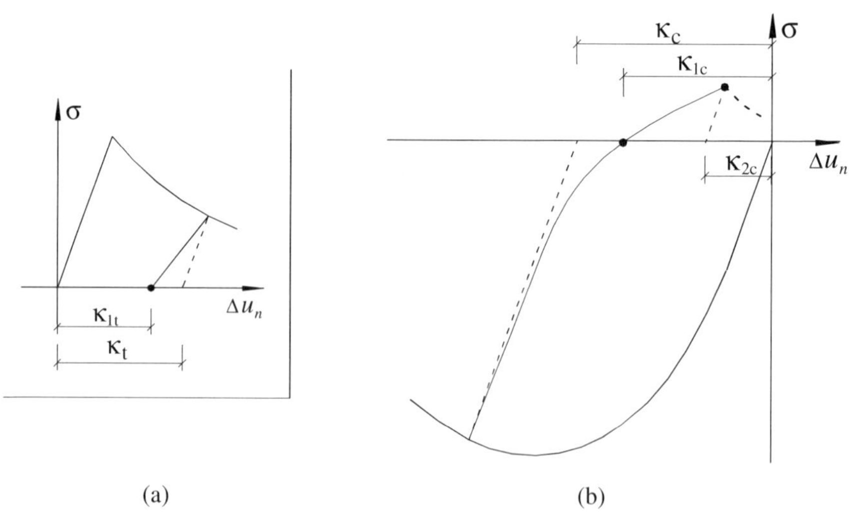

Fig. 6. Special points at the uniaxial $\sigma - \Delta u$ curve: (a) tensile loading and (b) compressive loading.

The integration of the non-linear rate equations over the finite step $(\cdot)_n \rightarrow (\cdot)_{n+1}$, by applying an implicit Euler backward integration scheme, allows obtaining the following discrete set of equations[23]:

$$
\begin{aligned}
\boldsymbol{\sigma}_{n+1} &= \mathbf{D}(\boldsymbol{\varepsilon}_{n+1} - \boldsymbol{\varepsilon}^p_{n+1}),\\
\boldsymbol{\varepsilon}^p_{n+1} &= \boldsymbol{\varepsilon}^p_n + \Delta\lambda_{U,n+1} \left.\frac{\partial g_U}{\partial \boldsymbol{\sigma}}\right|_{n+1},\\
\boldsymbol{\alpha}_{n+1} &= \boldsymbol{\alpha}_n + (1-\gamma)\Delta\lambda_{U,n+1} K_{ks} \mathbf{u}_{\alpha,n+1},\\
\boldsymbol{\kappa}_{U,n+1} &= \boldsymbol{\kappa}_{U,n} + \Delta\lambda_{U,n+1},\\
f_{U,n+1}(\boldsymbol{\sigma}_{n+1}, \boldsymbol{\alpha}_{n+1}, \boldsymbol{\kappa}_{U,n+1}) &= 0,
\end{aligned}
\tag{11}
$$

where $\boldsymbol{\varepsilon}^p$ is the plastic strain and K_{ks} is the kinematic secant hardening modulus defined as a function of the unloading hardening parameter and the kinematic effective stress. The discrete Kuhn–Tucker conditions at step $n+1$ are expressed as

$$
\begin{aligned}
\lambda_{U,n+1} &\geq 0,\\
f_{U,n+1}(\boldsymbol{\sigma}_{n+1}, \boldsymbol{\alpha}_{n+1}, \boldsymbol{\kappa}_{U,n+1}) &\leq 0,\\
\lambda_{U,n+1} f_{U,n+1}(\boldsymbol{\sigma}_{n+1}, \boldsymbol{\alpha}_{n+1}, \boldsymbol{\kappa}_{U,n+1}) &= 0.
\end{aligned}
\tag{12}
$$

Considering an auxiliary elastic trial state, where plastic flow is frozen during the finite step, Eqs. (11) can be reformulated and read as

$$
\begin{aligned}
\boldsymbol{\sigma}^{\text{trial}}_{n+1} &= \boldsymbol{\sigma}_n + \mathbf{D}\Delta\boldsymbol{\varepsilon}_{n+1},\\
\boldsymbol{\varepsilon}^{p,\text{trial}}_{n+1} &= \boldsymbol{\varepsilon}^p_n,\\
\boldsymbol{\alpha}^{\text{trial}}_{n+1} &= \boldsymbol{\alpha}_n,\\
\boldsymbol{\kappa}^{\text{trial}}_{U,n+1} &= \boldsymbol{\kappa}_{U,n},\\
f^{\text{trial}}_{U,n+1} &= f_{U,n+1}(\boldsymbol{\sigma}^{\text{trial}}_{n+1}, \boldsymbol{\alpha}^{\text{trial}}_{n+1}, \boldsymbol{\kappa}^{\text{trial}}_{U,n+1}).
\end{aligned}
\tag{13}
$$

A stress reversal occurrence is based on the elastic trial state. After a plastic process (monotonic or cyclic), a stress reversal case is established under the condition of a negative unloading yield function value. Within the notation inserted before, unloading movements CT or TC must be started from the respective monotonic envelope each time the following condition occurs, after a converged load step where $f_n(\boldsymbol{\sigma}_n, \boldsymbol{\kappa}_n) = 0$:

$$f_{n+1}^{\text{trial}} = f_{n+1}(\boldsymbol{\sigma}_{n+1}^{\text{trial}}, \boldsymbol{\kappa}_{n+1}^{\text{trial}}) < 0. \tag{14}$$

The remaining unloading hypotheses are triggered whenever, after a converged load step in which $f_{U,n}(\boldsymbol{\sigma}_n, \boldsymbol{\alpha}_n, \boldsymbol{\kappa}_{U,n}) = 0$, the following situation occurs:

$$f_{U,n+1}^{\text{trial}} = f_{U,n+1}(\boldsymbol{\sigma}_{n+1}^{\text{trial}}, \boldsymbol{\alpha}_{n+1}^{\text{trial}}, \boldsymbol{\kappa}_{U,n+1}^{\text{trial}}) < 0. \tag{15}$$

The system of non-linear equations expressed by Eqs. (11) can be significantly simplified because the variables $\boldsymbol{\sigma}_{n+1}$, $\boldsymbol{\alpha}_{n+1}$ and $\boldsymbol{\kappa}_{U,n+1}$ can be expressed as functions of $\Delta\lambda_{U,n+1}$, and therefore, Eq. (11)[5] is transformed into a non-linear equation of one single variable. The plastic corrector step consists of computing an admissible value of $\Delta\lambda_{U,n+1}$ that satisfies Eqs. (12), using the Newton–Rapshon method. The necessary derivative reads

$$\left.\frac{\partial f_U}{\partial \Delta\lambda_U}\right|_{n+1} = -\left(\frac{\partial f_U}{\partial \boldsymbol{\sigma}}\right)^{\mathrm{T}} \mathbf{H} \frac{\partial g_U}{\partial \boldsymbol{\sigma}} - h_U, \tag{16}$$

where

$$\mathbf{H} = \left[\mathbf{D}^{-1} + \Delta\lambda_{U,n+1} \frac{\partial^2 g_U}{\partial \boldsymbol{\sigma}^2}\right]^{-1};$$
$$h_U = (1-\gamma) K_t \left(\frac{\partial f_U}{\partial \boldsymbol{\sigma}}\right)^T \mathbf{u}_{\alpha,n+1} - \left.\frac{\partial f_U}{\partial \boldsymbol{\kappa}_U}\right|_{n+1}. \tag{17}$$

Figure 5 illustrates also that a composite yield criterion, composed by an unloading/shear corner, may occur. These two modes are assumed to be uncoupled, resulting in $\dot{\boldsymbol{\kappa}}_U = \dot{\lambda}_U$ and $\dot{\boldsymbol{\kappa}}_s = \dot{\lambda}_s$. Since all unknowns of the stress vector can be expressed as functions of $\Delta\lambda_{U,n+1}$ and $\Delta\lambda_{s,n+1}$, the system of non-linear equations to be solved can be reduced to

$$\begin{cases} f_s(\Delta\lambda_{U,n+1}, \Delta\lambda_{s,n+1}) = 0, \\ f_U(\Delta\lambda_{U,n+1}, \Delta\lambda_{s,n+1}) = 0. \end{cases} \tag{18}$$

The components of the Jacobian necessary for the iterative Newton–Raphson procedure to solve this system can be found in Ref. 23.

Each time a stress reversal takes place, a new unloading surface is activated, being deactivated when it reaches the monotonic envelope towards which it moves; thus, for the same load step, yielding may occur both on the unloading surface and on the monotonic surface. Therefore, a sub-incremental procedure must be used in order to split such load increment into two sub-increments, each one corresponding to a different yield surface. In a strain-driven process, in which the total strain vector is the only independent variable, the problem consists in the computation of the scalar

$$\varepsilon_{n+1} = \varepsilon_n + \beta\Delta\varepsilon_{n+1} + (1-\beta)\Delta\varepsilon_{n+1}, \tag{19}$$

for which the strain increment $\beta\Delta\varepsilon_{n+1}$ leads the unloading surface to touch the monotonic one. After the deactivation of the unloading surface, the remaining strain increment $(1-\beta)\Delta\varepsilon_{n+1}$ is used for the monotonic surface. In the present implementation, β is computed through the bisection method, where the monotonic yield function is evaluated at each iteration.

4.2. *A combined crack–shear–compression limit analysis model*

The limit analysis formulation for a rigid block assemblage presented here assumes standard hypotheses, which have been shown to be reasonable in normal applications: (a) the limit load occurs at small overall displacements; (b) masonry has no tensile strength; (c) shear failure at the joints is perfectly plastic; (d) the hinging failure mode at a joint occurs for a compressive force independent from the rotation.

The static variables, or generalised stresses, at an interface k are selected to be the shear force, V_k, the normal force, N_k, and the moment, M_k, all at the centre of the joint. Correspondingly, the kinematic variables, or generalised strains, are the relative tangential, normal and angular displacement rates, δn_k, δs_k and $\delta\theta_k$ at the interface centre, respectively. The degrees of freedom are the displacement rates in the x- and y-directions, and the angular change rate of the centroid of each block: δu_i, δv_i and $\delta\omega_i$ for the block i. In the same way, the external loads are described by the forces in x- and y-directions, as well as the moment at the centroid of the block. The loads are split in a constant part (with a subscript c) and a variable part (with a subscript v): f_{cxi}, f_{vxi}, for the forces in the x-direction, f_{cyi}, f_{vyi}, for the forces in the y-direction, and m_{ci}, m_{vi}, for the moments. These variables are collected in the vectors of generalised stresses $\boldsymbol{Q}$, generalised strains $\boldsymbol{\delta q}$, displacement rates $\boldsymbol{\delta u}$, constant (dead) loads $\mathbf{F}_c$ and variable (live) loads $\mathbf{F}_v$. Finally, the load factor α is defined, measuring the amount of the variable load vector applied to the structure.

The load factor is the limit (minimum) value that the analyst wants to determine and is associated with the collapse of the structure.

With the above notation, the total load vector $\mathbf{F}$ is given by

$$\mathbf{F} = \mathbf{F}_c + \alpha \mathbf{F}_v. \tag{20}$$

The yield function at each joint is rather complex for 3D problems due to the presence of torsion,[24,25] but rather simple for 2D problems, composed by the crushing–hinging criterion and the Coulomb criterion. For the crushing–hinging criterion, it is assumed that the normal force is equilibrated by a constant stress distribution near the edge of the joint (see Fig. 7(a)). Here, a is half of the length of a joint and w is the width of the joint normal to the plane of the block. The effective compressive stress value f_{cef} is given[26] by Eq. (21), where f_c is the compressive strength of the material expressed in N/mm^2:

$$f_{cef} = \left(0.7 - \frac{f_c}{200}\right) f_c. \tag{21}$$

The constant stress distribution hypothesis leads to the yield function φ given by Eq. (22), related to the equilibrium of moments; note that N_k represents a non-positive value. The Coulomb criterion is expressed

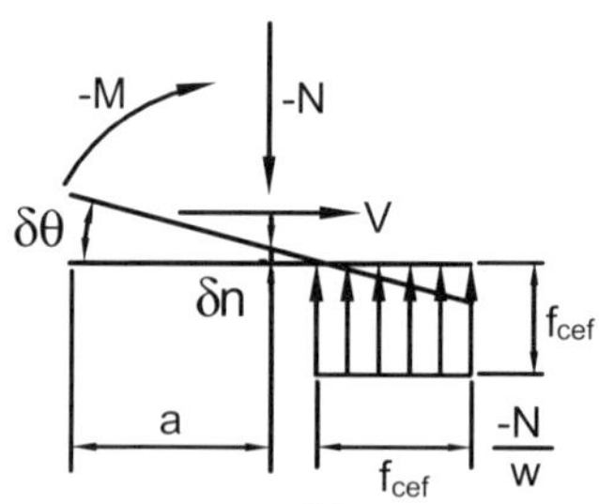

(a)

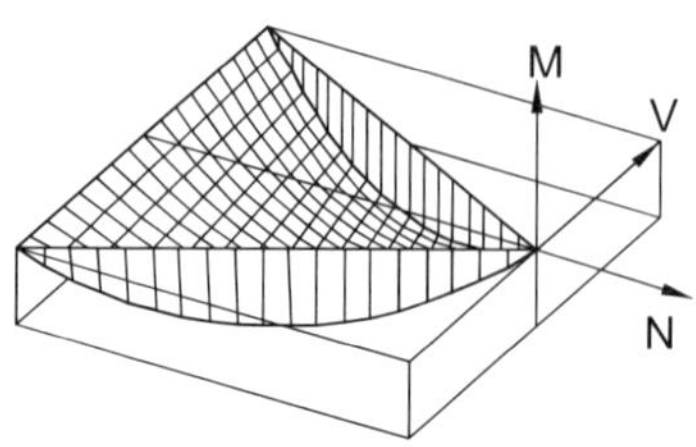

(b)

Fig. 7. Joint failure: (a) generalised stresses and strains for the crushing–hinging failure mode; (b) geometric representation of a half of the yield surface.

by Eq. (23), related to the equilibrium of tangential forces. Here, μ is the friction coefficient or the tangent of friction angle at the joint. The equilibrium of normal forces is automatically ensured by the rectangular distribution of normal stresses. It is noted that the complete yield function is composed by four surfaces, two surfaces given by Eq. (22) and two surfaces given by Eq. (23), in view of the use of the absolute value operator. Figure 7(b) represents half of the yield surface ($M < 0$), while the other half ($M > 0$) is symmetric to the part shown.

$$\varphi_{1,2} \equiv N_k \left(a_k + \frac{N_k}{2 f_{cef} w_k} \right) + |M_k| \leq 0 \tag{22}$$

$$\varphi_{3,4} \equiv \mu N_k + |V_k| \leq 0. \tag{23}$$

Figure 7(a) illustrates also the flow mode corresponding to crushing–hinging, in agreement with the normality rule. It is noted that, for the Coulomb criterion, the flow consists of a tangential displacement only. The flow rule at a joint can be written, in matrix form, as given by Eq. (24), and, in a component-wise form, as given by Eq. (25), in which the joint subscripts have been dropped for clarity. Here, $\mathbf{N}_{0k}$ is the flow rule matrix at joint k and $\delta\lambda_k$ is the vector of the flow multipliers, with each flow multiplier corresponding to a yield surface and satisfying Eqs. (26) and (27). These equations indicate that plastic flow must involve dissipation of energy (Eq. (26)), and that plastic flow cannot occur unless the stresses have reached the yield surface (Eq. (27)). For the entire structure, the flow rule results in Eq. (28), where the flow matrix $\mathbf{N}_0$ can be obtained by assembling all the joint matrices:

$$\delta\mathbf{q}_k = \mathbf{N}_{0k}\delta\boldsymbol{\lambda}_k, \tag{24}$$

$$\begin{bmatrix} \delta s \\ \delta n \\ \delta\theta \end{bmatrix} = \begin{bmatrix} 0 & 0 & -1 & 1 \\ a\left(1 - \dfrac{N}{f_{cef} w}\right) & a\left(1 - \dfrac{N}{f_{cef} w}\right) & 0 & 0 \\ -1 & 1 & 0 & 0 \end{bmatrix} \begin{bmatrix} \delta\lambda_1 \\ \delta\lambda_2 \\ \delta\lambda_3 \\ \delta\lambda_4 \end{bmatrix}, \tag{25}$$

$$\delta\boldsymbol{\lambda}_k \geq \mathbf{0}, \tag{26}$$

$$\varphi_k^{\mathrm{T}} \delta\boldsymbol{\lambda}_k = 0, \tag{27}$$

$$\delta\mathbf{q} = \mathbf{N}_0 \delta\boldsymbol{\lambda}. \tag{28}$$

Compatibility between joint k generalised strains and the displacement rates of the adjacent blocks i and j, is given in Eq. (29), the vector $\boldsymbol{\delta}\mathbf{u}_i$ being defined in Eq. (30) and the compatibility matrix $\mathbf{C}_{k,i}$, given in Eq. (31). Similarly, the vector $\boldsymbol{\delta}\mathbf{u}_j$ and the matrix $\mathbf{C}_{k,j}$ can be obtained. In this last equation γ_k, β_i, β_j, are the angles between the x-axis and, the direction of joint k, the line defined from the centroid of block i to the centre of joint k, and the line defined from the centroid of block j to the centre of joint k, respectively. Variables d_i, d_j, represent the distances from the centre of joint k to the centroid of the blocks i and j, respectively (Fig. 8):

$$\delta\mathbf{q}_k = \mathbf{C}_{k,j}\delta\mathbf{u}_j - \mathbf{C}_{k,i}\delta\mathbf{u}_i, \tag{29}$$

$$\delta\mathbf{u}_i^{\mathrm{T}} \equiv \begin{bmatrix} \delta u_i & \delta v_i & \delta\omega_i \end{bmatrix}, \tag{30}$$

$$\mathbf{C}_{k,i} = \begin{bmatrix} \cos(\gamma_k) & \sin(\gamma_k) & -d_i\sin(\beta_i - \gamma_k) \\ -\sin(\gamma_k) & \cos(\gamma_k) & d_i\cos(\beta_i - \gamma_k) \\ 0 & 0 & 1 \end{bmatrix}. \tag{31}$$

Compatibility for all the joints in the structure is given by Eq. (32), in which the compatibility matrix $\mathbf{C}$ is obtained by assembling the corresponding matrices for the joints of the structure:

$$\delta\mathbf{q} = \mathbf{C}\delta\mathbf{u}. \tag{32}$$

Applying the contragredience principle,[27] the equilibrium requirement is expressed by

$$\mathbf{F}_c + \alpha\mathbf{F}_v = \mathbf{C}^{\mathrm{T}}\mathbf{Q}. \tag{33}$$

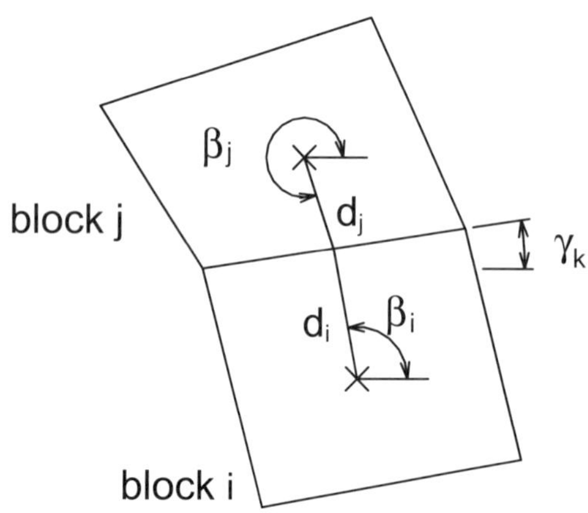

Fig. 8. Representation of main geometric parameters.

The solution to a limit analysis problem must fulfill the previously discussed principles. In the presence of non-associated flow, there is no unique solution satisfying these principles, and the actual failure load corresponds to the mechanism with a minimum load factor.[28] The proposed mathematical description results in the non-linear programming (NLP) problem expressed in Eqs. (34)–(40). Here, Eq. (34) is the objective function and Eq. (35) guarantees both compatibility and flow rule. Equation (36) is a scaling condition of the displacement rates that ensures the existence of non-zero values. This expression can be freely replaced by similar equations, as, at collapse, the displacement rates are undefined and it is only possible to determine their relative values. Equilibrium is given by Eq. (37), and Eq. (38) is the expression of the yield condition, which together with the flow rule, Eq. (39), must fulfill Eq. (40).

$$\text{Minimise: } \alpha, \quad \text{Subject to:} \tag{34}$$

$$\mathbf{N}_0 \delta \boldsymbol{\lambda} - \mathbf{C} \delta \mathbf{u} = \mathbf{0}, \tag{35}$$

$$\mathbf{F}_v^{\mathrm{T}} \delta \mathbf{u} - 1 = 0, \tag{36}$$

$$\mathbf{F}_c + \alpha \mathbf{F}_v = \mathbf{C}^{\mathrm{T}} \mathbf{Q}, \tag{37}$$

$$\varphi \leq \mathbf{0}, \tag{38}$$

$$\delta \boldsymbol{\lambda} \geq \mathbf{0}, \tag{39}$$

$$\varphi^{\mathrm{T}} \delta \boldsymbol{\lambda} = 0. \tag{40}$$

This set of equations represents a case known in the mathematical programming literature as a Mathematical Problem with Equilibrium Constraints (MPEQ).[29] This type of problems is hard to solve because of the complementarity constraint, Eq. (40). The solution adopted consists of two phases, in the first, a Mixed Complementarity Problem (MCP), constituted by Eqs. (35)–(40) is solved. This gives a feasible initial solution. In the second phase, the objective function (Eq. (34)), is reintroduced and Eq. (40) is substituted by Eq. (41). This equation provides a relaxation in the complementarity constraint, makes simpler the solution of the NLP, and allows to search for smaller values of the load factor. The relaxed NLP problem is solved for successively smaller values of ρ to force the complementarity term to approach zero:

$$-\varphi^{\mathrm{T}} \delta \boldsymbol{\lambda} \leq \rho. \tag{41}$$

It must be said that trying to solve a MPEQ as a NLP problem does not guarantee that the solution is a local minimum.[29] In addition, a load-path solution was developed in order not to reach incorrect over-conservative results.[25]

4.3. *Applications*

4.3.1. *Modelling masonry under compression*

The analysis of masonry assemblages under compression using detailed modelling strategies in which units and mortar are modelled separately is a challenging task. Sophisticated standard non-linear continuum models, based on plasticity and cracking, are widely available to represent the masonry components but such models overestimate the experimental strength of masonry prisms under compression.[30] Alternative modelling approaches are therefore needed.

A particle model consisting in a phenomenological discontinuum approach to represent the microstructure of units and mortar is shown here. The microstructure attributed to the masonry components is composed by linear elastic particles of polygonal shape separated by non-linear interface elements,[31] using the model detailed in Sec. 4.1. All the inelastic phenomena occur in the interfaces, and the process of fracturing consists of progressive bond-breakage.

Particle model simulations were carried out employing the same basic cell used for a traditional continuum model. The particle model is composed by approximately 13,000 linear triangular continuum elements, 6000 linear line interface elements and 15,000 nodes. The material parameters were defined by comparing the experimental and numerical responses of units and mortar considered separately.

Typical numerical results obtained for masonry prisms, together with experimental results, are shown in Fig. 9. The experimental collapse load seems to be overestimated by the particle and continuum models. However, a much better agreement with the experimental strength and peak strain has been achieved with the particle model, when compared to the continuum model. For the cases analysed, the numerical over experimental strength ratios ranged between 165% and 170% in the case of the continuum model, while in the case of the particle model, strength ratios ranging between 120% and 140% were found. The results obtained also show that the peak strain values are well reproduced by the

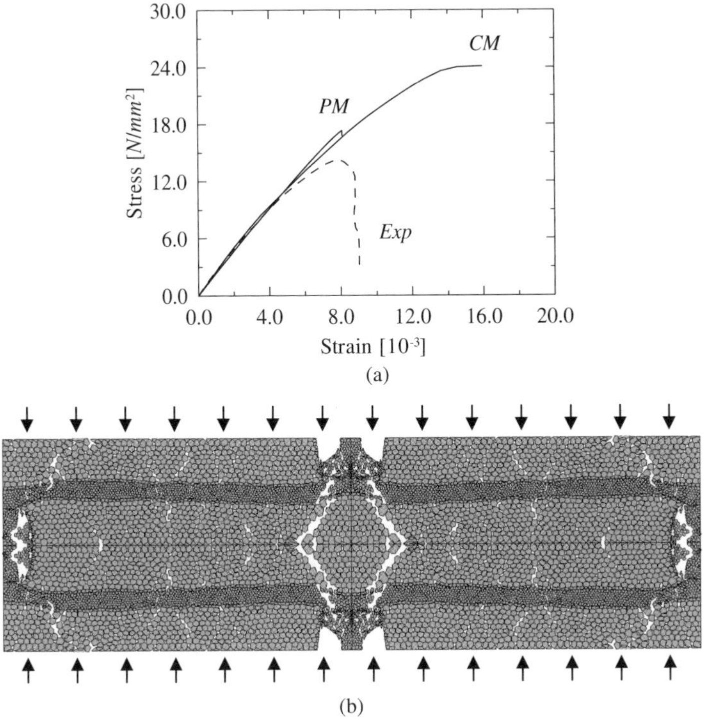

Fig. 9. Results for masonry compression: (a) experimental results, compared to a standard continuum model (CM) and a particulate model (PM); (b) incremental deformed mesh at failure for the particle model.

particle model but large overestimations are obtained with the continuum model.

In fact, for this last model, experimental over numerical peak strain ratios ranging between 190% and 510% were found.[30]

4.3.2. *Conventional micromodelling*

The ability of the model from Sec. 4.1 to reproduce the main features of structural masonry elements is now assessed through the numerical analysis of three masonry walls submitted to cyclic loads. In these simulations, the units were modelled using eight-node continuum plane stress elements with Gauss integration and, for the joints, six-node zero-thickness line interface elements with Lobatto integration were used. All the material parameters are discussed in Ref. 23.

Within the scope of the CUR project, several masonry shear walls were tested submitted to monotonic loads.[32,33] The walls were made of wire-cut solid clay bricks with dimensions of $210 \times 52 \times 10\,\text{mm}^3$ and 10 mm thick mortar joints and characterised by a height/width ratio of 1, with dimensions of $1000 \times 990\,\text{mm}^2$. The shear walls were built with 18 courses, from which only 16 were considered active, since the two extreme courses were clamped in steel beams.

During testing, different vertical uniform loads were initially applied to the walls. Then, for each level of vertical load, a horizontal displacement was imposed at the top steel beam, keeping the top and bottom steel beams horizontal and preventing any vertical movement of the top steel beam. The walls fail in a complex mode, starting from horizontal tensile cracks that develop at the bottom and top of the wall at an early loading stage. This is followed by a diagonal stepped crack that leads to collapse, simultaneously with cracks in the bricks and crushing of the compressed toes. Figure 10 presents the main results (see also Refs. 20 and 34).

Figure 10(a) presents the comparison between numerical and experimental load–displacement diagrams. The experimental behaviour is satisfactorily reproduced, and the collapse load can be estimated within a ~15% range of the experimental values. The sudden load drops are due to the opening of each complete crack across one brick. All the walls behave in a rather ductile manner, which seems to confirm the idea that confined masonry can withstand substantial post-peak deformation with reduced loss of strength, when subjected to in-plane loading.

Two horizontal tensile cracks develop at the bottom and top of the wall. A stepped diagonal crack through head and bed joints immediately follows. This crack starts in the middle of the wall and is accompanied by initiation of cracks in the bricks. Under increasing deformation, the crack progresses in the direction of the supports and, finally, a collapse mechanism is formed with crushing of the compressed toes and a complete diagonal crack through joints and bricks (Fig. 10(b)).

Initially also, the stress profiles are essentially "continuous". At this early stage, due to the different stiffness of joints and bricks, small struts are oriented parallel to the diagonal line defined by the centre of the bricks. This means that the direction of the principal stresses is mainly determined by the geometry of the bricks. After initiation of the diagonal crack the orientation of the compressive stresses gradually rotates. The diagonal crack prevents the formation of compressive struts parallel to the diagonal line defined by the centre of the bricks and, therefore, the internal force flow

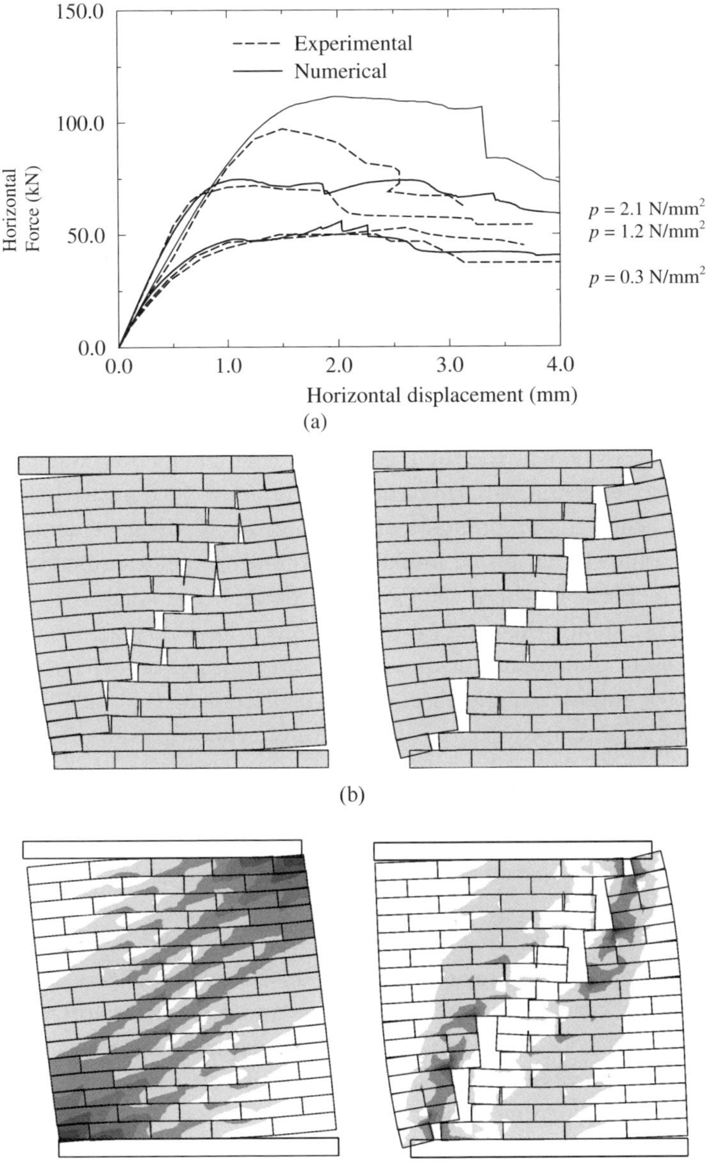

(a)

(b)

(c)

Fig. 10. Results from micromodelling of masonry shear walls: (a) force–displacement diagrams; (b) typical deformed mesh at peak and ultimate state; (c) minimum (compressive) principal stresses at early stage and ultimate state (darker regions indicate higher stresses).

between the two sides of the diagonal crack must be transmitted by shearing of the bed joints. Finally, when the diagonal crack is fully open, two distinct struts are formed, one at each side of the diagonal crack (Fig. 10(c)). The fact that the stress distribution at the supports is of "discontinuous" nature contributes further to the collapse of the wall due to compressive crushing.

For the purpose of investigating cyclic behaviour, a wall submitted to an average compressive stress value of $1.2\,\text{N/mm}^2$, without the possibility of cracking in the units for simplicity, is further considered here. The main purpose of this numerical analysis is to assess the qualitative ability of the model to simulate features related to cyclic behaviour, such as stiffness degradation and energy dissipation.

In order to investigate the cyclic behaviour,[21,23] it was decided to submit the wall to a set of loading–unloading cycles by imposing increasing horizontal displacements at the top steel beam, where unloading was performed at +1.0 mm, +2.0 mm, +3.0 mm and +4.0 mm, until a zero horizontal force value was achieved. The numerical horizontal load–displacement diagram, obtained using the proposed model and following the described procedure, is shown in Fig. 11, where the evolution of the total energy is also given. Figure 11(a) shows that the cyclic horizontal load–displacement diagram follows closely the monotonic one aside from the final branch, where failure occurs for a slightly smaller horizontal displacement (only 4% reduction). Unloading is performed in a quite linear fashion showing important stiffness degradation between cycles, while reloading presents initially high stiffness due to closing of diagonal cracks and then is followed by a progressive decrease of stiffness (reopening). From Fig. 11(b) it can also be observed that the energy dissipated in an unloading–reloading cycle is increased from cycle to cycle.

Figure 11(c) illustrates the incremental deformed meshes with the principal compressive stresses depicted on them, for imposed horizontal displacements corresponding to +4.0 mm and to a zero horizontal force after unloading from +4.0 mm. The initial structural response characterised by the formation of a single, large, compressive strut is quickly destroyed under loading–unloading. The development of the two struts, one at each side of the diagonal line, should be considered the normal condition with permanent residual opening of the head joints in the internal part of the wall.

The same wall is now analysed under load reversal (Fig. 12). It was found that the geometric asymmetry in the micro-structure (arrangements of the units) influenced significantly the structural behaviour of the wall.

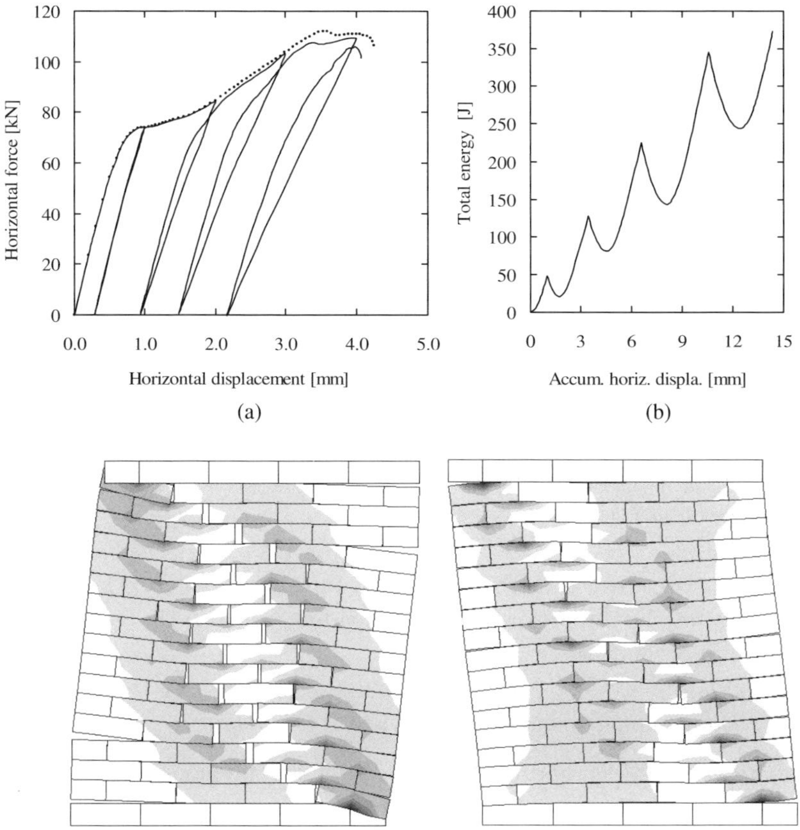

(c)

Fig. 11. Results of shear wall upon load–unloading cycles: (a) load–displacement diagram, where the dotted line represents the monotonic curve; (b) total energy evolution; (c) principal compressive stresses depicted on the incremental deformed mesh for a horizontal displacement of +4.0 mm; zero horizontal force after unloading from +4.0 mm.

Note that, depending on the loading direction, the masonry course starts either with a full unit or only with half unit. It is also clear from these analyses that masonry shear walls with diagonal zigzag cracks possess an appropriate seismic behaviour with respect to energy dissipation.

Figure 12 shows that the monotonic collapse load is 112.0 kN in the LR direction and 90.8 kN in the RL direction, where L indicates left and R indicates right. The cyclic collapse load is 78.7 kN, which represents a loss of ∼13% with respect to the minimum monotonic value but a loss

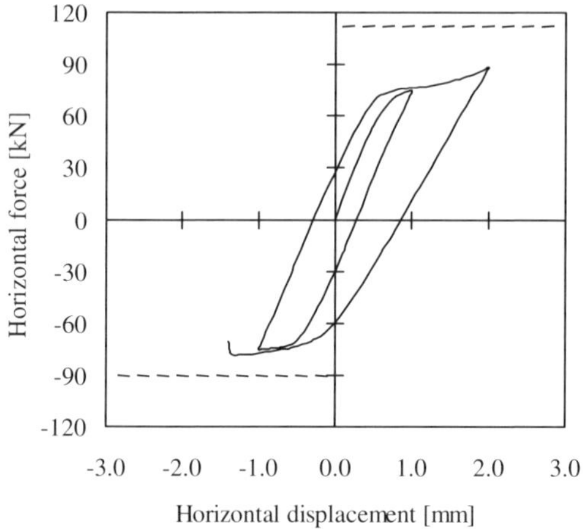

Fig. 12. Load–displacement diagram for shear wall upon load reversal, where the dashed lines represent the maximum monotonic loads.

of ~30% with respect to the maximum monotonic value. This not only demonstrates the importance of cyclic loading but also the importance of taking into account the microstructure.

4.3.3. *The macroblock approach for historical buildings*

The micromodelling approach as used in the previous sections is not practical for medium to large size or complex structure analysis. The use of macroblock models is becoming much popular in the last decades, and the tools discussed in the previous sections are directly applicable to this new application. Here, the model given in Sec. 4.2 is applied to a large-scale case study.

Knowledge about possible failure masonry mechanisms can be obtained by various ways: the engineer experience; the observation of the previous cracking patterns in the structure; and preferably, from studies about failure of structural elements and substructures performed through more detailed models and/or accurate approaches. There are two basic alternatives for developing a macroblock model for shear walls under seismic loading[35]: (1) to consider the wall as a single macroblock and to modify the yield functions for the joint at the base (and possibly top) of the wall on the basis of adequate formulas; and (2) to model each wall as two macroblocks as illustrated in Fig. 13. The latter approach is adopted here, being fully defined by the effective length B and the crack slope tan β_a.

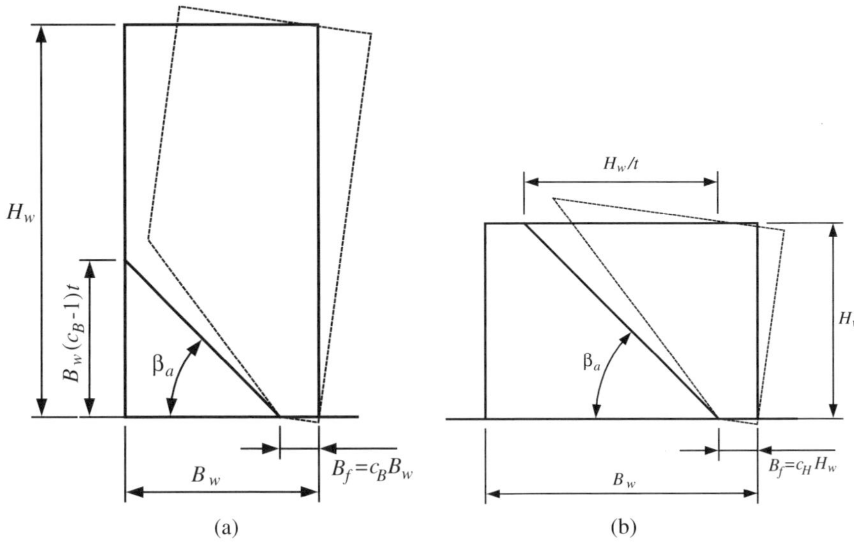

Fig. 13. Simplified model with limited compressive stress for (a) slender walls and (b) long walls.

The classification in slender or long wall depends on these parameters together with the overall wall dimensions.

It also well known that masonry buildings damaged by earthquake actions present cracks along the wall diagonals. So, the macroblock model of a wall can be constructed as illustrated in Fig. 14, where the potential diagonal crack goes from the base to the upper wall corner. The figure also illustrates the "window effect". This effect consists in the fact that the height of a wall contiguous to an opening depends on the load direction. The most critical example in the figure is the central wall: for the action directed to the right the wall height equals the door height, and for the action directed to the left the wall height is only the window height. The left and right walls have also different heights depending on whether the wall height includes or not the lintel height and the portion of the wall below the window. The rule to take into account the window effect can be stated as: *for a horizontal action directed to the right, the wall height is measured from the top of the left opening to the bottom of the right opening.*

For long walls, it is necessary to impose a lower limit to the crack slope due to the fact that for small unit aspect ratios it is probable that unit

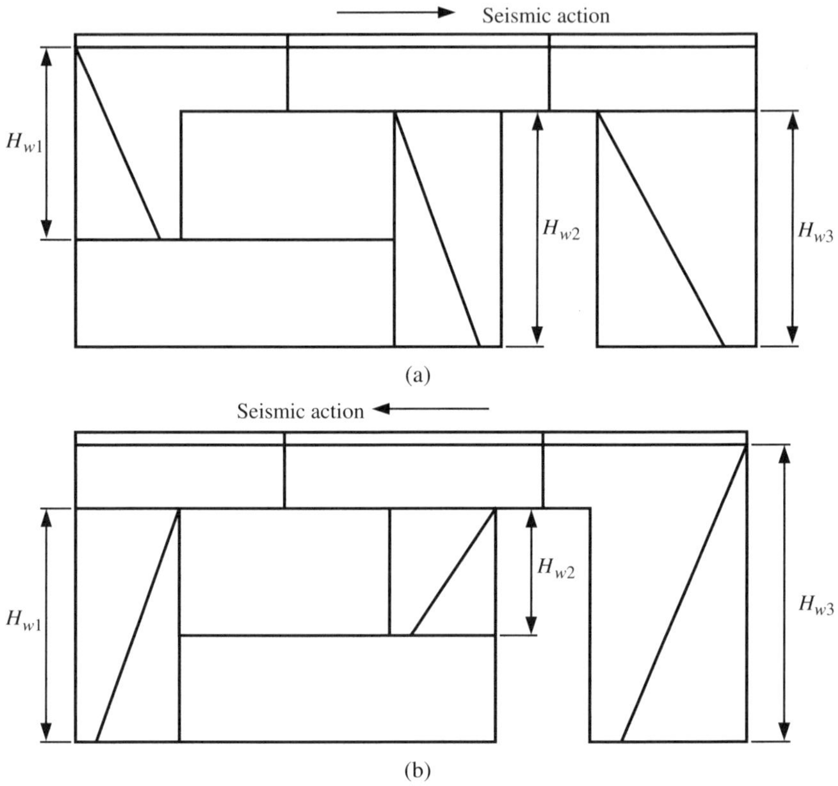

Fig. 14. Window effect for earthquake to the (a) right and (b) left.

cracking increases the wall crack slope. The limit $t \geq 1$, which represents a diagonal crack angle of 45°, is usually adopted.[36]

In the macroblock model, illustrated in Fig. 14, if the effective length of a wall is increased, then the crack slope also increases. Besides, in a multistorey building, the vertical load on the walls increases from the upper levels to the lower levels. Therefore, the effective lengths, which depend on the vertical loading, and, with them, the crack slopes also will increase from the upper levels to the lower levels. Furthermore, for slender and heavily loaded walls, or very slender walls, the model should consist only on a rectangle, with negligible effect on the lateral strength.

The lintels failure must also be considered in the analysis of shear walls with several openings. The normal forces transmitted by lintels are small because they depend, at failure, only on the relative strengths between

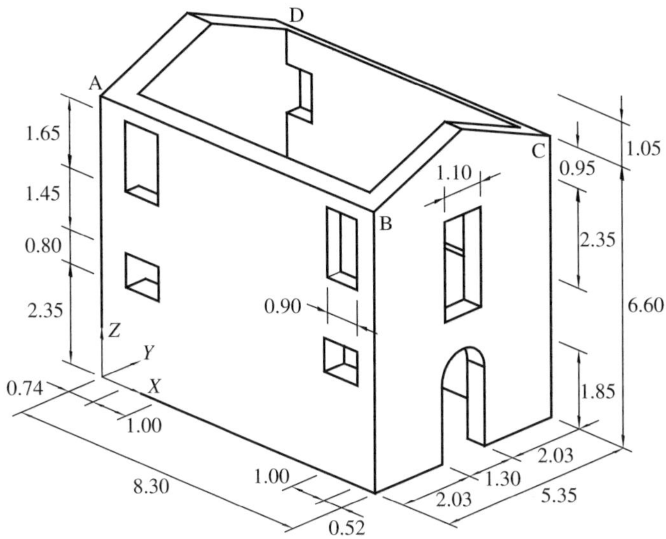

Fig. 15. Via Arizzi house model.

walls. The proposal is to include a vertical joint in the middle of the lintels, as already illustrated in Fig. 14, to allow the shear failure.

As an example, the seismic limit analysis of an ancient house before and after strengthening is presented. The house has two storeys and is located in a seismic area.[35,37] The plan measures are 8.30 m long and 5.35 m wide. Figure 15 presents a three-dimensional view of the main walls. Wall AD is shared with another house; so, walls AB and DC are continuous on that side. Due to this fact, the seismic action on the X-direction is taken as positive only for analysis purposes. The seismic action was considered both positive and negative in the Y-direction, although, due to the almost symmetry of the building, only the results for the positive direction are reported. The local construction code requires this structure to have a seismic coefficient equal or larger than 0.20. The seismic acceleration distribution is assumed constant through the height. The vertical, constant loads are the self-weight walls, as well as permanent and accidental loads on the floor and roof. The variable loads are the same but horizontally applied.

Two models for X and Y seismic action, respectively, were developed for the construction in its original state. It was assumed that no interlocking exists between perpendicular walls; this is a conservative assumption in the absence of better information. Figures 16(a) and 16(b) present the

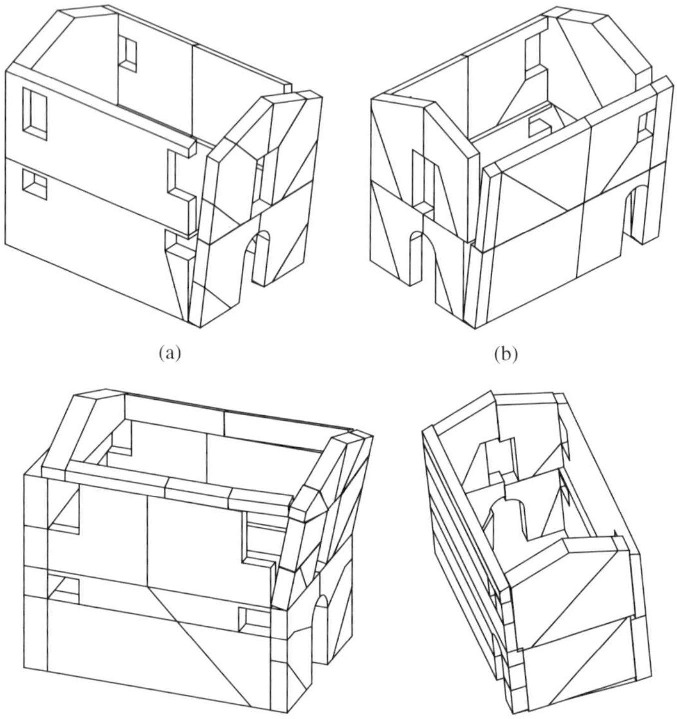

(a) (b) (c) (d)

Fig. 16. Via Arizzi house analysis: original state with earthquake in (a) X-direction ($\alpha = 0.050$) and (b) Y-direction ($\alpha = 0.068$); strengthened with earthquake in (c) X-direction ($\alpha = 0.38$) and (d) Y-direction ($\alpha = 0.28$).

failure mechanisms for the house subject to earthquakes in X- and Y-directions, respectively. Both failure mechanisms involve the overturning of the outmost wall, and the safety factors are sensibly lesser than the required seismic coefficient. These facts were expected since the horizontal load distribution capacity of the roof and floor was neglected, as well as the interlocking between perpendicular walls.

In order to improve the building seismic capacity, the following strengthening measures were proposed.[37] The roof and floor structures were strengthened in order to provide in-plane load distribution capacity. The construction of a concrete element at the top of the walls with an embedded steel bar was proposed. Also, installation of steel ties at floor level, two in the X-direction and three in the Y-direction was proposed. These elements tie the outmost walls each other in both directions.

Due to the deep structural changes introduced by the strengthening measures, the previously developed models were unable to reproduce the behaviour of the strengthened building. Therefore, it was necessary to develop two new models. Figures 16(c) and 16(d) show the failure mechanisms for the strengthened house subject to earthquakes in X- and Y-directions, respectively. For earthquake in the X-direction, the failure is again the overturning of the facade. Nevertheless, the embedded steel bars at the top of the walls drag the roof structure together with the facade. This increases significantly the safety factor to a value higher than the required seismic coefficient. For the earthquake in Y-direction, the strengthening modifies the failure mechanism. Now the failure occurs by shear in the AD and BC walls, increasing the safety factor to an acceptable level.

5. Homogenisation Approaches

The approach based on the use of averaged constitutive equations seems to be the only one suitable to be employed in a large-scale FE analysis.[38] Modelling strategies based on macromodelling,[39,40] have the drawback of requiring extensive laboratory testing of different unit and masonry geometries and arrangements. In this framework, homogenisation techniques can be used for the analysis of large-scale structures. Such techniques take into account at a cell level the mechanical properties of constituent materials and the geometry of the elementary cell, allowing the analysis of entire buildings through standard FE codes.

These two different approaches are illustrated in Fig. 17. A major difference is that homogenisation techniques provide continuum average results as a mathematical process that include the information on the microstructure. Average information, namely a continuum failure surface is not known, even if it can be calculated for different stress paths.

The complex geometry of the masonry representative volume, i.e. the geometrical pattern that repeats periodically in space, means that no closed form solution of the problems exists for running bond masonry.

One of the first ideas presented[41] was to substitute the complex geometry of the basic cell with a simplified geometry, so that a closed-form solution for the homogenisation problem was possible. This approach, rooted in geotechnical engineering applications, assumed masonry as a layered material and a so-called "two-step homogenisation". In the first step, a single row of masonry units and vertical mortar joints were taken

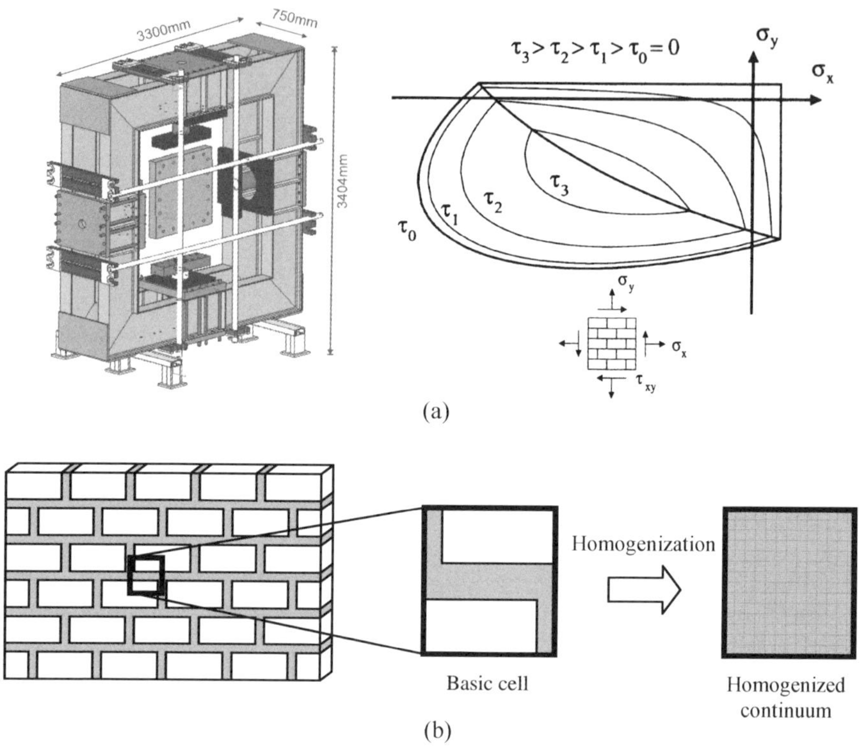

Fig. 17. Constitutive behaviour of materials with microstructure: (a) collating experimental data and defining failure surfaces; (b) a mathematical process that uses information on geometry and mechanics of components.

into consideration and homogenised as a layered system. In the second step, the "intermediate" homogenised material was further homogenised with horizontal joints in order to obtain the final material.

This simplification does not allow to include information on the arrangement of the masonry units with significant errors in the case of non-linear analysis. Moreover, the results depend on the sequence of homogenisation steps.

To overcome the limitations of the two-step homogenisation procedure, micromechanical homogenisation approaches that consider additional internal deformation mechanisms have been derived.[42–45] Other approaches[46,47] are based on the observation that, in general, masonry failure occurs with the damage of mortar joints, e.g. with cracking and shearing. In this way, masonry failure could occur as a combination of bed and head joints failure. The implementation of these approaches in standard macroscopic FE non-linear codes is simple, and the approaches can compete favourably with macroscopic approaches.

Here, a micromechanical model for the limit analysis of in- and out-of-plane loaded masonry walls is reviewed.[48,49] In the model, the elementary cell is subdivided along its thickness in several layers. For each layer, fully equilibrated stress fields are assumed, adopting polynomial expressions for the stress tensor components in a finite number of subdomains. The continuity of the stress vector on the interfaces between adjacent subdomains and suitable antiperiodicity conditions on the boundary surface is further imposed. In this way, linearised homogenised surfaces in six dimensions for masonry in- and out-of-plane loaded are obtained. Such surfaces are then implemented in a FE limit analysis code for simulation of entire 3D structures.

5.1. *Homogenised failure surfaces*

Figure 18 shows a masonry wall constituted by a periodic arrangement of bricks and mortar arranged in running bond. For a general rigid-plastic heterogeneous material, homogenisation techniques combined with limit analysis can be applied for the evaluation of the homogenised in- and out-of-plane strength domain Ω,[50] masonry being only a particular case of interest.

In the framework of perfect plasticity and associated flow rule for the constituent materials, and by means of the lower bound limit analysis theorem, S^{hom} can be derived by means of the following (non-linear) optimisation problem (see also Fig. 18):

$$S^{\mathrm{hom}} = \left\{ \max(\mathbf{M}, \mathbf{N}) \middle| \begin{cases} \mathbf{N} = \dfrac{1}{|Y|} \displaystyle\int_{Y \times h} \boldsymbol{\sigma}\, dV & \text{(a)} \\ \mathbf{M} = \dfrac{1}{|Y|} \displaystyle\int_{Y \times h} y_3 \boldsymbol{\sigma}\, dV & \text{(b)} \\ \operatorname{div} \boldsymbol{\sigma} = \mathbf{0} & \text{(c)} \\ [[\boldsymbol{\sigma}]]\mathbf{n}^{\mathrm{int}} = \mathbf{0} & \text{(d)} \\ \boldsymbol{\sigma}\mathbf{n} \text{ antiperiodic on } \partial Y_l & \text{(e)} \\ \boldsymbol{\sigma}(\mathbf{y}) \in S^m \quad \forall \mathbf{y} \in Y^m; \boldsymbol{\sigma}(\mathbf{y}) \in S^b \quad \forall \mathbf{y} \in Y^b & \text{(f)} \end{cases} \right\} \tag{42}$$

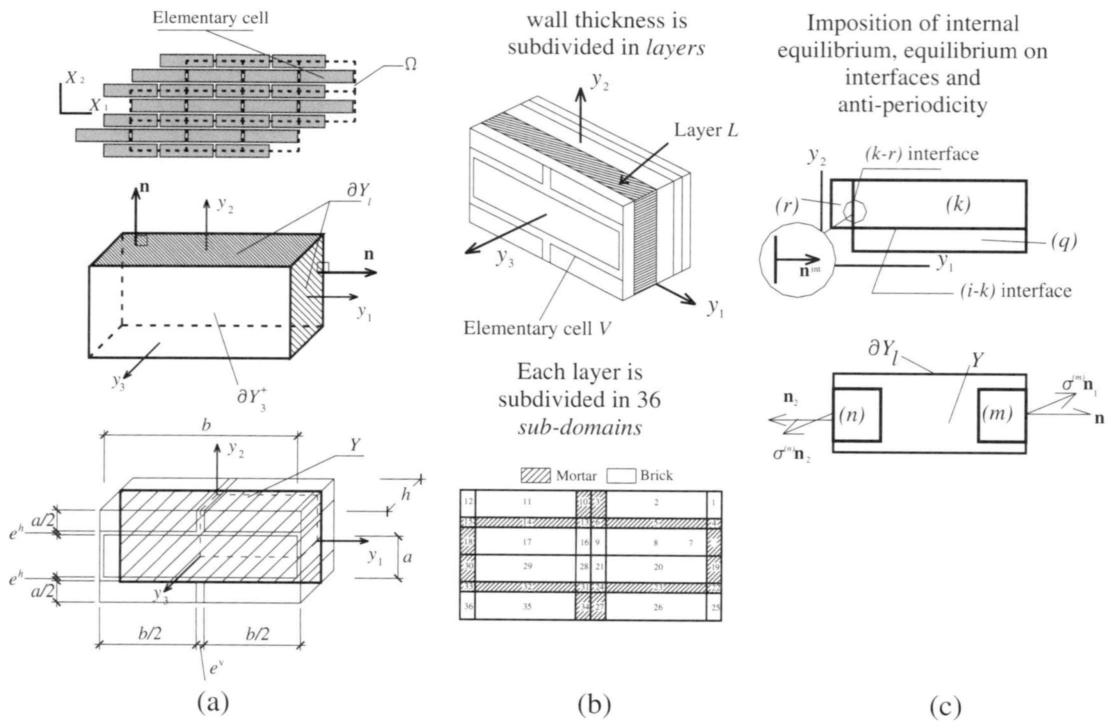

Fig. 18. Proposed micromechanical model: (a) elementary cell; (b) subdivision in layers along thickness and subdivision of each layer in subdomains; (c) imposition of internal equilibrium, equilibrium on interfaces and antiperiodicity.

where:

- $\mathbf{N}$ and $\mathbf{M}$ are the macroscopic in-plane (membrane forces) and out-of-plane (bending moments and torsion) tensors;
- $\boldsymbol{\sigma}$ denotes the microscopic stress tensor;
- $\mathbf{n}$ is the outward versor of ∂Y_l surface (see Fig. 18(a));
- $[[\boldsymbol{\sigma}]]$ is the jump of micro-stresses across any discontinuity surface of normal $\mathbf{n}^{\text{int}}$ (see Fig. 18(c));
- S^m and S^b denote respectively the strength domains of mortar and bricks;
- Y is the cross section of the 3D elementary cell with $y_3 = 0$ (see Fig. 18), $|Y|$ is its area, V is the elementary cell volume, h represents the wall thickness and $\mathbf{y} = (y_1 \quad y_2 \quad y_3)$ are the assumed material axes;
- Y^m and Y^b represent mortar joints and bricks, respectively (see Fig. 18).

It is worth noting that Eq. (42c) imposes the micro-equilibrium with zero body forces, usually neglected in the framework of the homogenisation theory and that antiperiodicity given by Eq. (42e) requires that stress vectors $\boldsymbol{\sigma}\mathbf{n}$ are opposite on opposite sides of ∂Y_l (Fig. 18(c)), i.e. $\sigma^{(m)}\mathbf{n}_1 = -\sigma^{(n)}\mathbf{n}_2$.

In order to solve Eqs. (5.1) numerically, an admissible and equilibrated micromechanical model is adopted.[48] The unit cell is subdivided into a fixed number of layers along its thickness, as shown in Fig. 18(b). For each layer, out-of-plane components σ_{i3} $(i = 1, 2, 3)$ of the microstress tensor $\boldsymbol{\sigma}$ are set to zero, so that only in-plane components σ_{ij} $(i, j = 1, 2)$ are considered active. Furthermore, σ_{ij} $(i, j = 1, 2)$ are kept constant along the Δ_L thickness of each layer, i.e. in each layer $\sigma_{ij} = \sigma_{ij}(y_1, y_2)$. For each layer in the wall thickness direction, one-fourth of the representative volume element is subdivided into nine geometrical elementary entities (subdomains), so that the entire elementary cell is subdivided into 36 subdomains.

For each subdomain (k) and layer (L), polynomial distributions of degree (m) in the variables (y_1, y_2) are *a priori* assumed for the stress components. Since stresses are polynomial expressions, the generic ijth component can be written as follows:

$$\sigma_{ij}^{(k,L)} = \mathbf{X}(\mathbf{y})\mathbf{S}_{ij}^{(k,L)T}, \quad \mathbf{y} \in Y^{(k,L)}, \tag{43}$$

where

- $\mathbf{X}(\mathbf{y}) = [1 \; y_1 \; y_2 \; y_1^2 \; y_1 y_2 \; y_2^2 \; \cdots]$;

— $\mathbf{S}_{ij}^{(k,L)} = [S_{ij}^{(k,L)(1)}\ S_{ij}^{(k,L)(2)}\ S_{ij}^{(k,L)(3)}\ S_{ij}^{(k,L)(4)}\ S_{ij}^{(k,L)(5)}\ S_{ij}^{(k,L)(6)}\ \cdots]$ is a vector representing the unknown stress parameters of subdomain (k) of layer (L);
— $Y^{(k,L)}$ represents the kth subdomain of layer (L).

The imposition of equilibrium inside each subdomain, the continuity of the stress vector on interfaces and the anti-periodicity of $\boldsymbol{\sigma}\mathbf{n}$ permit a reduction in the number of independent stress parameters.[48]

Assemblage operations on the local variables allow to write the stress vector $\tilde{\boldsymbol{\sigma}}^{(k,L)}$ of layer L inside each subdomain as

$$\tilde{\boldsymbol{\sigma}}^{(k,L)} = \tilde{\mathbf{X}}^{(k,L)}(\mathbf{y})\tilde{\mathbf{S}}^{(L)},$$
$$k = 1,\ldots,\ \text{no. of subdomains}\ L = 1,\ldots,\text{no. of layers}, \tag{44}$$

where $\tilde{\mathbf{S}}^{(L)}$ is a $N_{uk} \times 1$ (N_{uk} = number of unknowns per layer) vector of linearly independent unknown stress parameters of layer L, and $\tilde{\mathbf{X}}^{(k,L)}(\mathbf{y})$ is a $3 \times N_{uk}$ matrix depending only on the geometry of the elementary cell and on the position $\mathbf{y}$ of the point in which the microstress is evaluated.

For out-of-plane actions the proposed model requires a subdivision (n_L) of the wall thickness into several layers (see Fig. 18(b)), with a fixed constant thickness $\Delta_L = h/n_L$ for each layer. This allows to derive the following simple non-linear optimisation problem:

$$S^{\text{hom}} \equiv \begin{cases} \max\{\lambda\} \\ \text{such that} \begin{cases} \tilde{\mathbf{N}} = \displaystyle\int_{k,L} \tilde{\boldsymbol{\sigma}}^{(k,L)}\,dV & \text{(a)} \\ \tilde{\mathbf{M}} = \displaystyle\int_{k,L} y_3\tilde{\boldsymbol{\sigma}}^{(k,L)}\,dV & \text{(b)} \\ \boldsymbol{\Sigma} = \begin{bmatrix}\tilde{\mathbf{N}} & \tilde{\mathbf{M}}\end{bmatrix} = \lambda\mathbf{n}_\Sigma & \text{(c)} \\ \tilde{\boldsymbol{\sigma}}^{(k,L)} = \tilde{\mathbf{X}}^{(k,L)}(\mathbf{y})\tilde{\mathbf{S}} & \text{(d)} \\ \tilde{\boldsymbol{\sigma}}^{(k,L)} \in S^{(k,L)} & \text{(e)} \\ k = 1,\ldots,\text{number of } \textit{subdomains} & \text{(f)} \\ L = 1,\ldots,\text{number of } \textit{layers} & \text{(g)} \end{cases} \end{cases} \tag{45}$$

where

— λ is the load multiplier (ultimate moment, ultimate membrane action or a combination of moments and membrane actions) with fixed direction $\mathbf{n}_\Sigma$ in the six-dimensional space of membrane actions ($\tilde{\mathbf{N}} = [N_{xx}\ N_{xy}\ N_{yy}]$), together with bending and torsion moments ($\tilde{\mathbf{M}} = [M_{xx}\ M_{xy}\ M_{yy}]$);

— $S^{(k,L)}$ denotes the (non-linear) strength domain of the constituent material (mortar or brick) corresponding to the kth subdomain and Lth layer.
— $\tilde{\mathbf{S}}$ collects all the unknown polynomial coefficients (of each subdomain of each layer).

It is noted that the direction $\mathbf{n}_\Sigma$ is fixed arbitrarily in the six-dimensional space $[\tilde{\mathbf{N}}\ \tilde{\mathbf{M}}]$. As a rule, since $\mathbf{n}_\Sigma = [\alpha_1, \alpha_2, \ldots, \alpha_6]$ with $\Sigma\alpha_i^2 = 1$, the parameters α_i are chosen randomly between 0 and 1 satisfying the constraint $\Sigma\alpha_i^2 = 1$, so that a number of directions $\mathbf{n}_\Sigma$ are selected.

5.2. *Applications*

The homogenised failure surface obtained with the above approach has been coupled with FE limit analysis. Both upper and lower bound approaches have been developed, with the aim to provide a complete set of numerical data for the design and/or the structural assessment of complex structures. The FE lower bound analysis is based on an equilibrated triangular element,[51] while the upper bound is based on a triangular element with discontinuities of the velocity field in the interfaces.[52,53]

5.2.1. *Masonry shear wall*

Traditionally, experiments in shear walls have been adopted by the masonry community as the most common in-plane large test. The clay masonry shear walls tested at ETH Zurich[54] and analysed using non-linear analysis[39] are addressed next. These experiments are well suited for the validation of the model, not only because they are large and feature well-distributed cracking, but also because most of the parameters necessary to characterise the model are available from biaxial tests.

The walls consist of a masonry panel and two flanges, with two concrete slabs placed in the top and bottom of the specimen. Initially, the wall is subjected to a vertical load uniformly distributed, followed by the application of a horizontal force on the top slab. Experimental evidences show a very ductile response, justifying the use of limit analysis, with tensile and shear failure along diagonal stepped cracks.

In Figs. 19(a) and 19(b) the principal stress distribution at collapse from the lower bound analysis and the velocities at collapse from the upper bound analysis are reported. The results show the typical strut action and a combined shear-sliding mechanism for shear walls at collapse. Finally, in Fig. 19(c) a comparison between the numerical failure loads provided

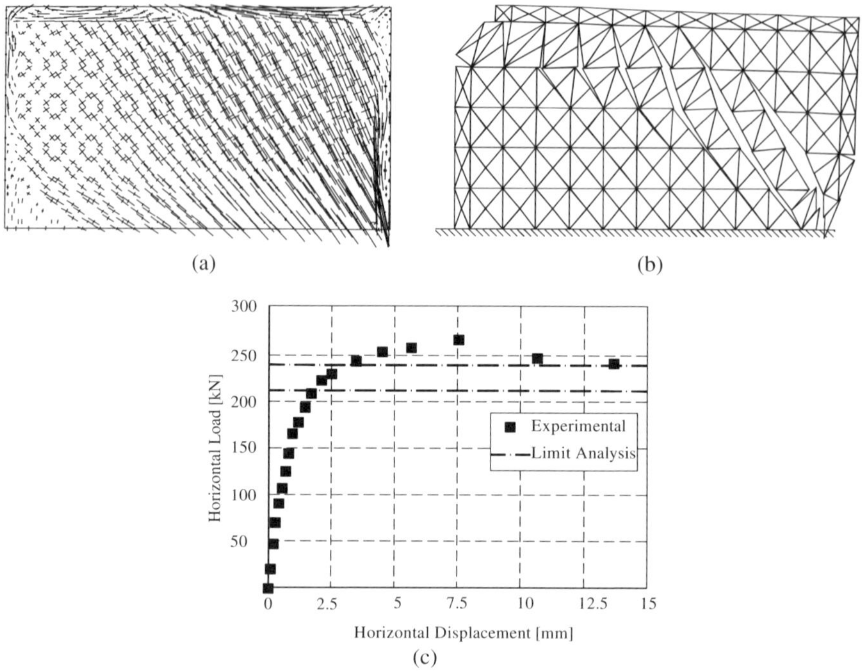

Fig. 19. Results from a masonry shear wall: (a) Principal stress distribution at collapse from the lower bound analysis; (b) Velocities at collapse from the upper bound analysis; (c) Comparison between experimental load–displacement diagram and the homogenised limit analysis (lower bound and upper bound approaches).

respectively by the lower and upper bound approaches and the experimental load–displacement diagram is reported. Collapse loads $P^- = 210\,\text{kN}$ and $P^+ = 245\,\text{kN}$ are numerically found using a model with 288 triangular elements, whereas the experimental failure shear load is approximately $P = 250\,\text{kN}$.

5.2.2. *Two-storeyed unreinforced masonry building*

Figure 20 presents a two-storeyed unreinforced masonry (URM) building tested[55] to reproduce some structural characteristics of typical existing buildings in the midwestern part of the United States. The dimensions of the structure are $7.32 \times 7.32\,\text{m}$ in plan with storey heights of 3.6 m for the first storey and 3.54 m for the second storey. The structure is constituted by four masonry walls labelled Walls A, B, 1 and 2, respectively. The walls have different thickness and opening ratios. Walls 1 and 2 are composed of

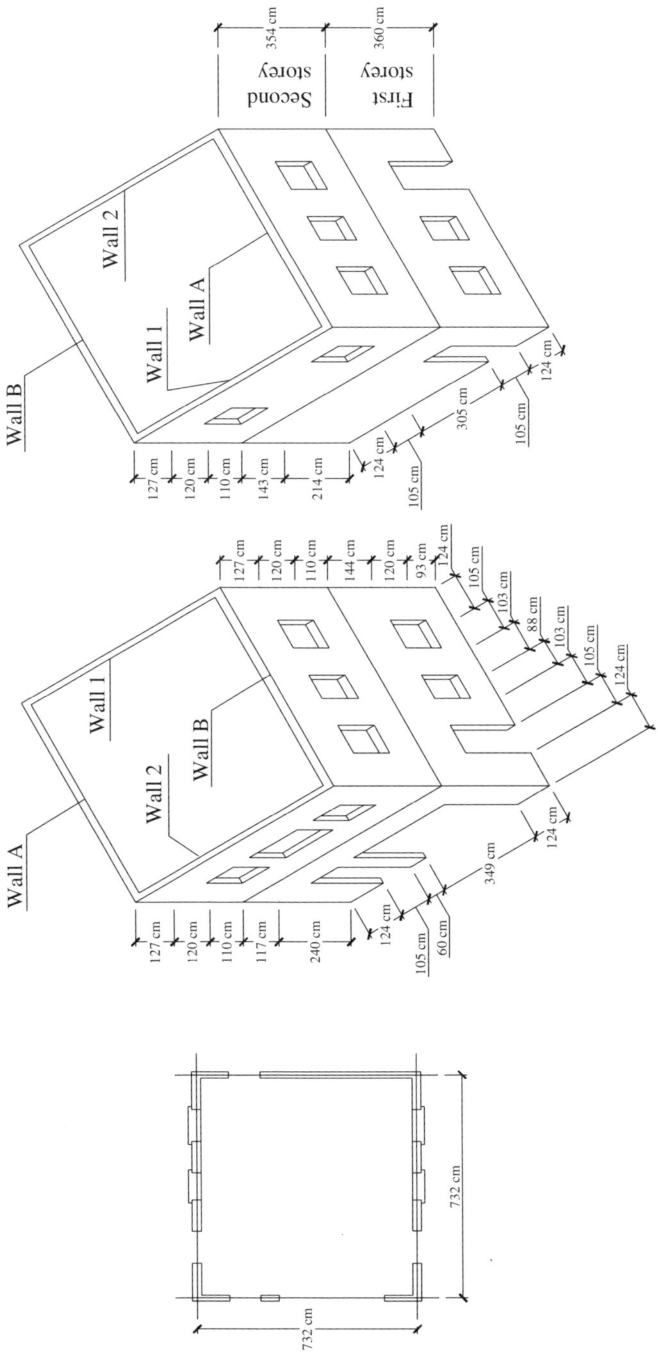

Fig. 20. Geometry of the unreinforced masonry building tested.[55]

brick masonry with thickness 20 cm. Wall 1 has relatively small openings, whereas Wall 2 contains a large door opening and larger window openings. The moderate opening ratios in these two walls are representative of many existing masonry buildings. The aspect ratios of piers range from 0.4 to 4.0. The four masonry walls are considered perfectly connected at the corners, a feature not always reproduced in the past URM tests. This allows to investigate also the contribution of transverse walls to the strength of the overall building.

A wood diaphragm and a timber roof are present in correspondence of the floors. Solid bricks and hollow cored bricks are employed in the structure. Vertical loading is constituted only by self-weight walls and permanent loads of the first floor and of the roof.

In order to numerically reproduce the actual experimental set-up, horizontal loads, depending on the limit multiplier, are applied in correspondence of first and second floor levels of Wall 1. The results obtained with the homogenised FE limit analysis model in terms of failure shear at the base are compared in Fig. 21(a), where total shear at the base of Walls A and B are reported. The kinematic FE homogenised limit analysis gives a total shear at the base for walls A and B of 183 kN, in excellent agreement with the results obtained experimentally. Figures 21(b) and 21(c) show the deformed shape of the model, which is also in agreement with the experimental results.[56] Failure involves torsion of the building, combining in-plane (damage in the piers and around openings) with out-of-plane mechanisms.

6. Conclusions

Constraints to be considered in the use of advanced modelling are the cost, the need of an experienced user/engineer, the level of accuracy required, the availability of input data, the need for validation and the use of the results.

As a rule, advanced modelling is a necessary means for understanding the behaviour and damage of (complex) historical masonry constructions, and examples have been addressed here. For this purpose, it is necessary to have reliable information on material data, and recommendations are provided in this contribution.

Micromodelling techniques for masonry structures allow a deep understanding of the mechanical phenomena involved. For large-scale applications, macroblock approaches or average continuum mechanics must be

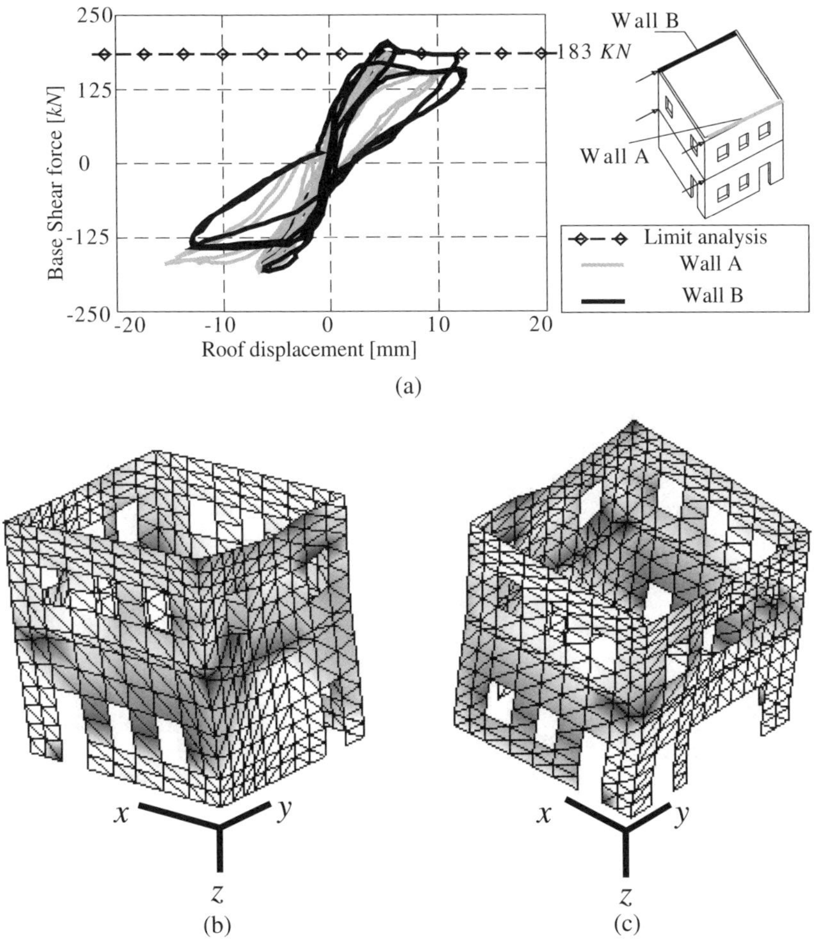

Fig. 21. Results for URM building: (a) Comparison between force–displacement experimental curves and numerical collapse load; Deformed shape at collapse for (b) Walls 1-B view and (c) Walls 2-A view. Darker areas indicate damage.

adopted, and homogenisation techniques represent a popular and active field in masonry research. Modern homogenisation techniques require a subdivision of the elementary cell in a number of different subdomains. A very simplified division of the elementary cell, such as layered approaches, is inadequate for the non-linear range. Examples of application of the micromodelling approach and the homogenisation approach are discussed, illustrating the power of modern numerical computations.

References

1. P. B. Lourenço, in *Structural Analysis of Historical Constructions II*, ed. P. Roca *et al.* (CIMNE, Barcelona, 1998), p. 57.
2. J. G. Rots (eds.), *Structural Masonry: An Experimental/Numerical Basis for Practical Design Rules* (Balkema, Rotterdam, 1997).
3. D. A. Hordijk, Local approach to fatigue of concrete, PhD thesis, Delft University of Technology, The Netherlands (1991).
4. R. van der Pluijm, Out-of-plane bending of masonry: Behaviour and strength, PhD thesis, Eindhoven University of Technology, The Netherlands (1999).
5. R. A. Vonk, Softening of concrete loaded in compression, PhD thesis, Eindhoven University of Technology, The Netherlands (1992).
6. P. B. Lourenço, J. C. Almeida and J. A. Barros, *Masonry Int.* **18**(1), 11 (2005).
7. G. Vasconcelos, Experimental investigations on the mechanics of stone masonry: Characterization of granites and behaviour of ancient masonry shear walls, PhD thesis, University of Minho, Portugal (2005). Available from www.civil.uminho.pt/masonry.
8. *CEB-FIP Model Code 90* (Thomas Telford Ltd., UK, 1993).
9. P. B. Lourenço, J. O. Barros and J. T. Oliveira, *Const. Bldng. Mat.* **18**, 125 (2004).
10. *Eurocode 6 — Design of masonry structures — Part 1-1: General rules for reinforced and unreinforced masonry structures* (European Committee for Standardization, Belgium, 2005).
11. J. G. Rots, *Heron* **36**(2), 49 (1991).
12. P. A. Cundall and O. D. L. Strack, *Geotechnique* **29**(1), 47 (1979).
13. G. Shi and R. E. Goodman, *Int. J. Numer. Anal. Meth. Geomech.* **9**, 541 (1985).
14. R. D. Hart, in *7th Congress on ISRM*, ed. W. Wittke (Balkema, Rotterdam, 1991), p. 1881.
15. J. Azevedo, G. Sincraian and J. V. Lemos, *Earthquake Spectra* **16**(2), 337 (2000).
16. D. Ngo and A. C. Scordelis, *J. Am. Concr. Inst.* **64**(3), 152 (1967).
17. R. E. Goodman, R. L. Taylor and T. L. Brekke, *J. Soil Mech. Found. Div. ASCE* **94**(3), 637 (1968).
18. A. W. Page, *J. Struct. Div. ASCE* **104**(8), 1267 (1978).
19. R. K. Livesley, *Int. J. Num. Meth. Eng.* **12**, 1853 (1978).
20. P. B. Lourenço and J. G. Rots, *J. Eng. Mech. ASCE* **123**(7), 660 (1997).
21. D. V. Oliveira and P. B. Lourenço, *Comp. Struct.* **82**(17–19), 1451 (2004).
22. P. H. Feenstra, Computational aspects of biaxial stress in plain and reinforced concrete, PhD thesis, Delft University of Technology, The Netherlands (1993).
23. D. V. Oliveira, Experimental and numerical analysis of blocky masonry structures under cyclic loading, PhD thesis, University of Minho, Portugal (2003). Available from www.civil.uminho.pt/masonry.

24. A. Orduña and P. B. Lourenço, *Int. J. Solids Struct.* **42**(18–19), 5140 (2005).
25. A. Orduña and P. B. Lourenço, *Int. J. Solids Struct.* **42**(18–19), 5161 (2005).
26. A. Orduña and P. B. Lourenço, *J. Struct. Eng. ASCE* **129**(10), 1367 (2003).
27. M. Mukhopadhyay, *Structures: Matrix and Finite Element* (A.A. Balkema, The Netherlands, 1993).
28. C. Baggio and P. Trovalusci, *Mech. Struct. Mach.* **26**(3), 287 (1998).
29. M. C. Ferris and F. Tin-Loi, *Int. J. Mech. Sci.* **43**, 209 (2001).
30. P. B. Lourenço and J. L. Pina-Henriques, *Comp. Struct.* **84**(29–30), 1977 (2006).
31. J. L. Pina-Henriques and P. B. Lourenço, *Eng. Comput.* **23**(4), 382 (2006).
32. T. M. J. Raijmakers and A. T. Vermeltfoort, Deformation controlled tests in masonry shear walls (in Dutch), Internal Report, Eindhoven University of Technology, The Netherlands (1992).
33. A. T. Vermeltfoort and T. M. J. Raijmakers, Deformation controlled tests in masonry shear walls, Part 2 (in Dutch), Internal Report, Eindhoven University of Technology, The Netherlands (1993).
34. P. B. Lourenço, Computational strategies for masonry structures, PhD thesis, Delft University of Technology, The Netherlands (1996). Available from www.civil.uminho.pt/masonry.
35. A. Orduña, Seismic assessment of ancient masonry structures by rigid blocks limit analysis, PhD thesis, University of Minho, Portugal (2003). Available from www.civil.uminho.pt/masonry.
36. A. Giuffrè, *Safety and Conservation of Historical Centres: The Ortigia Case* (in Italian), Guide to the seismic retrofit project (Editori Laterza, Italy, 1991), Chap. 8, p. 151.
37. R. de Benedictis, G. de Felice and A. Giuffrè, *Safety and Conservation of Historical Centres: The Ortigia Case* (in Italian), Seismic retrofit of a building (Editori Laterza, Italy, 1991), Chap. 9, p. 189.
38. P. B. Lourenço, R. de Borst and J. G. Rots, *Int. J. Num. Meth. Eng.* **40**, 4033 (1997).
39. P. B. Lourenço, J. G. Rots and J. Blaauwendraad, *J. Struct. Eng. ASCE* **124**(6), 642 (1998).
40. P. B. Lourenço, *J. Struct. Eng. ASCE* **126**(9), 1008 (2000).
41. G. N. Pande, J. X. Liang and J. Middleton, *Comp. Geotech.* **8**, 243 (1989).
42. J. Lopez, S. Oller, E. Oñate and J. Lubliner, *Int. J. Num. Meth. Eng.* **46**, 1651 (1999).
43. A. Zucchini and P. B. Lourenço, *Int. J. Sol. Struct.* **39**, 3233 (2002).
44. A. Zucchini and P. B. Lourenço, *Comp. Struct.* **82**, 917 (2004).
45. A. Zucchini and P. B. Lourenço, *Comp. Struct.* **85**, 193 (2007).
46. L. Gambarotta and S. Lagomarsino, *Earth Eng. Struct. Dyn.* **26**, 423 (1997).
47. C. Calderini and S. Lagomarsino, *J. Earth Eng.* **10**, 453 (2006).
48. G. Milani, P. B. Lourenço, and A. Tralli, *Comp. Struct.* **84**, 166 (2006).
49. G. Milani, P. B. Lourenço, and A. Tralli, *J. Struct. Eng. ASCE* **132**(10), 1650 (2006).
50. P. Suquet, Comptes Rendus de l'Academie des Sciences — Series IIB — *Mechanics* (in French) **296**, 1355 (1983).

51. S. W. Sloan, *Int. J. Num. Anal. Meth. Geomech.* **12**, 61 (1988).
52. S. W. Sloan and P. W. Kleeman. *Comp. Meth. Appl. Mech. Eng.* **127**(1–4), 293 (1995).
53. J. Munro and A. M. A. da Fonseca, *J. Struct. Eng. ASCE* **56B**, 37 (1978).
54. H. R. Ganz and B. Thürlimann, Tests on masonry walls under normal and shear loading (in German), Internal Report, Institute of Structural Engineering, Switzerland (1984).
55. Y. Tianyi, F. L. Moon, R. T. Leon and L. F. Khan, *J. Struct. Eng. ASCE* **132**(5), 643 (2006).
56. G. Milani, P. B. Lourenço and A. Tralli, 3D homogenized limit analysis of masonry buildings under horizontal loads, *Eng. Struct.* (in press, 2007).

MECHANICS OF MATERIALS WITH SELF-SIMILAR HIERARCHICAL MICROSTRUCTURE

R. C. Picu* and M. A. Soare†

*Department of Mechanical, Aerospace and Nuclear Engineering
Rensselaer Polytechnic Institute, Troy, NY 12180, USA
picuc@rpi.edu

†Division of Engineering, Brown University
Providence, RI, 02912, USA

Many natural materials have hierarchical microstructure that extends over a broad range of length scales. Examples include the trabecular bone, aerogels, filled polymers, etc. Performing efficient design of structures made from such materials requires the ability to integrate the governing equations of the respective physics, with the support of complex geometry.

Traditional homogenisation methods apply when scales are decoupled and when the microstructure has certain translational symmetry. A microstructure that is self-similar to a scaling operation lacks both these features. Several efforts have been made recently to develop new formulations of mechanics that include information about the geometry in the governing equations. This new concept is based on the idea that the geometric complexity of the domain can be incorporated in the governing equations, rather than in the definition of the boundary conditions, as done within classical continuum mechanics. In this chapter we review the progress made to date in this direction. We discuss elements of fractal geometry, the geometry that best describes the type of microstructure considered, and of fractional calculus. A detailed review of the various works performed to date in this area of research is presented.

1. Introduction

Many natural materials have hierarchical microstructure extending over length scales that cover many orders of magnitude. By this we understand microstructures formed by assembling blocks, which in turn, are made from smaller blocks. This hierarchy may be limited to two scales or may be composed of multiple levels. If the structure observed within blocks belonging to various scales is geometrically similar, the hierarchical microstructure is denoted as "self-similar".

Examples of such materials include the trabecular bone, muscles and tendons, the sticky foot of the Gecko lizard, the structural support of various plants and algae, etc. In general, most biological materials, which are made by controlled assembly of molecular components, exhibit hierarchical structures. Self-similarity is usually observed over a limited range of scales and may be deterministic or stochastic in nature. Stochastic self-similarity is generally the rule. The stochastic nature of the structure is related either to both the dimensions of the building blocks that are replicated from scale to scale, or to their relative position. Figure 1(a) shows a schematic representation of a tendon composed mainly from collagen fibers arranged in a hierarchical manner in a non-collagenous matrix.[1] Figure 1(b) shows an image of the skeleton of a marine algae displaying a branched hierarchical structure.

Man is just beginning, mostly by biomimetics, to discover the benefits of hierarchical designs. Man-made materials belonging to this category include aerogels, filled polymers in which fillers form fractal aggregates as in tire rubber and some dendritic structures. In all these materials the amount of geometric detail observed when zooming in increases with decreasing scale of observation. The micro (and nano)-structure may be self-similar from scale to scale (either in a deterministic or a stochastic fashion) or not. Aerogels are obtained by the aggregation of colloidal particles (e.g. base-catalysed hydrolysis and condensation of silicon tetramethoxide (TMOS) or tetraethoxysilane (TEOS) in alcohol[2,3]), followed by removal of the solvent through evaporation. These materials usually have low density (as low as $0.09\,\text{g/cm}^3$) and high porosity, displaying a fractal mass distribution. Tire rubber gains desirable properties (stiffness and wear resistance) only after

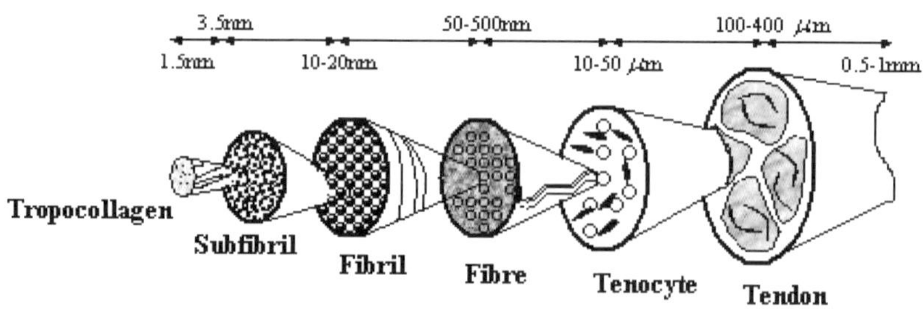

Fig. 1(a). Schematic representation of the hierarchical organisation of a tendon (adapted from Kastelic *et al.*[1]).

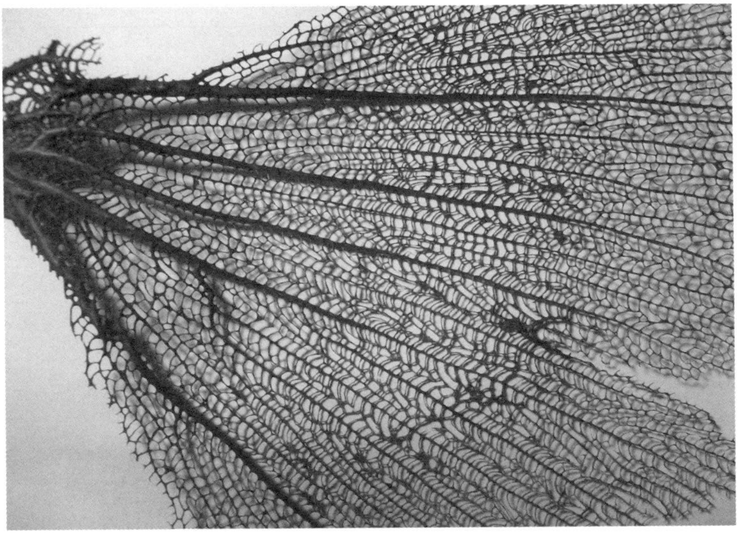

Fig. 1(b). Image of the skeleton of a marine alga.

mixing with carbon black nanoparticles. These nanoscale inclusions are known to aggregate in structures that percolate the material and which have fractal geometry over a range of scales. This microstructure is the result of a long optimisation effort performed mostly by experimental trial and error. Clearly, design assisted by multiscale modelling technologies able to capture the behaviour of such network materials would have been highly desirable as it would have reduced the time to market and the associate cost of the product.

The fractal microstructure may be evidenced by scattering experiments. Figure 2 shows a typical scattering pattern (the scattering intensity versus the scattering vector in units of inverse wavelength of the probing radiation) obtained from a silica aerogel.[4] At long wavelengths, the scattering intensity is independent of the wave-vector indicating that at these length scales the material responds as a continuum. At smaller wave vectors, the diagram becomes linear (in log–log coordinates) indicating fractal microstructure. The intensity scales as $I(k) \sim k^{-q}$, where q is the fractal dimension. Scattering diagrams obtained from other materials may exhibit multiple straight lines with different slopes, indicating a multifractal structure. The Porod regime (slope -4) is observed at even smaller length scales.

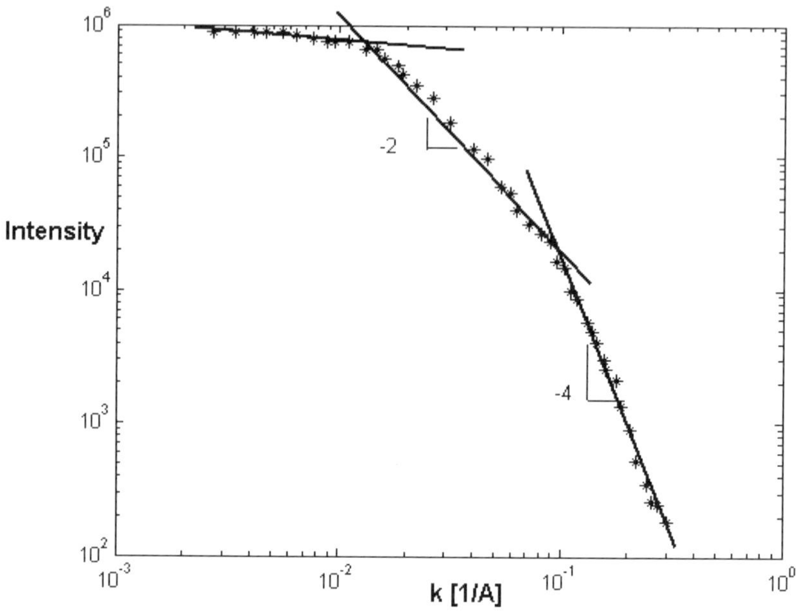

Fig. 2. Small angle X-ray and light scattering from silica aerogels. Three regimes can be distinguished. For large wavelengths (small k), the material behaves as a continuum and the intensity is essentially independent of k (Guinier regime), For intermediate k, the fractal structure of the material in this range of scales leads to a slope -2. The Porod regime with a slope of -4 is seen at the largest k.

Performing efficient design of structures made from such materials requires representing the mechanics of the microstructure. This can be done in a number of ways. One may employ, as traditionally done, a constitutive law obtained by adequate homogenisation on the system scale. Of course, this is pending on the possibility of deriving such constitutive law starting from the multiscale structure. As discussed in this chapter, this approach is practically and conceptually difficult for these types of structures. In principle, one may also fit such equation to the macroscopic (system scale) response obtained experimentally. This method suffers from the usual disadvantages of experimentally-derived constitutive laws: the resulting equation is valid only over the range of experimental conditions considered in experiments and its use outside this range is problematic.

The difficulties preventing the application of usual homogenisation techniques to modelling the deformation of hierarchical materials with self-similar multiscale structures are related to the nature of the geometry.

Traditional homogenisation methods apply when scales are decoupled and when the microstructure has translational symmetry. Then, the microstructure of the underlying scales may be smeared out into a continuum on the scale of observation and usual periodic boundary conditions may be used. A self-similar microstructure constructed by a scaling operation lacks both these features. Deterministic fractals have scaling but no translational symmetry. Stochastic fractals do not have exact translational symmetry either, the situation being identical to that of any other random media.

The concurrent multiscale methods are a new class of methods designed to address problems with no scale decoupling. In these methods, various models representing the behaviour of the material on various scales are used simultaneously in the problem domain. An example is the use of atomistic models in regions of the model where large field gradients exist, while various forms of continuum are used elsewhere, as in Quasicontinuum.[5,6] Other types of hybrid discrete–continuum model have been developed to date. In principle, one may employ this technique when dealing with deterministic fractal microstructures, however, due to the presence of a large number of scales (e.g. very large and very small inclusions are simultaneously present) its advantage becomes marginal. In essence, one would largely recover the efficiency of a model defined on the smallest relevant scale, which is also the most accurate but the most expensive model.

A completely new approach for such problems began to be developed recently. The key idea is to include information about the complex geometry in the governing equations. This is opposed to the traditional method of representing the complexity through complicated boundary conditions. To clarify the discussion, let us consider a composite with a very large number of strongly interacting inclusions of various dimensions. Consider also that a continuum description of the material is adequate both for the matrix and the inclusion materials. To solve boundary value problems on such domain, one would integrate the governing equations (equilibrium, compatibility and the constitutive equations) while imposing displacement and traction continuity across all matrix–inclusion interfaces. Hence, explicit representation of the interfaces is required. The solution will be defined over subdomains, each subdomain being made from a single material, either matrix or inclusion material. Clearly, when a large number of such interfaces are present, the cost of using this method becomes prohibitive. The new concept discussed here is based on the idea that the geometric complexity of the domain may be incorporated in the governing equations, rather than in the definition of the boundary conditions. This is a revolutionary idea. However, as with any such attempt, reaching a form that is broadly accepted and effectively useful in practice is not straightforward.

In this chapter we review the progress made to date in this direction. A brief review of fractal geometry, the geometry that best describes the type of microstructure considered here, is presented first. Few mathematical results relevant for the various formulations described in the chapter are discussed. A detailed review of various works performed to date is then presented, underlying the advances made and their respective limitations.

2. Elements of Fractal Geometry

The fractal geometry developed based on the ideas of Mandelbrot[7] appears adequate to describe certain types of multiscale hierarchical microstructures. We begin with a brief overview of few relevant notions.

Let us consider a one-dimensional example: the Cantor set. This set is a fractal embedded in one-dimension, say in the interval $A = [a, b]$, and is generated by the following geometric iterative procedure: the interval is divided into three segments of equal length, and the middle segment is excluded from the set. The procedure is repeated with the end segments. Each such iteration, n, is identified with a "scale". The sets generated from the first three iterations are shown in Fig. 3. In the following we will denote by F the domains belonging to the fractal and by $A - F$ their embedding complement.

It is useful to observe that the Cantor set is neither discrete, nor continuum. It has the following properties that set it in a class of itself[8,9]: it is compact, i.e. it is bounded and closed (the limits of all sets of points from F are included in F), it is perfect, i.e. any point from F is the limit of a set of points from F, and is disconnected, i.e. between any two points of F exists at least a point from $A - F$. The property of being disconnected may be lost for fractals embedded in Euclidean spaces of dimension larger

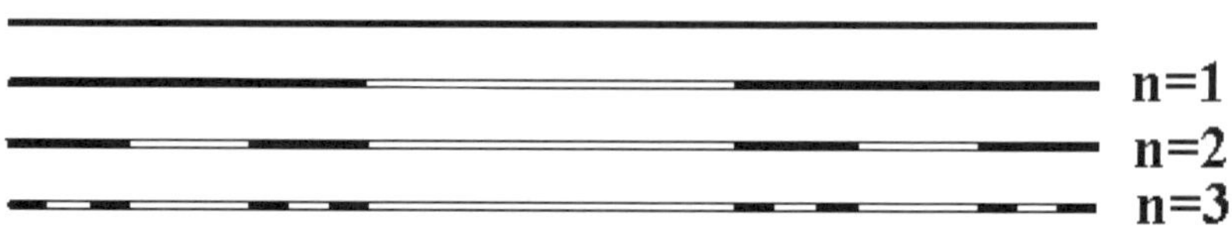

Fig. 3. Three steps of the iteration leading to a deterministic Cantor fractal set.

than one, however, they still have unusual properties that situate them between the discrete and the continuum. They are compact and perfect (a characteristic property of a continuum), but any open set from A includes at least an open set containing only points from $A - F$, and the Euclidean dimension of F is zero (which is a characteristic of discrete systems).

A stochastic fractal may be generated by selecting at random the segment eliminated from the structure at each step, while preserving the ratio of segment lengths of the deterministic procedure. An example of a generalised Cantor set embedded in two dimensions is shown in Fig. 4 (first three steps/scales only). The domain is divided into M equal parts of which P are preserved in the next iteration. Here $M = 4$ and $P = 2$. The P parts that are retained are selected at random in each iteration from the M subdomains. At iteration n, the initial domain is divided in M^n cells of characteristic dimension ε_n of which P^n are occupied by the fractal material. The number of possible configurations at iteration/scale n is $[M!/(P!(M-P)!)]^{p^{n-1}+\cdots+p+1}$.

Fractals are usually characterised by their fractal dimensions. Many such measures have been proposed. One of the most used is the box counting dimension which is determined by covering the set with segments of length ε_n, where n is a natural number representing the iteration step. In the first iteration of the set in Fig. 3 one needs 2 segments of length $(b-a)/3$. In the nth iteration, $M_n = 2^n$ segments of length $(b-a)/3^n$ are needed. This leads to $\varepsilon_n = (b-a)/3^n$. Denoting by $\varepsilon_n^0 = \varepsilon_n/(b-a)$ a non-dimensional coefficient that decreases to zero as the iteration order increases, the box counting dimension, q, results from the identity $M_n = (\varepsilon_n^0)^{-q}$. Specifically,

$$q = \log(M_n)/\log(1/\varepsilon_n^0) = \log(M_n)/\log[(b-a)/\varepsilon_n]. \tag{1}$$

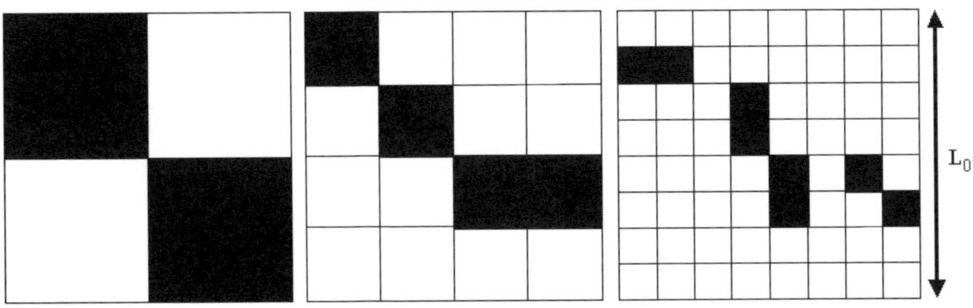

Fig. 4. Three steps of the iteration leading to a generalised stochastic Cantor set embedded in two dimensions.

In the Cantor set (Fig. 3) case, $q = \log 2/\log 3$, with $q < 1$, i.e. the fractal dimension is smaller than the dimension of the Euclidean embedding space. Note that the total length corresponding to scale "n" is $L_n = M_n\varepsilon_n = (\varepsilon_n^0)^{-q}\varepsilon_n = (b-a)(\varepsilon_n^0)^{-q}$ (e.g. Refs. 9 and 10). Since the parameter ε_n^0 is non-dimensional, L_n has the usual units (e.g. metre) for any n. Generally, if the object is embedded in a space of dimension d, and the topological dimension of the object is D_T, the object is a fractal if its dimension, q, has the property $D_T < q < d$.

For the generalised Cantor set of Fig. 4, the dimension of the resulting object is $q = \log(P^n)/\log(1/\varepsilon_n^0)$, where the non-dimensional quantity ε_n^0 results from the scaling of the characteristic length $\varepsilon_n = \varepsilon_n^0 \text{ Vol}(A)^{1/d}$. For this structure one obtains $q = 1$.

It should be noted that the fractal dimension does not fully characterise the geometry and is only an indication of the irregularity of the object. Other measures (e.g. the lacunarity) were developed to provide additional information on the multiscale geometry; however, a unique set of quantities of this type probably does not exist, as some problem specificity should always be present.

3. Elements of Fractional Calculus

As discussed by several authors,[11] fractional calculus appears to be well suited for operation on domains with fractal geometry. However, this relationship has been identified only recently despite the fact that fractional calculus has its origins more than 100 years ago. Two equivalent expressions for the integral/differential operators were initially proposed. One such set is known as the Grunwald–Letnikov operators and is based on the notion of fractional finite differences.[12,13] Let us consider a function f defined on a one-dimensional domain $[a,b]$, and a real number $q \in (0,1)$. The differential of order q is given by

$$\frac{d^q}{d(x-a)^q} f = \lim_{h\to 0} \frac{1}{h^q} \sum_{i=0,N} (-1)^i \frac{\Gamma(q+1)}{i!\Gamma(q-i+1)} f(x-ih),$$

$$N = [(x-a)/h], \tag{2a}$$

and the integral of order q over $[a,b]$ is given by

$$I^q([a,x],f) = \lim_{h\to 0} h^q \sum_{i=0,N} \frac{\Gamma(q+i)}{i!\Gamma(q)} f(x-ih). \tag{2b}$$

The second set of operators is based on an extension of the multiple integral and is denoted as the Riemann–Liouville operators.[14,15] The fractional integral of order q is given by

$$I^q([a,x],f) = \frac{1}{\Gamma(q)} \int_a^x \frac{f(x')}{(x-x')^{1-q}} dx', \tag{3a}$$

and the differential of order q results from Eq. (3a) as

$$\begin{aligned} \frac{d^q}{d(x-a)^q} f &= \left(\frac{d}{dx}\right) I^{1-q}([a,x],f) \\ &= \frac{1}{\Gamma(1-q)} \left(\frac{d}{dx}\right) \int_a^x \frac{f(x')}{(x-x')^q} dx'. \end{aligned} \tag{3b}$$

Although the operators come as a natural generalisation of the classical ones, they have the peculiarity that the derivative of constant functions is not zero: $d^q C/d(x-a)^q = Cx^{-q}/\Gamma(1-q)$. In addition, all these operators are non-local (derivative at x depends on the choice of the left end of the interval, a, which makes their interpretation difficult).

The non-local fractional operators were extensively used for modelling various physical phenomena as diffusion and transport on porous media,[16] relaxation processes of polymers,[17,18] turbulent flows,[19] viscous fingering and diffusion limited aggregation,[20] for the characterisation of the rheologic (viscoelastic) behaviour of materials[21–23] and in fracture mechanics.[24,25]

3.1. *Local fractional differential operators*

In a series of recent publications, Kolwankar and Gangal[26–28] introduced a local version of the Riemann–Liouville derivative as

$$D^{qK} f(x_0) = \lim_{x \to x_0} \frac{1}{\Gamma(1-q)} \left(\frac{d}{dx}\right) \int_{x_0}^x \frac{f(x') - f(x_0)}{(x-x')^q} dx'. \tag{4}$$

An alternate form of (4), described in Ref. 29 has the expression:

$$D^q f(x_0) = \lim_{n \to \infty} L^{q-1} \frac{f(x_n) - f(x_0)}{|x_n - x_0|^q \operatorname{sign}(x_n - x_0)}. \tag{5}$$

The physical meaning of Eq. (5) is transparent: the derivative of order q is computed in a manner similar to the classical derivative, except that the length of the segment is measured on the fractal set, rather than in the embedding space. $(x_n - x_0)^q / L^{q-1}$ is the distance from x_n to x_0 measured on the fractal set, provided $L = \varepsilon_n$. Expressions (4) and (5) differ through a multiplicative constant: $D^q f(x_0) = (L^{q-1}/\Gamma(1+q))D^{qK} f(x_0)$. It is also noted that the units of the derivative (5) are similar to those of the classical derivative, i.e. 1/length (due to the introduction of parameter L), while those of expression (4) are 1/length^q.

The elementary function $f : [0,1] \to R$, $f(x) = x^q$ with $0 < q < 1$ is not differentiable classically in $x_0 = 0$ but it is fractional differentiable of order q and $D^q x^q|_{x=0} = L^{q-1}$ (with definition (5)). On the other hand, at all other points of [0, 1], where f is classically differentiable, the local fractional derivative (5) vanishes.

It is useful to give an example of a function fractional differentiable at an infinite number of points in an embedding domain. This will also demonstrate, by means of an example, the relationship between the fractal support and the fractional operators. This function is an extension of the "devil staircase" function defined on interval [0, 1]. It is constructed starting from the deterministic Cantor set. Specifically, consider first the nth step of the iteration leading to the Cantor set F in Fig. 3. A continuous function is constructed having linear variation over each segment from $A - F_n$ and power law variation over segments from F_n (Fig. 5):

$$f_n(x) = \begin{cases} f_n(x_i) + \beta(x - x_i) & \text{if } x \in (x_i, x_{i+1}] \subset A - F_n, \\ f_n(x_i) + L^{1-q}(x - x_i)^q \gamma & \text{if } x \in (x_i, x_{i+1}] \subset F_n. \end{cases} \tag{6}$$

The function of interest here, f, results by taking the limit $n \to \infty$, so that $f(x) = \lim_{n\to\infty} f_n(x)$. The set of non-derivability points (in the classical sense) coincides with the points of$F$. This function has the property:

$$\begin{cases} Df(x) = \beta & \text{if } x \in A - F \\ D^q f(x) = \gamma & \text{if } x \in F \end{cases}, \quad A = [0,1], \tag{7}$$

i.e. it is a linear function on F and $A - F$ (note that F is defined in the limit $n \to \infty$). By using this procedure, one may define an entire family of power law functions of higher order, starting from Eq. (7).

An expression for the derivative of a function upon a change of variable can be derived. If g is a continuous, differentiable function $g : A \to A$ whose inverse exists,

$$D_y^q f(y_0) = D_x^q f(x_0)[g'^{-1}(y_0)]^q = D_x^q f(x_0)\frac{1}{(g'(x_0))^q}, \tag{8}$$

where x_0 is a point from A where f is differentiable of order q, $y_0 = g(x_0)$ and $g'(x_0) \neq 0$.

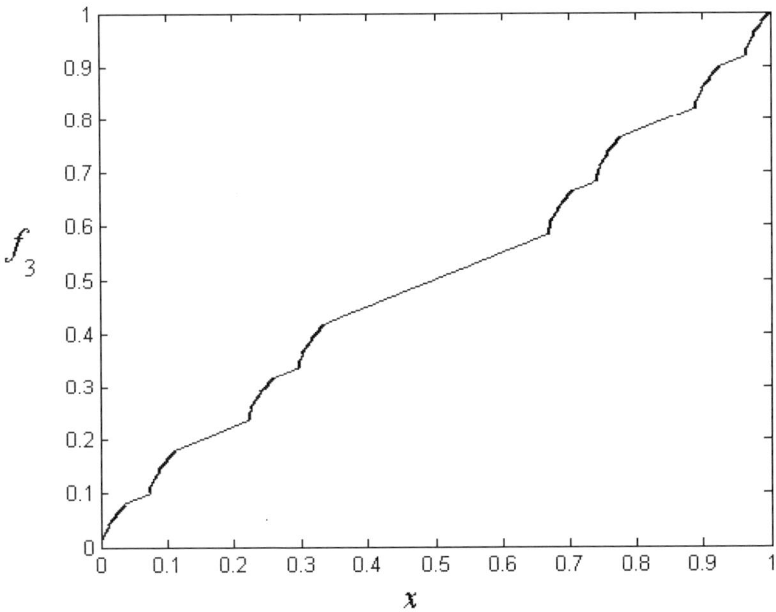

Fig. 5. The third order approximation of a function fractionally differentiable of order $q = \log 2/\log 3$ at the points of a Cantor set.

The directional fractional derivative along vector $\mathbf{v}$ defined in the embedding Euclidean space can be defined based on Eq. (5). Thus, f is fractional differentiable of order q in $\mathbf{x}_0$ in direction $\mathbf{v}$ if there exists a set of points taken in the respective direction, $\mathbf{x}_n - \mathbf{x}_0 = \alpha_n \mathbf{v}$, converging to $\mathbf{x}_0$ and the following limit is finite:

$$D_{\mathbf{v}}^q f(\mathbf{x}_0) = \lim_{n \to \infty} L^{1-q} \frac{f(\mathbf{x}_n) - f(\mathbf{x}_0)}{||\mathbf{x}_n - \mathbf{x}_0||^q \operatorname{sign}(\alpha_n)}, \tag{9}$$

where $||.||$ is the Euclidean norm (e.g. L^2 norm) of the embedding space. The fractional derivatives in the frame directions are denoted here by $D_{\mathbf{e}_i}^q$ or simply D_i^q.

3.2. *Fractional integral operators*

3.2.1. *The 1D case*

A formulation based on the extension of Riemann sums proposed in Ref. 29 is presented here. Let $A = [a, b]$ be an arbitrary one-dimensional domain containing a fractal set F and f a real-valued function defined on A. A point

$x \in [a,b]$ is taken in A and the subdomain $[a,x]$ is partitioned by a set of points $\{x^m_{i=0\cdots m}\}_{m\geq 1}$. In particular, the partition can be uniform, with the length $\varepsilon_m = x^m_{i+1} - x^m_i$ being independent of i. Obviously, $\varepsilon_m \to 0$ as $m \to \infty$. For any order m of the partition, one may distinguish intervals $[x^m_i, x^m_{i+1}] \subset A - F$ and intervals that contain at least one point of F. The following sum can be evaluated for any m:

$$\begin{aligned}\Theta_m(f) = &\sum_i [x^m_{i+1} - x^m_i] f(x^{*m}_i) \\ &+ \sum_{\substack{i| \\ [x^m_i, x^m_{i+1}] \cap F \neq \Phi}} L^{1-q}[x^m_{i+1} - x^m_i]^q f(x^{**m}_i) \\ &- \sum_{\substack{i| \\ [x^m_i, x^m_{i+1}] \cap F \neq \Phi}} L^{1-q}[x^m_{i+1} - x^m_i]^q f(x^{*m}_i).\end{aligned} \tag{10}$$

The point x^{*m}_i is a point from $A - F$ in the interval $[x^m_i, x^m_{i+1}]$ and x^{**m}_i is a point from F. As with the differential operator, L is a parameter in this discussion.

If the limit of the above sum for $m \to \infty$ exists and is finite for any sequence of partitions, $\{x^m_{i=0\cdots m}\}_{m\geq 1}$ of the interval $A = [a,x]$ with the division norms going to zero, the fractional integral of the function f over $[a,x]$ is defined as

$$\int_A f(x')\, d^{\mathrm{Fr}}x' = \lim_{m\to\infty} \Theta_m(f). \tag{11}$$

In particular, if $F = \Phi$ the Riemann integral is recovered (first term in (10)). If the integrand is defined strictly on the fractal set F, only the second term appears in the sum (10), and the fractional integral operator defined by Kolwankar and Gangal[26] is recovered up to a multiplicative constant. When the integral is considered on the entire domain A containing a fractal F, besides the Riemann sum (first term in (10)) the integral on the fractal (the second term in (10)) is added. Hence, intervals containing F are counted twice. The third term in (10) provides a correction. Note that this third term is well defined since, due to the property of F being disconnected, any closed interval contains points x^{*m}_i from $A - F$.

The integral and differential operators defined with Eqs. (5) and (11) can be shown to be inverse to each other, in the sense that

$$\int_A D^{q(x')} f(x')\, d^{\mathrm{Fr}} x' = f(x) - f(a),$$

where $q(x')$ is the order of differentiability at x'.

3.2.2. *Approximate relationship between classical and fractional integrals*

The relationship between the integral operator discussed above and a fractal set embedded in one dimension is relatively obvious. To better define it, one may partition A in segments of length $\varepsilon_n = |x_{i+1}^n - x_i^n|$ such that each segment becomes the size of the fractal box in the nth step of the fractal generation. Then, the integrals become

$$\int_F f(x')\, d^{\mathrm{Fr}} x' = \lim_{n\to\infty} \left\{ \sum_{F_n} L^{1-q} \varepsilon_n^q f(x_i^{**n}) \right\}, \tag{12a}$$

$$\int_{A-F} f(x')\, d^{\mathrm{Fr}} x' = \lim_{n\to\infty} \left\{ \sum_{A-F_n} \varepsilon_n f(x_i^{*n}) - \sum_{F_n} L^{1-q} \varepsilon_n^q f(x_i^{*n}) \right\}, \tag{12b}$$

which are defined only in the limit $n \to \infty$ (limit in which the fractal exists).

If one limits attention to a given iteration/scale of the fractal generation procedure (n is given), the integrals in Eq. (12) become the classical Riemann sums. This happens for a particular choice of $L (L = \varepsilon_n)$.[29] It is noted that for this L the expression $L^{1-q}\varepsilon_n^q$ becomes the length of the respective segment measured *on the fractal set* and hence, the sums containing such terms in (12) recover the meaning of the equivalent terms in the classical Riemann sum. Specifically, one may write

$$\int_{F_n} f_n(x)\, dx \approx \int_F f(x)\, d^{\mathrm{Fr}} x|_{L=\varepsilon_n}, \tag{13a}$$

$$\int_{A-F_n} f_n(x)\, dx \approx \int_{A-F} f(x)\, d^{\mathrm{Fr}} x|_{L=\varepsilon_n}. \tag{13b}$$

The approximation becomes better as the order n increases. The expressions can be used to great advantage in order to predict the solution of a boundary value problem (BVP) defined on a structure with a finite number of scales, n, from a fictitious solution of the same BVP defined for the infinite fractal, $n \to \infty$. This is discussed further in Sec. 4.1.2.2.

4. Mechanics BVPs on Materials with Fractal Microstructure

As discussed in the Introduction, solving BVPs of deformation on materials with fractal (hierarchical and self-similar) microstructure is notoriously difficult. Using methods commonly employed to homogenise composite materials is not feasible. This is primarily due to the presence of inclusions with a broad range of sizes (size distribution is represented by a power law), which are located close to each other. Scale decoupling is not possible under these circumstances. Within the classical theory, the only alternative for addressing these problems is numerical. However, applying usual numerical methods is also difficult since the discretization employed must be on the scale of the finest object that needs to be resolved, which becomes impractical very quickly with increasing the number of scales present in the hierarchical microstructure.

Some of the methods overviewed in this section provide an alternate approach to this problem. They are based on the concept that the operators derived from the balance equations have to operate on the field defined over the entire problem domain containing heterogeneities. This is opposed to the "patching" method commonly used in mechanics, in which the solution is sought over each homogeneous subdomain, under the condition of traction and displacement continuity across the boundaries of the subdomain. In finite elements (FEs), this requires placing element boundaries along each interface of the microstructure. Certainly, the concept is rather revolutionary, but it comes for a price. The geometry of the microstructure has to be known and the equations that need to be solved become non-standard. Specifically, fractional calculus has to replace the usual mathematical apparatus. The main advantage of fractional calculus in this context is the fact that it can operate on functions which are not classically differentiable everywhere. If one considers the deformation problem of a composite with dense inclusions forming a fractal structure, the number of points where the fields (displacements, stresses) are not differentiable (interfaces) diverges. Hence, fractional calculus appears to be optimal in addressing such problems.

Other BVPs, such as transport,[30,31] defined on fractal supports have been posed and solved in the past, with or without using concepts from fractional calculus. As it turns out, such problems are significantly simpler than mechanics problems. This is due to the fact that the transport process takes place on the fractal support, and hence the solution is sought with the metric of and within the space defined by the structure. With mechanics problems the situation is different since deformation takes place in the embedding space and hence, one has to take into account the fractal geometry (with reduced dimensionality) while employing fields defined in the embedding space.

Wave propagation, although a mechanics problem, has been treated in the spirit of transport problems. Strichartz *et al.*[32] proposed solutions of the wave and heat equations for a particular class of fractals that can be approximated by a sequence of finite graphs (called post-critical finite). Such an example is the Sierpinsky gasket represented in Fig. 6. In their approach, the Laplacian operator is redefined based on the Lagrangian (or intrinsic) metric on the fractal space.[33] As inferred earlier by Kigami[34] this operator results as a renormalised limit of graph Laplacians. Details of this construction and of the eigenfunctions of the Laplacian on Sierpinsky gasket-like structures can be found in Refs. 32 and 35. Vibrations on fractals were studied by Alexander and Orbach[36,37] and by others,[38] and a new class of localised vibration modes (fractons) has been identified.

Fig. 6. The Sierpinky gasket.

In the following, we limit attention to quasi-static problems in mechanics, in particular, to the deformation of a body containing a fractal microstructure under specified tractions and/or displacements applied on its boundary. The boundary is defined at the macroscale, i.e. on the scale of the entire modelled structure. We review theoretical results obtained by Panagiatopoulos,[39,40] Carpinteri and collaborators,[41,42] Tarasov[43,44] and by the present authors[29,45] and numerical results obtained by several other groups.[46,47]

4.1. *Deterministic fractal microstructures*

We divide the approaches used to date to address mechanics problems on objects with fractal microstructure in two classes: iterative approaches and approaches based on modifying the governing equations to account for the geometric complexity. We review them below.

4.1.1. *Iterative approaches*

Oshmyan *et al.*[46] considered the problem of finding the effective moduli for composite structures for which the matrix material is linear elastic, while the inclusions are either rigid or voids and are organised in a Sierpinski carpet. The first three generations of this deterministic geometry are shown in Fig. 7.

Although the matrix material is isotropic, the composite is expected to have cubic symmetry. The effective moduli $\mathbf{C}_{1111}$, $\mathbf{C}_{1122}$, $\mathbf{C}_{1212}$ are computed for the first generation using the classical finite element method (FEM). The procedure is repeated and the effective elastic constants $\{\mathbf{C}^{(n)}\}$ are determined for the next n generations. Due to computational

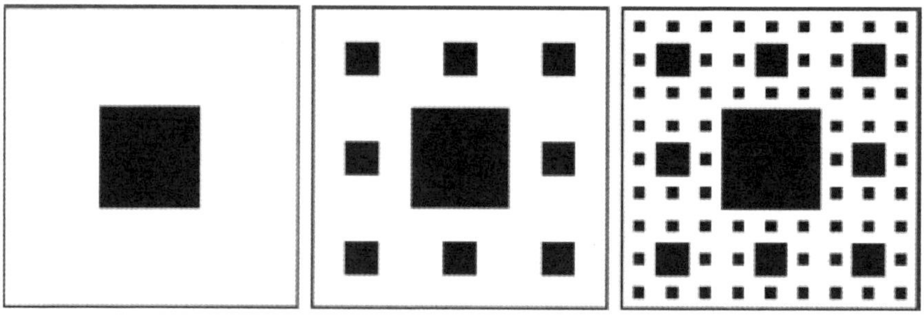

Fig. 7. The first three generations of the Sierpinki gasket.

limitations, the analysis was performed only up to the fourth generation ($n = 4$). Renormalisation group techniques and a fixed-point theorem were used to extrapolate this information for larger n. The fixed-point theorem was formulated in more general terms by Panagiatopoulos *et al.*[39,40] Let us recall this theorem.

Let F (of boundary ∂F) be a fractal structure expressed as $F = \lim_{n\to\infty} F_n$ (in the Hausdorff metric). For example F may represent an attractor for an iterated function system (IFS).[8] Because for each iteration step (scale) F_n is an Euclidean set, one may consider a *classical* deformation problem formulated for this geometry. In general terms, the solution of this problem formulated at each step n, $\mathbf{X}_n$, verifies an equation of the form

$$\mathbf{\Psi}(F_n)\mathbf{X}_n = \mathbf{t}(F_n). \tag{14a}$$

$\mathbf{X}_n$ stands for the displacement or the stress field in an elasticity problem, the temperature field for a heat conduction problem, etc. In this equation, $\mathbf{\Psi}(F_n)$ is an appropriate operator (e.g. Lame operator for the linear elasticity equations, the Laplacian operator for the heat equation, etc.) and $\mathbf{t}$ is the applied perturbation.

The operator $\mathbf{\Psi}$ is defined on the admissible Hilbert space V and has the following properties:

1. It is linear, bonded, symmetric and coercive ($\langle \mathbf{\Psi X}, \mathbf{X}\rangle \geq c\|\mathbf{X}\|$ $\forall \mathbf{X} \in V$).
2. $\|\mathbf{\Psi}(F_n)\mathbf{X} - \mathbf{\Psi}(F)\mathbf{X}\| \to 0$ $\forall \mathbf{X} \in V$ and $\|\mathbf{t}(F_n) - \mathbf{t}(F)\| \to 0$.

Then, the existence of the solution for the problem formulated for the fractal structure $F = \lim_{n\to\infty} F_n$,

$$\mathbf{\Psi}(F)\mathbf{X} = \mathbf{t}(F). \tag{14b}$$

can be determined as the limit $\mathbf{X} = \lim_{n\to\infty} \mathbf{X}_n$.

Aiming to use this theorem to study the elastic constants of the Sierpinski carpet, Oshmyan *et al.* identified the mapping $\mathbf{f}_n\colon \mathbf{C}^{(0)} \to \mathbf{C}^{(n)}$ between the elastic moduli of the host and the normalised effective moduli for the nth generation of the Sierpinski carpet. Thus, the effective elastic moduli for the mnth generation of the carpet can be obtained iterating the mapping m times: $\mathbf{C}^{(mn)} = \underbrace{(\mathbf{f}_n \circ \cdots \circ \mathbf{f}_n)}_{m} \mathbf{C}^{(0)}$.

Showing that the mapping $\mathbf{f}_n$ is a contraction, the authors find the elastic moduli of the ideal fractal structure as

$$\mathbf{C}^{(\infty)} = \lim_{n\to\infty} (\mathbf{f}_n \circ \cdots \circ \mathbf{f}_n)\mathbf{C}^{(0)} \tag{15}$$

(these also represent the fixed point of the map $\mathbf{f}_n$ as $\mathbf{C}^{(\infty)} = \mathbf{f}_n(\mathbf{C}^{(\infty)})$).

The Poisson ratio $\nu = \mathbf{C}_{1122}/\mathbf{C}_{1111}$ and the coefficient of anisotropy $\alpha = (\mathbf{C}_{1111} - \mathbf{C}_{1122})/2\mathbf{C}_{1212}$ converge to 0.065 and 4.43, respectively, for carpets with voids, and to 0.063 and 3.74, for carpets with rigid inclusions. As expected for both voids and rigid inclusions the scaling of the elastic constants is given by a power law:

$$\mathbf{C}^{(n)} = \mathbf{C}^{(0)}\mathbf{L}^{\beta(n)}, \tag{16}$$

where L is a normalised characteristic parameter of the structure at step n, $\mathbf{L} = 3^n$ and β is an exponent depending on the fractal dimension of the microstructure. At each step n the exponent can be expressed as a function of ν and α; so it converges to a finite value. The variation of the elastic constants with the characteristic length of the structure at step n for porous carpets is represented in Fig. 8. The results are in agreement with those previously obtained by Poutet *et al.*[47]

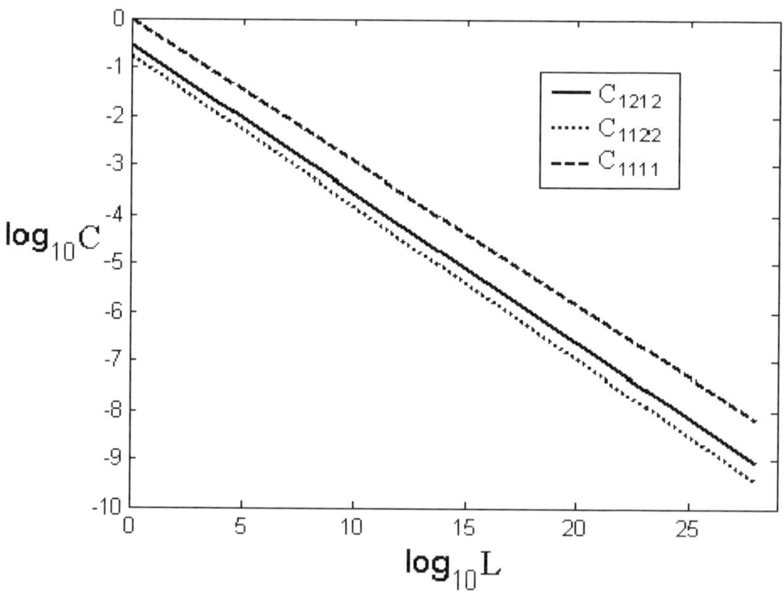

Fig. 8. Variation of the homogenised elastic constants with the characteristic length L, adapted from Ref. 46.

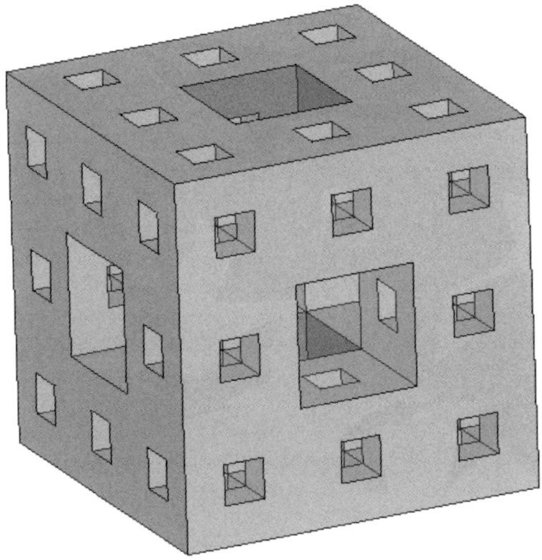

Fig. 9. The second generation of the Menger sponge.

Poutet *et al.*[47] were also interested in finding the elastic properties of porous fractals embedded in the 3D space, such as a fractal foam (with a fractal box dimension of log(26)/log(3)) and the Menger sponge shown in Fig. 9 (with a fractal box dimension of log(20)/log(3)).

Due to the complexity of the geometries and to limited computational resources, only the first two generations could be numerically simulated. In both cases, although cubic symmetry was expected, the elastic moduli results almost isotropic (the approximation is less accurate for the Menger sponge). A power-law scaling of the equivalent Young modulus of the form $E^{(n)} = E_0(25/27)^n$ was found for fractal foams and $E^{(n)} = E_0(2/3)^n$ for the Menger sponge. The results are extrapolated for large n using renormalisation arguments starting from the numerical values for the first two iterations.

Dyskin[48] used the differential self-consistent method[49] for media containing self-similar distributions of spherical/ellipsoidal pores or cracks to find the homogenised elastic constants. The author proposes to model such materials by a sequence of continua defined by homogenising over a sequence of volume elements of various sizes. Specifically, the constitutive behaviour of the continuum on scale ε is obtained based on averaged stresses and strains over volume elements of size ε. Under the restrictive hypothesis that at each scale inhomogeneities (pores/inclusions) of equal size do not interact, the elastic constants on scale ε are only functions of the volume fraction of the voids/inclusions of smaller size. The inhomogeneities defined on scale ε are embedded in this effective continuum. The procedure is applied iteratively on larger scales.

Not surprisingly, the elastic constant scaling is described by a power law:

$$E(\varepsilon) \prec \varepsilon^{\beta}, \tag{17}$$

where E is the effective modulus of interest and β is an exponent depending on the fractal dimension of the microstructure. This is expected as if one disregards the interaction of inclusions, the moduli should scale in a manner similar to the scaling of the volume fraction.

4.1.2. *Approaches based on the reformulation of governing equations*

These are attempts to incorporate information about the geometry into the governing equations of elastostatics. The critical physical aspect that needs to be accommodated is the large (infinite) number density of interfaces. Hence, the fields are not classically differentiable at an infinite number of points in the problem domain. This suggested that one may use fractional calculus to render the fields differentiable in the entire domain. We divide the few attempts that use this concept in methods based on non-local fractional operators, and methods that employ local operators.

4.1.2.1. Non-local fractional operators-based approach

Tarasov[43,44] studied porous materials with fractional mass dimensionality. The material contains pores with a broad range of sizes and the mass enclosed in a volume of characteristic dimension ε scales as $m(\varepsilon) \prec \varepsilon^{q_m}$. Here q_m is a non-integer number indicating the fractal mass dimension. The author replaces the fractal body with an equivalent continuum having a "fractional measure". The fractional measure (or fractal volume) $\mu_{q_m}(W)$ of a ball W of radius ε is defined such that the mass it contains scales as

$$m(W) = \rho_0 \mu_{q_m}(W) \prec \varepsilon^{q_m}, \tag{18}$$

where ρ_0 is the density of the matrix material. The fractional integral of an arbitrary function f on a fractal volume W of dimension q_m embedded in the 3D Euclidean space is defined using the new measure as (Riesz form)

$$I^{q_m} f = \int_W f(\varepsilon)\, d\mu_{q_m}(W), \tag{19}$$

where the relationship between the differential fractal volume/measure and the differential Euclidean volume is given by

$$d\mu_{q_m}(W) = \frac{2^{3-q_m}\Gamma(3/2)}{|\boldsymbol{\varepsilon}|^{3-q_m}\Gamma(q_m/2)} d^3\boldsymbol{\varepsilon}. \tag{20}$$

Thus, if the ball W of radius $\varepsilon = |\boldsymbol{\varepsilon}|$ contains a fractal set, its fractal volume/measure is $\mu_{q_m}(W) = I^{q_m} 1_W = 4\pi\varepsilon^{q_m}/q_m$. If it does not contain a fractal, the classical volume is recovered $I^{q_m} 1_W = 4\pi\varepsilon^3/3$.

The balance equations for mass, linear and angular momentum conservation are reformulated for the equivalent continuum using this measure.

The fractional operators proposed by Tarasov are a generalisation of the classical Riemann–Liouville operators to embedding spaces of arbitrary dimension. They are non-local and depend on the origin of the coordinate system. Thus, the field equations are rewritten for the equivalent "homogenised" continuum and the solutions are obtained in an average sense, without distinguishing between a material point and a pore point. Furthemore, the formulation is limited to homogeneous fractals in the sense that the mass of the material contained in a certain Euclidean volume is independent of the translation or rotation of the respective volume.

4.1.2.2. Local fractional operators-based approach

In the framework of solid mechanics, a local version of the fractional operators were initially used by Carpinteri *et al.*[24,41,42] in an attempt to explain size effects in deformation and fracture processes of heterogeneous materials. Using renormalisation group procedures for the fractal-like structures, these authors defined new mechanical quantities (such as fractal strain, fractal stress and the corresponding work), which are scale-invariant but have unconventional physical dimensions that depend on the fractal dimensions of the structure.

The kinematics equations and the principle of virtual work for fractal media embedded in Euclidean spaces were formally presented. They use local fractional operators previously developed by Kolwankar and Gangal,[26–28] which allow writing (formally) the balance equations in a local form. As discussed below, this is not always warranted. These authors went further by postulating the existence of an elastic potential for the fractal microstructure which is then used to define the fractal stress.

This formulation was employed to date only in the rather trivial case of a 1D rigid bar which is allowed to deform only at a set of points that form a Cantor set. In this case, the displacement is represented by the Devil staircase function. This function is piecewise constant (no deformation for the rigid part of the bar) and discontinuous at the Cantor set points.

For structures embedded in multidimensional spaces, the authors suggested a variational formulation of the elasticity problem using their fractional operators and a method to approximate the solution using the Devil's staircase function. Nevertheless, the formulation has not been used to solve any problem in multidimensions so far.

A limitation of this description is that deformation is allowed to take place on the fractal support only. Obviously, this is not the case in most practical situations.

A new formulation of mechanics on composite bodies containing fractal inclusions was presented in Ref. 29. This work makes the following advances:

- it describes the deformation in the embedding space rather than in the space of the fractal object;
- the material is treated as a composite, with both fractal inclusions and the matrix complement deforming;
- goes all the way from the formulation of the governing equations to implementation and to solving example problems;
- presents a new type of FE (in two dimensions) that includes information about the fractal geometry of the underlying material, i.e. it does not require partitioning the problem domain in elements of size comparable with the smallest inclusion present;
- devises a procedure by which the solution for the composite with fractal inclusions having a lower scale cut-off is obtained from a parametric solution of the same problem with an infinite number of scales. This implies that once one solves the parametric problem, solutions for an entire family of problems (with variable number of scales present) are obtained at no extra cost.

These advances make possible obtaining solutions for realistic problems defined on bodies embedded in Euclidean spaces of dimensionality larger than one. The formulation is briefly outlined below.

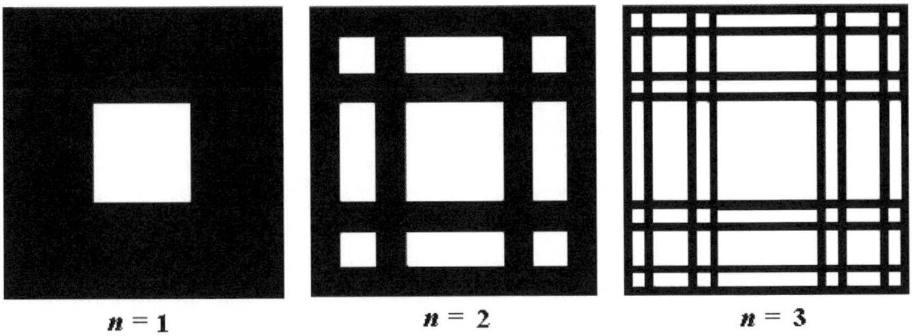

Fig. 10. The first three iterations of a fractal plate.

A 2D composite domain of the type shown in Fig. 10 is considered as an example. The embedding material (the matrix) in the subdomain $A - F$ is shown in white. The fractal geometry is defined by partitioning the two axes (reference frame) in Cantor sets. The subdomain F that results through this procedure is shown in black. The fractal box counting dimension of this structure results $q = \log 2/\log 3 + 1$. The two materials are considered linear elastic with elastic constants E, ν for F and E_0, ν for $A - F$. Both cases of stiffer ($E_0 > E$) and more compliant matrix material ($E_0 < E$) are studied.

The kinematics is represented using strains, which within the assumption of small deformation, are given by

$$\varepsilon_{ij}(\mathbf{u}) = \begin{cases} (D_i u_j(\mathbf{X}) + D_j u_i(\mathbf{X}))/2 & \text{if } \mathbf{X} \in A - F, \\ (D_i^{q_i(\mathbf{X})} u_j(\mathbf{X}) + D_j^{q_j(\mathbf{X})} u_i(\mathbf{X}))/2 & \text{if } \mathbf{X} \in F, \end{cases} \tag{21}$$

where q_i and q_j are the fractal dimensions in the two principal directions $\{\mathbf{e}_1, \mathbf{e}_2\}$, used to describe the geometry.

This expression is similar to that proposed by Carpinteri and collaborators,[41,42] except that the fractional derivatives used here are given by Eq. (9), i.e. include a parameter L (with physical dimensions of length), and have units similar to those of the classical derivative. Also, the matrix $A-F$ is allowed to deform and its response is modelled with classical continuum mechanics.

The strain of Eq. (21) does not rotate according to the usual tensor rule and therefore is called a "pseudo-strain". This limitation stems from the fact that the definitions of the geometry and of the strain are relative to a fixed coordinate system $\{\mathbf{e}_i\}$. If this reference frame is modified, the description of the fractal object changes (e.g. the fractal dimensions in the two principal directions change). This situation is inherent once the "material point" of the usual continuum is replaced by an entire structure with no isotropy.

The balance of linear momentum leads to the equation

$$\int_P \rho(a_k(\mathbf{x},t) - b_k(\mathbf{x},t))\, d^{\mathrm{Fr}}\mathbf{x} = \int_P D_i^{q_i(\mathbf{x})} T_{ki}(\mathbf{x},t)\, d^{\mathrm{Fr}}\mathbf{x}, \qquad (22)$$

where $\mathbf{a}$ and $\mathbf{b}$ are the acceleration and the body forces, while the "pseudo-stress" $T_{ki}(\mathbf{x},t) = t_k(\mathbf{x},\mathbf{e}_i,t)$ is defined, as usual, based on the tractions acting on a plane of normal $\mathbf{e}_i$. In continuum mechanics, when the traction vector is everywhere differentiable and the domain boundaries are smooth, these are the components of the Cauchy stress tensor. In the present case neither of these conditions is fulfilled. In addition, since the reference frame is kept fixed, $\mathbf{T}$ does not rotate. A similar expression was also proposed in Ref. 42, but using the fractional operators defined in Refs. 26–28.

It is noted that the weak form (22) cannot be localised, except under rather restrictive conditions related to the continuity of the integrand over A.

A constitutive relation similar to the linear elasticity is postulated to exist between the pseudo-stress and the pseudo-strain:

$$\varepsilon_{ij}(\mathbf{x}) = L_{ijkl}(\mathbf{x}) T_{kl}(\mathbf{x}). \qquad (23)$$

As with any constitutive relation, this expression is postulated. The issue is discussed further in Ref. 29.

A mixed boundary value problem is defined on the outer boundary of the domain A. This boundary is smooth since in general, it has no relation with the interface between matrix and inclusions (Fig. 10).

The solution is sought by reformulating the problem in the FE framework. The mixed variational formulation (Hellinger–Reissner principle) is used, which leads to the solution by seeking the stationary points of the functional

$$\begin{aligned}\mathbf{U}(\mathbf{v},\mathbf{P}) &= \frac{1}{2}\int_A P_{ij} L_{ijkl} P_{kl}\, d^{\mathrm{Fr}}\mathbf{x} - \int_A P_{ij}\varepsilon_{ij}(\mathbf{v})\, d^{\mathrm{Fr}}\mathbf{x} \\ &\quad + \int_A \varepsilon_{ij}(\mathbf{v}) L^{-1}_{ijkl} \varepsilon_{kl}(\mathbf{v})\, d^{\mathrm{Fr}}\mathbf{x} - \int_{\Gamma_t} u_i t^0\, d^{\mathrm{Fr}}\Gamma, \qquad (24)\end{aligned}$$

where $\mathbf{v}$ and $\mathbf{P}$ are the probing displacement and pseudo-stress fields used to render $\mathbf{U}$ stationary. The variation of $\mathbf{U}$ with $\mathbf{P}$ leads to the constitutive equation, while the variation of the functional $\mathbf{U}$ with respect to $\mathbf{v}$ leads to the equilibrium equation.

The solution is approximated using a set of generic shape functions $N_m(\xi), m = 1 \cdots M_u$ for the displacement field $\mathbf{u} = \sum_{m=1,M_u} \mathbf{u}^m N_m$ and a corresponding set of shape functions $M_m(\xi), m = 1 \cdots M_T$ for the pseudo-stress field $\mathbf{T} = \sum_{m=1,M_T} \mathbf{T}^m M_m$. As usual, this transforms the minimisation problem stated above into a system of equations for the unknown coefficients $\{\mathbf{u}^m, \mathbf{T}^m\}$.

The solution results in the reference frame with respect to which both pseudo-stress and pseudo-strains are defined. This coordinate system spans the embedding space and is selected at the beginning of the analysis. Defining a given deterministic fractal structure with respect to multiple coordinate systems of the embedding space (i.e. rules for the reference frame rotation) is still an open issue in fractal geometry. Therefore, the method discussed here can be used at this time only with respect to a single frame in which the geometry is defined.

The shape functions are selected to reflect the complexity of the geometry and hence must be developed separately for each problem of known geometry and for the given reference frame in which the geometry is described. Shape functions of the type shown in Fig. 5, and derived from Eq. (7) are used. These are equivalent to the common linear shape functions used in FEM. Higher order functions can be derived from Eqs. (5) and (7). These functions contain two parameters, β and γ. They are related by the normalisation condition requiring that the shape function takes the value 1 at one of the nodes. The other parameter remains and is carried over in the variational formulation. The value of this internal parameter of the element results as part of the solution (it is solved for while the stationary points of the variational form (24) are sought). Explicit forms for the two sets of shape functions $N_m(\xi)$ and $M_m(\xi)$ are presented in Ref. 29. It is also noted that the continuity of the interpolation functions used for the stress field insures the continuity of tractions across all interfaces in the problem.

As noted above, the solution is obtained for the plate with an infinite number of scales. This is a fictitious problem since at infinite refinement, the fractal inclusions disappear and one recovers the homogenous plate. However, the solution is given in a parametric form, in terms of L, which vanishes for $n \to \infty$. So, this physically meaningful result (the homogeneous plate) is recovered. When a lower scale cut-off exists, the solution is approximated using Eq. (13) that provides approximations of the integral operators. In essence, one replaces in the solution for the infinite number of scales $L = \varepsilon_n$, where ε_n is the characteristic length scale of the fractal structure on the finest scale n.

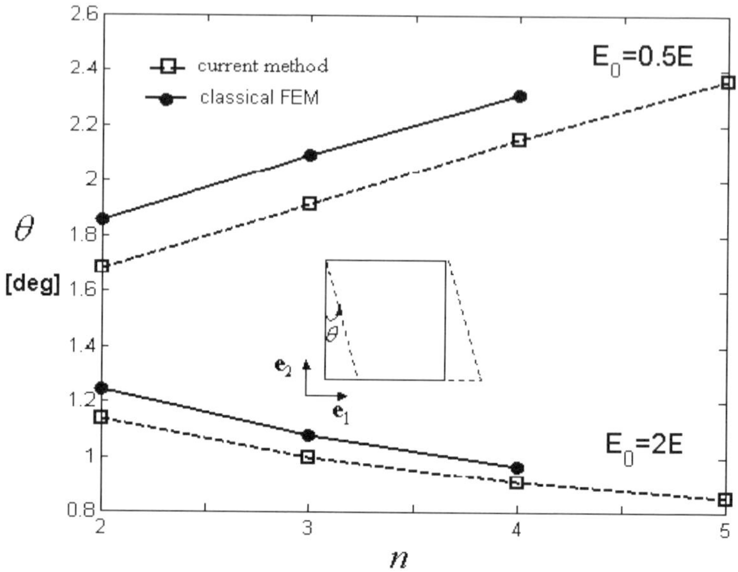

Fig. 11. Deformation of the fractal plate in Fig. 10 subjected to shear. Here n represents the number of scales in the hierarchical microstructure. The continuous lines marked by filled symbols represent the solution obtained with usual FEs and classical continuum mechanics. The dashed lines marked by open symbols show the predictions of the method discussed here. A single boundary value problem is solved, for the structure with an infinite number of scales. All solutions for finite n result by proper particularisations of the parameter$L$ that enters the definition of the fractional operators.

The problem results non-linear and is solved using standard numerical procedures. To demonstrate the method, an example is shown below. The plate in Fig. 10 is subjected to simple shear. The shear stress is applied in the frame $\{\mathbf{e}_1, \mathbf{e}_2\}$ and the resulting change of angle of the plate is computed (see inset to Fig. 11). Values are shown for various plate geometries corresponding to $n > 2$, for the case when the fractal material is two times more compliant and two times stiffer than its complement. These data are compared with the results obtained using regular FEs and a fine discretisation (continuous lines marked by filled symbols).

When using regular FEM, a separate problem is solved for each scale n. The mesh has to be refined such that the smallest element of the structure is at most of equal size with the finest inclusion. Therefore, the number of elements increases very fast with n. In contrast, the solution obtained using the formulation discussed here is obtained with one element (element with internal microstructure and special shape functions). Furthermore, as discussed before, once the solution for the plate with an infinite number of scales is obtained, solutions for all finite n (dashed lines in Fig. 11) result at no extra cost.

Various boundary value problems formulated for finite generations of the fractal geometry presented in Fig. 10 are described and solved in Ref. 45.

4.2. *Stochastic fractal microstructures*

Deterministic fractals are rarely (if ever) found in nature. Hierarchical self-similar structures with stochastic characteristics are widespread. There are multiple ways in which such geometries can be generated and hence these structures can be classified in two categories. In the first case, the scaling properties are fulfilled exactly for all scales, and the resulting structure has a well-defined fractal dimension. The stochastic nature comes from the way the material is distributed in the problem domain at each scale. In the second, the scaling is followed only in average. Combinations of the two types are conceivable.

Solving boundary value problems over domains with this type of microstructure received very little attention. In this section we summarise the method used and the results obtained in Ref. 50. for the quasistatic deformation of a composite domain containing a stochastic self-similar structure of the first type (see above). The structure is embedded in two dimensions and is generated according to the rule discussed in the Introduction. Specifically, the domain is divided into M equal parts of which P are preserved in the next iteration. In Fig. 4, $M = 4$ and $P = 2$. The P parts that are retained are selected at random in each iteration from the M subdomains. The number of possible configurations at iteration (or scale) n is $[M!/(P!(M-P)!)]^{P^{n-1}+\cdots+P+1}$. Since the scaling is exactly fulfilled in each iteration, the fractal dimension is well defined and is given by $q = \log(P)/\log(M^{1/d})$, i.e. for the structure in Fig. 4 ($d = 2$), one obtains $q = 1$.

The randomness of the geometry reflects in the randomness of the distribution of material properties (elastic constants) in the problem domain. It is assumed that the two phases forming the composite are each homogeneous and linear elastic materials of compliance $L_{ijkl}(\mathbf{x}) = L_{ijkl}$ if $\mathbf{x}$ belongs to the fractal inclusions, and $L_{ijkl}(\mathbf{x}) = L^0_{ijkl}$ otherwise. A mixed boundary value problem is defined on the boundary of the body. This contour is smooth since it is defined in the embedding space. The solution fields (stresses, $\mathbf{T}(\mathbf{x}, \boldsymbol{\omega})$ and displacements, $\mathbf{u}(\mathbf{x}, \boldsymbol{\omega})$) are functions of a deterministic variable representing the spatial position, $\mathbf{x}$, and a stochastic variable, $\boldsymbol{\omega}$, which accounts for the variability associated with the elastic constants at $\mathbf{x}$. Hence, one is interested in the statistical properties of the solution only. The values of the imposed tractions and displacements along the boundary are deterministic and identical for all realisations of the structure.

The problem is solved using the stochastic (spectral) finite element method (SFEM) developed by Ghanem and Spanos.[51] The formulation of Sec. 4.1 based on fractional calculus is not used in this study. This is imposed by the fact that the geometry is known only in the statistical sense and cannot be described with the tools employed in the previous section. One may generate realisations of the structure that are compatible with the required scaling and for which the geometry would be exactly known; however, this is not desirable since it would require solving a large number of replicas separately. In contrast, in the SFEM method one solves directly for the mean and standard deviation of the solution (stresses and displacements). The higher order moments of the solution cannot be obtained with this version of SFEM.

The elasticity problem defined on structures similar to those in Fig. 4 with traction imposed boundary conditions:

$$\mathbf{Tn} = \mathbf{t}^0 \quad \text{on } \Gamma_t \quad \text{and} \quad \mathbf{u} = \mathbf{0} \quad \text{on } \Gamma_u = \partial A_n - \Gamma_t \tag{25}$$

can be written in the variational form, in terms of the displacement field $\mathbf{u}(\mathbf{x}, \boldsymbol{\omega})$, as

$$\left\langle \int_{A_n} u_{i,j}(\mathbf{x}, \boldsymbol{\omega}) L^{-1}_{ijkl}(\mathbf{x}, \boldsymbol{\omega}) \nu_{k,l}(\mathbf{x}, \boldsymbol{\omega})\, d\mathbf{x} \right\rangle = \left\langle \int_{\Gamma_t} t^0_i \nu_i(s, \boldsymbol{\omega})\, ds \right\rangle, \tag{26}$$

where $\mathbf{v}(\mathbf{x}, \boldsymbol{\omega})$ is the probing field. The sign $\langle\rangle$ stands here for ensemble averaging.

This equation is solved using a procedure similar to that of the usual FEs. The problem domain is discretized in a number of elements $A_n = \bigcup_{p=1,N} A_{(n)p}$ with a total number of M_u nodes. The deterministic part of the solution is approximated using a set of generic shape functions $N_m(\mathbf{x})$, $m = 1 \cdots M_u$ such that $\int_{A_n} N_m(\mathbf{x}) N_p(\mathbf{x})\, d\mathbf{x} = \delta_{mp}$. These shape functions are identical to those commonly used in FEM.

The stochastic component of the solution is approximated by a superposition of chaos polynomials[52,53] $\psi_1(\boldsymbol{\omega}), \ldots, \psi_{M_\xi}(\boldsymbol{\omega})$ having the orthogonality property $\langle \psi_i(\boldsymbol{\omega}) \psi_j(\boldsymbol{\omega}) \rangle = \delta_{ij}$. In the probabilistic space, the chaos polynomials play the role of Hermite polynomials in the deterministic space and can be used as a decomposition base.[51,54]

The approximation of the displacement field is written in the separable form:

$$\mathbf{u}(\mathbf{x},\boldsymbol{\omega}) = \sum_{q=1\cdots M_\xi} \sum_{m=1,M_u} \mathbf{u}^{mq} N_m(\mathbf{x})\psi_q(\boldsymbol{\omega}). \tag{27}$$

Finding the solution amounts to identifying the $M_\xi M_u$ coefficients $\mathbf{u}^{mq}$.

Substituting the discrete solution (27) in Eq. (26) and probing with test functions $\nu_k(\mathbf{x},\boldsymbol{\omega}) = N_a^{(k)}(\mathbf{x})\psi_b(\boldsymbol{\omega})$ for $a = 1\cdots M_u$, $b = 1\cdots M_\xi$, $k = 1\cdots d$, leads to the system

$$\sum_{q=1\cdots M_\xi} \sum_{m=1\cdots M_u} \sum_{i=1\cdots d} \langle A_{ima}(\boldsymbol{\omega})\psi_q(\boldsymbol{\omega})\psi_b(\boldsymbol{\omega})\rangle u_i^{mq} = \langle B_a\psi_b(\boldsymbol{\omega})\rangle, \tag{28}$$

where

$$A_{ima}(\boldsymbol{\omega}) = \sum_{j=1\cdots d} \sum_{k=1\cdots d} \sum_{l=1\cdots d} \int_{A_n} N_{m,j}^{(i)}(\mathbf{x}) L_{ijkl}^{-1}(\mathbf{x},\boldsymbol{\omega}) N_{a,l}^{(k)}(\mathbf{x})\, d\mathbf{x}$$

and

$$B_a = \sum_{i=1\cdots d} \int_{\Gamma_u} t_i^0 N_a^{(i)}\, ds.$$

The objective of the analysis is to determine the mean and variance of the solution that result directly as

$$\langle u_i(\mathbf{x},\boldsymbol{\omega})\rangle = \sum_{m=1,M_u} u_i^{m1} N_m(\mathbf{x}) \tag{29a}$$

$$\langle u_i^2(\mathbf{x})\rangle - \langle u_i(\mathbf{x})\rangle^2 = \sum_{q=2\cdots M_\xi} \sum_{m=1,M_u} (u_i^{mq})^2 N_m^2(\mathbf{x}) i = 1\cdots d. \tag{29b}$$

To evaluate the stiffness matrix out of expression (28) it is necessary to specify the stiffness constants. The major step forward in SFEM is to decompose this function in a Karhunen–Loève[55] form, such that the average in (28) can be explicitly evaluated.

The Karhunen–Loève decomposition of a function $f_n(\mathbf{x},\xi)$, which depends on a deterministic and a stochastic variable is given by

$$f_n(\mathbf{x},\boldsymbol{\omega}) = \langle f_n(\mathbf{x})\rangle + \sum_{k=1,M_n} \sqrt{\alpha_n^{(k)}}\, \boldsymbol{\xi}_n^{(k)}(\boldsymbol{\omega}) F_n^{(k)}(\mathbf{x}). \tag{30}$$

The fluctuating part of the function is written in terms of a set of random uncorrelated variables $\boldsymbol{\xi}_n$, an othonormal set of deterministic functions $\{F_n^{(k)}\}$ and a set of constants $\{\alpha_n^{(k)}\}$, which are obtained as the eigenfunctions and eigenvalues of the covariance of function f_n, respectively. The spectral decomposition of the covariance can be written formally as

$$\mathrm{Cov}(f_n(\mathbf{x}, \mathbf{y})) = \sum_{i=1,\infty} \alpha_n^{(i)} F_n^{(i)}(\mathbf{x}) F_n^{(i)}(\mathbf{y}). \tag{31}$$

The Karhunen–Loève decomposition separates the correlation information (which is represented by $\{F_n^{(k)}\}$) in a deterministic fashion and reconstructs the initial function using uncorrelated random variables. This separation of the stochastic variable is used to great advantage in expression (28) as shown below. It should be noted that the random variables $\boldsymbol{\xi}_n$ may not be independent and, in principle, one may expand each of them in a series of chaos polynomials in a manner similar to the approximation of the unknown displacement field in (27).[54,56]

Considering that both the fractal and base materials are linear elastic, the compliance matrix is expressed only in terms of Young's moduls (equal with E on F_n and E_0 on $A - F_n$) and Poisson coefficient (considered the same for both phases). The tensor $\mathbf{A}$ in (28) can now be written as

$$\begin{aligned} A_{ma}(\boldsymbol{\omega}) = \int_{A_n} \langle E_n(\mathbf{x})\rangle + (E - E_0) \sum_{k=1,M_\alpha} \\ \times \sqrt{\alpha_n^{(k)}}\, \boldsymbol{\xi}_n^{(k)}(\boldsymbol{\omega}) F_n^{(k)}(\mathbf{x}))[(\nabla \mathbf{N}(\mathbf{x}))^{\mathrm{T}} \mathbf{C} \nabla \mathbf{N}(\mathbf{x})]_{ma}\, d\mathbf{x}, \end{aligned} \tag{32}$$

where it was made explicit that $\boldsymbol{\xi}_n$ is a function of the set of independent uncorrelated random variables $\boldsymbol{\omega}$. This allows the evaluation of the stochastic stiffness matrix entering the equilibrium equation

$$\sum_{q=1\cdots M_\xi} \sum_{m=1\cdots M_u} K_{maqb} X^{mq} = B_a \delta_{b1}$$

as

$$K_{maqb} = \int_{A_n} G_{qb} D_{ma}(\mathbf{x})\, d\mathbf{x}. \tag{33}$$

Here $\mathbf{D}(\mathbf{x}) = (\nabla \mathbf{N}(\mathbf{x}))^{\mathrm{T}} \mathbf{C} \nabla \mathbf{N}(\mathbf{x})$ is deterministic, the constant matrix depending only on the Poisson ratio is

$$\mathbf{C} = (\mathbf{e}_1 \otimes \mathbf{e}_1 + \mathbf{e}_2 \otimes \mathbf{e}_2 + \nu \mathbf{e}_1 \otimes \mathbf{e}_2)/(1 - \nu^2) + \mathbf{e}_3 \otimes \mathbf{e}_3/(2 + 2\nu)$$

and

$$G_{qb} = [\langle E_n(\mathbf{x})\rangle \delta_{qb} + (E - E_0) \sum_{k=1,M\xi} \sqrt{\alpha_n^{(k)}} \langle \boldsymbol{\xi}_n^{(k)}(\boldsymbol{\omega})\psi_q(\boldsymbol{\omega})\psi_b(\boldsymbol{\omega})\rangle F_n^{(k)}(\mathbf{x})]$$

$$q, b = 1 \cdots M_\xi \tag{34}$$

is stochastic.

The Karhunen–Loève decomposition of the elastic constant field over fractals generated with the rule considered here (and a variant of it) has been presented in Refs. 50 and 57. It was observed that the self-similarity of the structure leads to an interesting structure of the set of eigenfunctions $\{F_n^{(k)}\}$. The eigenfunction set at generation n of the fractal microstructure contains the eigenfunctions corresponding to all generations of index smaller than n. Then, the decomposition (30) written for example for Young's modulus, can be rewritten as

$$\begin{aligned} E_n(\mathbf{x}, \boldsymbol{\omega}) = \langle E_n(\mathbf{x})\rangle &+ \sum_{k=1,M} \sqrt{\alpha_n^{(k)}} \boldsymbol{\xi}_1^{(k)}(\boldsymbol{\omega}) F_1^{(k)}(\mathbf{x}) \\ &+ \sum_{k=M+1,M^2} \sqrt{\alpha_n^{(k)}} \boldsymbol{\xi}_2^{(k)}(\boldsymbol{\omega}) F_2^{(k)}(\mathbf{x}) + \cdots \\ &+ \sum_{k=M^{n-1}+1,M^n} \sqrt{\alpha_n^{(k)}} \boldsymbol{\xi}_n^{(k)}(\boldsymbol{\omega}) F_n^{(k)}(\mathbf{x}), \end{aligned} \tag{35}$$

where $\{F_1^{(k)}\}_{k=1\cdots M}$ are the eigenfunctions corresponding to the first generation, $\{\{F_1^{(k)}\}_{k=1\cdots M}, \{F_2^{(k)}\}_{k=M+1\cdots M^2}\}$ are eigenfunctions corresponding to the second generation, etc. Note that the decomposition has a finite number of terms, which depends on the number of scales present in the structure as M^n.

Similarly, the eigenvalues can be evaluated analytically as

$$\alpha_n^{(1)} = 0; \quad \alpha_n^{(k)} = \frac{1}{P}\left(\frac{P}{M}\right)^{2n} \frac{M-P}{M-1} \quad \text{for } k = 2 \cdots M, \tag{36a}$$

$$\alpha_n^{(k)} = \frac{1}{P^2}\left(\frac{P}{M}\right)^{2n} \frac{M-P}{M-1} \quad \text{for } k = M+1 \cdots M^2, \tag{36b}$$

$$\alpha_n^{(k)} = \frac{1}{P^n}\left(\frac{P}{M}\right)^{2n} \frac{M-P}{M-1} \quad \text{for } k = M^{n-1}+1 \cdots M^n. \tag{36c}$$

As mentioned above, the self-similar nature of the structure allows for important simplifications in performing the decomposition. If n is large, this decomposition has a large number of terms. This leads to difficulties when attempting to use the result in SFEM. In order to keep the analysis manageable, it is important to work with a small number of eigenfunctions. For this purpose, the Karhunen–Loève decomposition is truncated using the eigenfunctions of the first n_0 generations ($n_0 < n$), and (35) becomes

$$\begin{aligned} E_n(\mathbf{x}, \boldsymbol{\omega}) = \langle E_n(\mathbf{x})\rangle &+ \sum_{k=1,M} \sqrt{\alpha_n^{(k)}}\boldsymbol{\xi}_1^{(k)}(\boldsymbol{\omega})F_1^{(k)}(\mathbf{x}) \\ &+ \sum_{k=M+1,M^2} \sqrt{\alpha_n^{(k)}}\boldsymbol{\xi}_2^{(k)}(\boldsymbol{\omega})F_2^{(k)}(\mathbf{x}) + \cdots \\ &+ \sum_{k=M^{n_0-1}+1,M^{n_0}} \sqrt{\alpha_n^{(k)}}\boldsymbol{\xi}_{n_0}^{(k)}(\boldsymbol{\omega})F_{n_0}^{(k)}(\mathbf{x}). \end{aligned} \tag{37}$$

As mentioned above, in principle, one can approximate each of the functions $\boldsymbol{\xi}_n^{(k)}$ appearing in the decomposition (37) using chaos polynomials. This increases the number of unknowns in the problem which can be solved for as part of the general solution. On the other hand, it is shown in Ref. 50 that for this particular field, selecting $\boldsymbol{\xi}_n^{(k)}(\boldsymbol{\omega}) = \boldsymbol{\omega}_n^{(k)}$ in (37) leads to the same mean and covariance of the elastic constant field. Hence, considering the stochastic variables in the Karhunen–Loève decomposition uncorrelated *and* independent, the input field remains unchanged at least up to the second moment of the respective probability distribution function. On the other hand, the computational cost is significantly reduced. Another benefit of this observation is that the average in the second term in (34) can be evaluated explicitly for all pairs (q, b) of chaos polynomials, based on the moments of the probability distribution function for ω. Therefore, G_{qb} in (33) does not have a stochastic nature, and the resulting problem to solve is deterministic.

These issues and the approximation related to the truncation transforming (35) into (37) are discussed in detail in Ref. 50.

To demonstrate the method, in Ref. 50, the plate in Fig. 4 was loaded uniaxially by applying a specified distributed force in the vertical direction on the upper and lower sides of the square plate. The mean and variance of the solution were evaluated. Figure 12 shows these quantities for the displacement of the upper side of the plate, i.e. the displacement component work conjugated with the applied force.

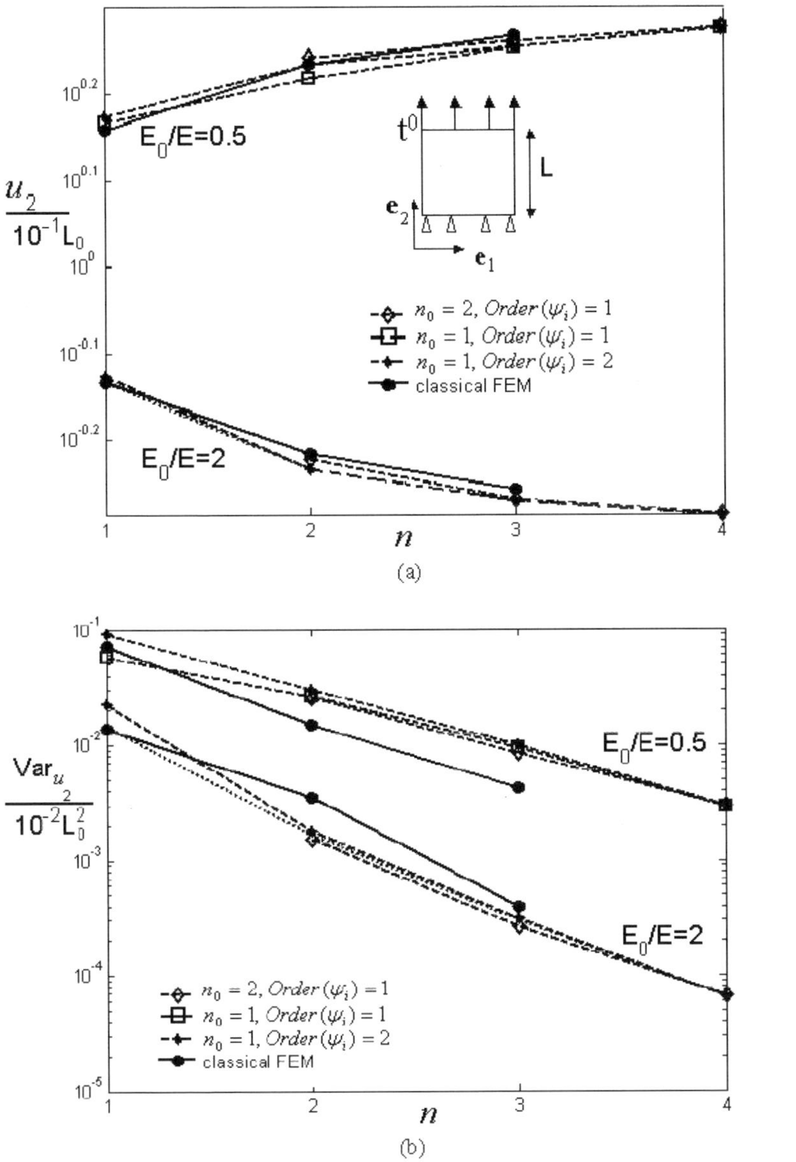

Fig. 12. The variation of the mean and variance of the displacement of the upper edge of the plate with the generation (scale) index n (logarithmic plots). The dashed curves correspond to the various approximations described in text for the Karhunen–Loève expansion and for the chaos polynomials used to approximate the solution. The continuous lines marked by filed circles correspond to averages of results from a large number of deterministic FEM simulations.

The solution obtained using the SFEM method is compared with the equivalent one obtained using the standard (deterministic) FEM. To obtain the desired information with usual FEM, a large number of simulations were performed for each generation of the structure in order to allow for a meaningful statistical analysis. Because the number of relevant realisations of the structure increases very fast with n, this "brute force" evaluation could be performed only for $n \leq 3$. The total number of realisations of the structure for $n = 1$, 2 and 3 are 6, 216 and 6^7 respectively. For the two lower n values, statistics was collected by sampling all possible configurations, while for $n = 3, 4500$ replicas were solved explicitly.

Obviously, obtaining the solution using the SFEM method is significantly less expensive: a single simulation is needed for each generation. Furthermore, because of the truncation of the series (35) at n_0, the size of the model does not change with n although many scales are present in the structure; predictions can be made with minimal effort for any desired n. This is not the case with the "brute force" method.

Two types of approximations are made in order to obtain the solution with minimal effort: the Karhunen–Loève expansion of Young's modulus distribution (36) is truncated retaining only the eigenfunctions corresponding to the first n_0 functions, and the order of the chaos polynomials used to approximate the solution is limited to two.

The results obtained for $n_0 = 1$ (first-order approximation in the Karhunen–Loève decomposition) are represented with dashed lines and marked by squares if first-order chaos polynomials are used, and with stars if second-order chaos polynomials are used. The results obtained for $n_0 = 2$ (second-order approximation of the Karhunen–Loève decomposition) and first-order chaos polynomials are represented by dashed lines marked with diamonds. The continuous lines marked by filled circles represent results obtained with the classical FEM (deterministic models) and a high number of sample configurations.

The method is compared favourably in terms of accuracy with the results obtained by "brute force" simulations. Its performance depends on the two approximations made: the approximation of the stochastic component of the displacement field with chaos polynomials and the approximation of the elastic constant distribution with the truncated Karhunen–Loève decomposition. It results in that the error (relative to the deterministic FEM solution) is much more sensitive to the number of terms considered in the Karhunen–Loève decomposition, than it is to the order of chaos polynomials used to express the displacement field. This is expected since the decomposition reproduces the spatial correlations of the random field, which in turn, represent the scaling and the geometrical details of the fractal microstructure.

In terms of computational efficiency, the method is clearly superior to any other method that requires averaging over deterministic replicas. As the iteration index n increases, the number of replicas increases very fast and "brute force" calculations are simply impossible. Furthermore, the self-similar nature of the geometry makes possible obtaining the solution for any n without actually discretizing the structure with a characteristic length scale comparable with the finest feature of the geometry. Of course, this is a peculiarity of the type of geometries considered here and is independent of the SFEM method. These two observations make the method flexible enough for use in applications that involve other types of self-similar geometries.

5. Conclusions

Materials and structures with hierarchical microstructures/substructures are ubiquitous. They have interesting optical, magnetic, transport and mechanical properties. To a large extent, these structures were developed by living organisms to perform multiple functions and were optimised over millennia of evolution. The geometric complexity observed in such materials is large and increases with decreasing scale of observation. Some of them have self-similar geometric characteristics.

Integrating field equations on such supports, while taking into account all scales of the structure, is difficult. The approaches presented in this chapter represent just the beginning of a long path towards a consistent framework that permits addressing complexity in its entirety, rather than attempting to "divide and conquer". Much future work is needed to achieve this goal, both on describing the geometric complexity and on the representation of physics in such environments.

References

1. J. Kastelic, A. Galeski and E. Baer, *Connective Tissue Res.* **6**, 11 (1978).
2. G. Beaucage, *J. Appl. Cryst.* **29**, 134 (1996).
3. C. Marliere, F. Despetis, P. Etienne, T. Woignier, P. Dieudonne and J. Phalippou, *J. Non-Cryst. Solids* **285**, 148 (2001).

4. D. W. Schaefer and K. D. Keefer, *Phys. Rev. Lett.* **56**(20), 2199 (1986).
5. E. B. Tadmor, M. Ortiz and R. Phillips, *Phil. Mag. A* **6**(73), 1529 (1996).
6. R. Miller, E. B. Tadmor, R. Phillips and M. Ortiz, *Model. Simul. Mater. Sci. Eng.* **6**, 607 (1998).
7. B. B. Mandelbrot, *The Fractal Geometry of Nature* (Freeman, New York, 1983).
8. M. F. Barnsley, *Fractals Everywhere* (Academic Press, Cambridge, MA, 1993).
9. K. J. Falconer, *The Geometry of Fractal Sets* (Cambridge University Press, Cambridge, 1985).
10. J. Feder, *Fractals* (Plenum Press, New York, 1988).
11. F. B. Tatom, *Fractals* **3**(1), 217 (1995).
12. A. K. Grundwald, *Zeitschrift fur Mathematik und Physik* **XII**(6), 441 (1967).
13. A. V. Letnikov, *Mat. Sb.* **3**, 1 (1868).
14. R. Gorenflo and F. Mainardi, in *CISM Lecture Notes Fractals and Fractional Calculus in Continuum Mechanics*, eds. A. Carpinteri and F. Mainardi (Springer Verlag, Wien, NY, 1978), p. 223.
15. R. Hilfer, *Applications of Fractional Calculus in Physics* (World Scientific, Singapore, 2000).
16. M. Giona and H. E. Roman, *J. Phys. A. Math. Gen.* **25**, 2093 (1992).
17. W. G. Glocke and T. F. Nonnemacher, *J. Stat. Phys.* **71**, 741 (1992).
18. T. F. Nonnemacher, *J. Phys.* **23A**, L697 (1990).
19. D. Del-Castillo-Negrete and B. A. Carreras, *Phys. Rev. Lett.* **94**, 065003-1 (2005).
20. T. A. Witten and L. M. Sander, *Phys. Rev. Lett.* **47**, 1400 (1981).
21. F. Mainardi, in *CISM Lecture Notes Fractals and Fractional Calculus in Continuum Mechanics*, eds. A. Carpinteri and F. Mainardi (Springer Verlag, Wien, NY, 1998), p. 291.
22. V. D. Djordjevic, J. Jaric, B. Fabry, J. J. Fredberg and D. Stamenovic, *Annals Biomed. Eng.* **31**, 692 (2003).
23. R. L. Bagley and P. J. Torvik, *J. Rheol.* **30**, 133 (1986).
24. A. Carpinteri and B. Chiaia, *Chaos, Solitons & Fractals* **9**, 1343 (1996).
25. G. P. Cherepanov, A. S. Balankin and V. S. Ivanova, *Eng. Frac. Mech.* **51**(6), 997 (1995).
26. K. M. Kolwankar and A. D. Gangal, *Chaos* **6**, 505 (1996).
27. K. M. Kolwankar and A. D. Gangal, in *Proc. Conf. Fractals in Engineering* (Archanon, 1997).
28. K. M. Kolwankar, Studies of fractal structures and processes using methods of fractional calculus, PhD thesis, University of Pune, India (1998).
29. M. A. Soare, Mechanics of materials with hierarchical fractal structure, PhD thesis, Rensselaer Polytechnic Institute, Troy, NY (2006).
30. M. Meerschaert, D. Benson, H. P. Scheffler and B. Baeumer, *Phys. Rev. E* **65**, 1103 (2002).
31. A. I. Saichev and G. M. Zaslavsky, *Chaos* **7**, 753 (1997).
32. R. S. Strichartz and M. Usher, *Math. Proc. Cambridge Phil. Soc.* **129**, 331 (2000).

33. U. Mosco, *Phys Rev. Lett.* **79**(21), 4067 (1997).
34. J. Kigami,*Jpn J. Appl. Math.* **8**, 259 (1989).
35. R. S. Strichartz, *Notices AMS* **46**(10), 1199 (1999).
36. S. Alexander and R. Orbach, *J. Phys. Lett.* **43**, L625 (1982).
37. R. Orbach, *Science* **231**(4740), 814 (1986).
38. T. Nakayama, K. Yakubo and R. Orbach, *Rev. Mod. Phys.* **66**, 381 (1994).
39. P. D. Panagiatopoulos, E. S. Mistakidis and O. K. Panagouli, *Comp. Meth. Appl. Mech. Eng.* **99**, 395 (1992).
40. P. D. Panagiatopoulos, *Int. J. Sol. Struct.* **29**(17), 2159 (1992).
41. A. Carpinteri, B. Chiaia and P. Cornetti, *Comput. Meth. Appl. Mech. Eng.* **191**, 3 (2001).
42. A. Carpinteri, B. Chiaia and P. Cornetti, *Mater. Sci. Eng. A* **365**, 235 (2004).
43. V. E. Tarasov, *Ann. Phys.* **318**(2), 286 (2005).
44. V. E. Tarasov, *Phys. Lett. A* **336**, 167 (2005).
45. M. A. Soare and R. C. Picu, *Int. J. Sol. Struct.* **44**(24), 7877 (2007).
46. V. G. Oshmyan, S. A. Patlashan and S. A. Timan, *Phys. Rev. E* **64**, 056108: 1 (2001).
47. J. Poutet, D. Manzoni, F. H. Chehade, C. J. Jacquin, M. J. Bouteca, J. F. Thovert and P. M. Adler, *J. Mech. Phys. Solids* **44**(10), 1587 (1996).
48. A. V. Dyskin, *Int. J. Sol. Struct.* **42**, 477 (2005).
49. R. L. Salganik, *Mech. Solids* **8**, 135 (1973).
50. M. A. Soare and R. C. Picu, *Int. J. Num. Meth. Eng.* **74**, 668 (2008).
51. G. Ghanem and P. D. Spanos, *Stochastic Finite Elements: A Spectral Approach* (Springer-Verlag, New York, 1991).
52. R. H. Cameron and W. T. Martin, *Ann. Math.* **48**, 385 (1947).
53. N. Wiener, *Nonlinear Problems in Random Theory* (Technology Press of the Massachussetts Institute of Technology and John Wiley and Sons Inc., New York, 1958).
54. H. G. Matthies, C. E. Brenner, C. G. Bucher and C.G. Soares, *Struct. Safety* **19**, 283 (1997).
55. K. Karhunen, *Am. Acad. Sci. Fennicade, Ser.A.I.* **37**, 3 (1947) (Translation RAND Corporation, Santa Monica, California, Rep. T-131, 1960).
56. H. G. Matthies and A. Keese, *Comp. Meth. Appl. Mech. Eng.* **194**, 1295 (2005).
57. M. A. Soare and R. C. Picu, *Chaos, Solitons & Fractals* **37**, 566 (2008).

Index

apparent properties, 31
Artificial Neural Network (ANN), 228, 231, 234, 235, 237, 244
asymptotic homogenisation, 3, 46
averaging theorems, 11, 12

beams, 34
BEM/FEM comparison, 141
boundary conditions, 5, 11, 113
boundary element, 101
boundary element method, 102
bounds, 209

Cantor set, 300
cohesive surfaces, 103
cohesive zone, 117, 119
compliance tensor, 110, 111
computational homogenisation, 1, 4, 30
computational plasticity, 258
constant subparametric elements, 116

"deformation driven" procedure, 8
deformation gradient tensor, 12, 14
deterministic fractal microstructures, 310
deterministic fractals, 299
discontinuity in the displacement field, 70
discontinuous finite element method, 257
discrete element method, 257
displacement compatible elements, 65
displacement incompatible element, 72

effective material coefficients, 218
effective material properties, 160
effective medium approximation, 3
effective properties, 43
element-free Galerkin method, 68
energy averaging theorem, 14
enhanced-strain element, 80

FEM, 258
finite element code, 188
first Piola–Kirchhoff stress tensor, 13, 14
First- and Higher-Order, 46
first-order computational homogenisation, 9
fractal (hierarchical and self-similar) microstructure, 308
fractal geometry, 295
fractals, 301
fractional calculus, 302
fracture, 101
fully prescribed boundary displacements, 19
functional materials, 2
functionally graded materials, 7

global length scale, 47
global–local analysis, 4
grain boundary interface, 118

hierarchical structures, 207
Hill–Mandel condition, 14
homogenisation, 160, 162
homogenisation for heat conduction, 36
homogenised failure surfaces, 283
hybrid stress elements, 76

interfaces with discontinuous first-order derivatives, 70

Karhunen–Loève decomposition, 177,323–326, 328
Karhunen–Loève procedure, 196

LAn, 258
limit analysis, 257, 265
local length scale, 47
localisation, 190, 208
localisation tensors, 178

macro-to-micro transition, 10
macroblock models, 276
macroscopic behaviour, 2
macroscopic loading paths, 1
macroscopic scale, 162
macroscopic tangent, 23
macroscopic tangent stiffness, 17, 21
masonry, 251, 270
materials with fractal microstructure, 308
micro-stress field, 44
microscopic boundary conditions, 4
microscopic scale, 162
microstructure, 295
microstructure generation, 107
molecular dimensions, 7

non-local approach, 126
non-uniform transformation field analysis, 159, 162, 171

parallel processing formulation, 138
period of the structure, 211
periodic boundary condition, 6, 16, 19, 22, 59, 84, 104, 126
periodicity, 213
periodicity conditions, 15
polycrystalline material, 110, 118
prescribed boundary displacements, 22
prescribed displacements, 11, 15
prescribed periodicity, 11
prescribed tractions, 11, 15
principal component analysis, 177
principle of local action, 33
proper orthogonal decomposition, 177

re-localisation, 216
recovering method, 208
representative unit cell (RUC), 44
representative volume element (RVE), 4, 7, 9, 16, 25, 30, 44, 103, 128
rule of mixtures, 2
RVE boundary conditions, 132

Sachs (or Reuss) assumption, 10
second-order homogenisation, 33
self-consistent approach, 3
self-learning FE model, 230
self-similar, 295
self-similar microstructure, 299
shells, 34
statistically representative, 30
stiffness tensor, 110
stochastic finite element method (SFEM), 322, 323, 326, 328
stochastic fractal microstructures, 321
stochastic fractals, 299
"stress-driven" procedure, 8
superconvergent patch recovery (SPR), 85

Taylor (or Voigt) assumption, 10
through-thickness RVE, 35
Transformation Field Analysis (TFA), 159, 161, 171
two length scales, 47
two-scale asymptotic homogenisation, 43
two-scale expansion, 49, 162

unit cell methods, 3
unit-cell V, 167
unsmearing, 208, 216

variational bounding methods, 3
Voronoi tessellation, 107

Y-periodicity, 48

Computational and Experimental Methods in Structures

Editorial Board